Assessment
Practices
in
Undergraduate
Mathematics

©1999 by The Mathematical Association of America

ISBN 0-88385-161-X

Library of Congress Catalog Number 99-63047

Printed in the United States of America

Current Printing

10 9 8 7 6 5 4 3 2 1

MAA NOTES NUMBER 49

Assessment Practices in Undergraduate Mathematics

BONNIE GOLD

SANDRA Z. KEITH

WILLIAM A. MARION

EDITORS

Published by
The Mathematical Association of America

The MAA Notes Series, started in 1982, addresses a broad range of topics and themes of interest to all who are involved with undergraduate mathematics. The volumes in this series are readable, informative, and useful, and help the mathematical community keep up with developments of importance to mathematics.

MAA Notes

11. Keys to Improved Instruction by Teaching Assistants and Part-Time Instructors, *Committee on Teaching Assistants and Part-Time Instructors, Bettye Anne Case,* Editor.

13. Reshaping College Mathematics, *Committee on the Undergraduate Program in Mathematics, Lynn A. Steen,* Editor.

14. Mathematical Writing, by *Donald E. Knuth, Tracy Larrabee, and Paul M. Roberts.*

16. Using Writing to Teach Mathematics, *Andrew Sterrett,* Editor.

17. Priming the Calculus Pump: Innovations and Resources, *Committee on Calculus Reform and the First Two Years,* a subcomittee of the Committee on the Undergraduate Program in Mathematics, *Thomas W. Tucker,* Editor.

18. Models for Undergraduate Research in Mathematics, *Lester Senechal,* Editor.

19. Visualization in Teaching and Learning Mathematics, *Committee on Computers in Mathematics Education, Steve Cunningham and Walter S. Zimmermann,* Editors.

20. The Laboratory Approach to Teaching Calculus, *L. Carl Leinbach et al.,* Editors.

21. Perspectives on Contemporary Statistics, *David C. Hoaglin and David S. Moore,* Editors.

22. Heeding the Call for Change: Suggestions for Curricular Action, *Lynn A. Steen,* Editor.

24. Symbolic Computation in Undergraduate Mathematics Education, *Zaven A. Karian,* Editor.

25. The Concept of Function: Aspects of Epistemology and Pedagogy, *Guershon Harel and Ed Dubinsky,* Editors.

26. Statistics for the Twenty-First Century, *Florence and Sheldon Gordon,* Editors.

27. Resources for Calculus Collection, Volume 1: Learning by Discovery: A Lab Manual for Calculus, *Anita E. Solow,* Editor.

28. Resources for Calculus Collection, Volume 2: Calculus Problems for a New Century, *Robert Fraga,* Editor.

29. Resources for Calculus Collection, Volume 3: Applications of Calculus, *Philip Straffin,* Editor.

30. Resources for Calculus Collection, Volume 4: Problems for Student Investigation, *Michael B. Jackson and John R. Ramsay,* Editors.

31. Resources for Calculus Collection, Volume 5: Readings for Calculus, *Underwood Dudley,* Editor.

32. Essays in Humanistic Mathematics, *Alvin White,* Editor.

33. Research Issues in Undergraduate Mathematics Learning: Preliminary Analyses and Results, *James J. Kaput and Ed Dubinsky,* Editors.

34. In Eves' Circles, *Joby Milo Anthony,* Editor.

35. You're the Professor, What Next? Ideas and Resources for Preparing College Teachers, *The Committee on Preparation for College Teaching, Bettye Anne Case,* Editor.

36. Preparing for a New Calculus: Conference Proceedings, *Anita E. Solow,* Editor.

37. A Practical Guide to Cooperative Learning in Collegiate Mathematics, \ *Nancy L. Hagelgans, Barbara E. Reynolds, SDS, Keith Schwingendorf, Draga Vidakovic, Ed Dubinsky, Mazen Shahin, G. Joseph Wimbish, Jr.*

38. Models That Work: Case Studies in Effective Undergraduate Mathematics Programs, *Alan C. Tucker,* Editor.

39. Calculus: The Dynamics of Change, *CUPM Subcommittee on Calculus Reform and the First Two Years, A. Wayne Roberts,* Editor.

These volumes can be ordered from:
MAA Service Center
P.O. Box 91112
Washington, DC 20090-1112
800-331-1MAA   FAX: 301-206-9789

This book is dedicated to the memory of **James R. C. Leitzel,** who chaired both the Committee on the Teaching of Undergraduate Mathematics (CTUM) (on which Bonnie Gold served) and the Committee on the Undergraduate Program in Mathematics (on which Bill Marion served), and who also co-authored an article with Sandy Keith. Jim dedicated his life to improving the teaching of undergraduate mathematics, and worked on it tirelessly, giving generously of his time to those he worked with, until his untimely death. We hope this book will continue his work.

Preface

This book grew out of a report by a subcommittee of the Committee on the Undergraduate Program in Mathematics, chaired by Bernie Madison, and out of a concern by Sandy Keith that mathematicians take assessment seriously. The introductions by Lynn Steen and Bernie Madison set the context more fully.

Acknowledgements

We would like to acknowledge the support of the institutions at which this book was edited, and of the staff who helped us with the details. In particular, at Wabash College (Bonnie Gold's former institution), Marcia Labbe wrote all the initial letters inviting contributions and the computer center staff, particularly Kelly Pfledderer and Carl Muma, helped with translating electronic versions of articles in many different formats into usable form. At Valparaiso University, Paula Scott, secretary of the Department of Mathematics and Computer Science, provided Bill Marion with significant word processing help in corresponding with the authors of the articles in the section on assessing the major, especially during the revision stage. We would also like to acknowledge the very helpful suggestions of the Notes subcommittee which worked with us, Nancy Baxter Hastings, Mary Parker, and Anita Solow.

Table of Contents

ASSESSING ASSESSMENT

Lynn Arthur Steen
St. Olaf College

We open letters from the city assessor with trepidation since we expect to learn that our taxes are about to go up. Mathematicians typically view academic assessment with similar emotion. Some react with indifference and apathy, others with suspicion and hostility. Virtually no one greets a request for assessment with enthusiasm. Assessment, it often seems, is the academic equivalent of death and taxes: an unavoidable waste.

In ancient times, an assessor (from *ad + sedere*) was one who sat beside the sovereign to provide technical advice on the value of things that were to be taxed. Only tax collectors welcomed assessors. Tradition, self-interest, and common sense compel faculty to resist assessment for many of the same reasons that citizens resist taxes.

Yet academic sovereigns (read: administrators) insist on assessment. For many reasons, both wise and foolish, administrators feel compelled to determine the value of things. Are students learning what they should? Do they receive the education they have been promised? Do our institutions serve well the needs of all students? Are parents and the public receiving value for their investment in education? Are educational programs well suited to the needs of students? Do program benefits justify costs? Academic sovereigns ask these questions not to impose taxes but to determine institutional priorities and allocate future resources.

What we assess defines what we value [22]. Students' irreverent questions ("Will it be on the test?") signal their understanding of this basic truth. They know, for example, that faculty who assess only calculation do not really value understanding. In this respect, mathematics faculty are not unlike their students: while giving lip service to higher goals, both faculty and students are generally satisfied with evidence of routine performance. Mathematics departments commonly claim to want their majors to be capable of solving real-world problems and communicating mathematically. Yet these goals ring hollow unless students are evaluated by their ability to identify and analyze problems in real-world settings and communicate their conclusions to a variety of audiences. Assessment not only places value on things, but also identifies the things we value.

In this era of accountability, the constituencies of educational assessment are not just students, faculty, and administrators, but also parents, legislators, journalists, and the public. For these broader audiences, simple numerical indicators of student performance take on totemic significance. Test acronyms (SAT, TIMSS, NAEP, AP, ACT, GRE) compete with academic subjects (mathematics, science, history) as the public vocabulary of educational discourse. Never mind that GPA is more a measure of student compliance than of useful knowledge, or that SAT scores reflect relatively narrow test-taking abilities. These national assessments have become, in the public eye, surrogate definitions of education. In today's assessment-saturated environment, mathematics *is* the mathematics that is tested.

College Mathematics

In most colleges and universities, mathematics is arguably the most critical academic program. Since students in a large majority of degree programs and majors are required to take (or test out of) courses in the mathematical sciences, on most campuses mathematics enrollments are among the highest of any subject. Yet for many reasons, the withdrawal and failure rates in mathematics courses are higher than in most other courses. The combination of large enrollments and high failure rates makes mathematics departments responsible for more student frustration — and dropout — than any other single department.

What's more, in most colleges and universities mathematics is the most elementary academic program. Despite mathematics' reputation as an advanced and esoteric subject, the average mathematics course offered by most postsecondary institutions is at the high-school level. Traditional postsecondary level mathematics — calculus and above — accounts for less than 30% of the 3.3 million mathematical science enrollments in American higher education [8].

Finally, in most colleges and universities, mathematics is the program that serves the most diverse student needs. In addition to satisfying ordinary obligations of providing courses for general education and for mathematics majors, departments of mathematical sciences are also responsible for developmental courses for students with weak mathematics backgrounds; for service courses for programs ranging from agriculture to engineering and from business to biochemistry; for the mathematical preparation of prospective teachers in elementary, middle, and secondary schools; for research experiences to prepare interested students for graduate programs in the mathematical sciences; and, in smaller institutions, for courses and majors in statistics, computer science, and operations research.

Thus the spotlight of educational improvement often falls first and brightest on mathematics. In the last ten years alone, new expectations have been advanced for school mathematics [13], for college mathematics below calculus [1], for calculus [3, 14, 16], for statistics [6], for undergraduate mathematics [17], for departmental goals [10] and for faculty rewards [7]. Collectively, these reports convey new values for mathematics education that focus departments more on student learning than on course coverage; more on student engagement than on faculty presentation; more on broad scholarship than on narrow research; more on context than on techniques; more on communication than on calculation. In short, these reports stress mathematics for all rather than mathematics for the few, or (to adopt the slogan of calculus reform) mathematics as "a pump, not a filter."

Principles of Assessment

Assessment serves many purposes. It is used, among other things, to diagnose student needs, to monitor student progress, to give students grades, to judge teaching effectiveness, to determine raises and promotions, to evaluate curricula and programs, and to decide on allocation of resources. During planning (of courses, programs, curricula, majors) assessment addresses the basic questions of why, who, what, how, and when. In the thick of things (in midcourse or mid-project) so-called formative assessment monitors implementation (is the plan going as expected?) and progress (are students advancing adequately?). At the summative stage — which may be at the end of a class period, or of a course, or of a special project — assessment seeks to record impact (both intended and unintended), to compare outcomes with goals, to rank students, and to stimulate action either to modify, extend, or replicate.

Several years ago a committee of the Mathematical Association of America undertook one of the very first efforts in higher education to comprehend the role of assessment in a single academic discipline [2, 9]. Although this committee focused on assessing the mathematics major, its findings and analyses apply to most forms of assessment. The committee's key finding is that assessment, broadly defined, must be a cyclic process of setting goals, selecting methods, gathering evidence, drawing inferences, taking action, and then reexamining goals and methods. Assessment is the feedback loop of education. As the system of thermostat, furnace, and radiators can heat a house, so a similar assessment system of planning, instruction, and evaluation can help faculty develop and provide effective instructional programs. Thus the first principle: *Assessment is not a single event, but a continuous cycle.*

The assessment cycle begins with goals. If you want heat, then you must measure temperature. On the other hand, if it is humidity that is needed, then a thermostat won't be of much use. Thus one of the benefits of an assessment program is that it fosters — indeed, necessitates — reflection on program and course goals. In his influential study of scholarship for the Carnegie Foundation, Ernest Boyer identified reflective critique as one of the key principles underlying assessment practices of students, faculty, programs, and higher education [5]. Indeed, unless linked to an effective process of reflection, assessment can easily become what many faculty fear: a waste of time and effort.

But what if the faculty want more heat and the students need more humidity? How do we find that out if we only measure the temperature? It is not uncommon for mathematics faculty to measure success in terms of the number of majors or the number of graduates who go to graduate school, while students, parents, and administrators may look more to the support mathematics provides for other subjects such as business and engineering. To ensure that goals are appropriate and that faculty expectations match those of others with stakes in the outcome, the assessment cycle must from the beginning involve many constituencies in helping set goals. Principle two: *Assessment must be an open process.*

Almost certainly, a goal-setting process that involves diverse constituencies will yield different and sometimes incompatible goals. It is important to recognize the value of this variety and not expect (much less force) too much uniformity. The individual backgrounds and needs of students make it clear that uniform objectives are not an important goal of mathematics assessment programs. Indeed, consensus does not necessarily yield strength if it masks important diversity of goals.

The purpose of assessment is to gather evidence in order to make improvements. If the temperature is too low, the thermostat turns on the heat. The attribution of cause (lack of heat) from evidence (low temperature) is one of the most important and most vexing aspects of assessment. Perhaps the cause of the drop in temperature is an open window or door, not lack of heat from the furnace. Perhaps the cause of students' inability to apply calculus in their economics courses is that they don't recognize it when the setting has changed, not that they have forgotten the repertoire of algorithms. The effectiveness of actions taken in response to evidence depends on the validity of inferences drawn about causes of observed effects. Yet in assessment, as in other events, the more distant the effect, the more difficult the attribution. Thus principle three: *Assessment must promote valid inferences.*

Compared to assessing the quality of education, taking the temperature of a home is trivial. Even though temperature does vary slightly from floor to ceiling and feels lower in moving air, it is fundamentally easy to measure. Temperature is one-dimensional, it changes slowly, and common measuring instruments are relatively accurate. None of this is true of mathematics. Mathematical performance is highly multidimensional and varies enormously from one context to another. Known means of measuring mathematical performance are relatively crude — either simple but misleading, or insightful but forbiddingly complex.

Objective tests, the favorite of politicians and parents, atomize knowledge and ignore the interrelatedness of concepts. Few questions on such tests address higher level thinking and contextual problem solving — the ostensible goals of education. Although authentic assessments that replicate real challenges are widely used to assess performance in music, athletics, and drama, they are rarely used to assess mathematics performance. To be sure, performance testing is expensive. But the deeper reason such tests are used less for formal assessment in mathematics is that they are perceived to be less objective and more subject to manipulation.

The quality of evidence in an assessment process is of fundamental importance to its value and credibility. The integrity of assessment data must be commensurate with the possible consequences of their use. For example, informal comments from students at the end of each class may help an instructor refine the next class, but such comments have no place in an evaluation process for tenure or promotion. Similarly, standardized diagnostic tests are helpful to advise students about appropriate courses, but are inappropriate if used to block access to career programs. There are very few generalizations about assessment that hold up under virtually all conditions but this fourth principle is one of them: *Assessment that matters should always employ multiple measures of performance.*

Mathematics assessment is of no value if it does not measure appropriate goals — the mathematics that is important for today and tomorrow [11, 12]. It needs to penetrate the common facade of thoughtless mastery and inert ideas. Rhetorical skill with borrowed ideas is not evidence of understanding, nor is facility with symbolic manipulation evidence of useful performance [21]. Assessment instruments in mathematics need to measure all standards, including those that call for higher order skills and contextual problem solving. Thus the content principle: *Assessment should measure what is worth learning, not just what is easy to measure.*

The goal of mathematics education is not to equip all students with identical mathematical tool kits but to amplify the multiplicity of student interests and forms of mathematical talent. As mathematical ability is diverse, so must be mathematics instruction and assessment. Any assessment must pass muster in terms of its impact on various subpopulations — not only for ethnic groups, women, and social classes, but also for students of different ages, aspirations (science, education, business) and educational backgrounds (recent or remote, weak or strong).

As the continuing national debate about the role of the SAT exam illustrates, the impact of high stakes assessments is a continuing source of deep anxiety and anger over issues of fairness and appropriate use. Exams whose items are psychometrically unbiased can nevertheless result in unbalanced impact because of the context in which they are given (e.g., to students of uneven preparation) or the way they are used (e.g., to award admissions or scholarships). Inappropriate use can and does amplify bias arising from other sources. Thus a final principle, perhaps the most important of all, echoing recommendations put forward by both the Mathematical Sciences Education Board [11] and the National Council of Teachers of Mathematics [12]: *Assessment should support every student's opportunity to learn important mathematics.*

Implementations of Assessment

In earlier times, mathematics assessment meant mostly written examinations — often just multiple choice tests. It still means just that for high-stakes school mathematics assessment (e.g., NAEP, SAT), although the public focus on standardized exams is much less visible (but not entirely absent) in higher education. A plethora of other methods, well illustrated in this volume, enhance the options for assessment of students and programs at the postsecondary level:

- *Capstone courses* that tie together different parts of mathematics;
- *Comprehensive exams* that examine advanced parts of a student's major;
- *Core exams* that cover what all mathematics majors have in common;
- *Diagnostics exams* that help identify students' strengths and weaknesses;

- *External examiners* who offer independent assessments of student work;
- *Employer advisors* to ensure compatibility of courses with business needs;
- *Feedback* from graduates concerning the benefits of their major program;
- *Focus groups* that help faculty identify patterns in student reactions;
- *Group projects* that engage student teams in complex tasks;
- *Individual projects* which lead to written papers or oral presentations;
- *Interviews* with students to elicit their beliefs, understandings, and concerns;
- *Journals* that reveal students' reactions to their mathematics studies;
- *Oral examinations* in which faculty can probe students' understanding;
- *Performance tasks* that require students to use mathematics in context;
- *Portfolios* in which students present examples of their best work;
- *Research projects* in which students employ methods from different courses;
- *Samples* of student work performed as part of regular course assignments;
- *Senior seminars* in which students take turns presenting advanced topics;
- *Senior theses* in which students prepare a substantial written paper in their major;
- *Surveys* of seniors to reveal how they feel about their studies;
- *Visiting committees* to periodically assess program strengths and weaknesses.

These multitude means of assessment provide options for many purposes — from student placement and grading to course revisions and program review. Tests and evaluations are central to instruction and inevitably shine a spotlight (or cast a shadow) on students' work. Broader assessments provide summative judgments about a students' major and about departmental (or institutional) effectiveness. Since assessments are often preludes to decisions, they not only monitor standards, but also set them.

Yet for many reasons, assessment systems often distort the reality they claim to reflect. Institutional policies and funding patterns often reward delaying tactics (e.g., by supporting late summative evaluation in preference to timely formative evaluation) or encourage a facade of accountability (e.g., by delegating assessment to individuals who bear no responsibility for instruction). Moreover, instructors or project directors often unwittingly disguise advocacy as assessment by slanting the selection of evaluation criteria. Even external evaluators often succumb to promotional pressure to produce overly favorable evaluations.

Other traps arise when the means of assessment do not reflect the intended ends. Follow-up ratings (e.g., course evaluations) measure primarily student satisfaction, not course effectiveness; statements of needs (from employers or client departments) measure primarily what people think they need, not what they really need; written examinations reveal primarily what students can do with well-posed problems, not whether they can use mathematics in external contexts. More than almost anything else a mathematician engages in, assessment provides virtually unlimited opportunities for meaningless numbers, self-delusion, and unsubstantiated inferences. Several reports [e.g., 15, 18, 19] offer informative maps for navigating these uncharted waters.

Assessment is sometimes said to be a search for footprints, for identifying marks that remain visible for some time [4]. Like detectives seeking evidence, assessors attempt to determine where evidence can be found, what marks were made, who made them, and how they were made. Impressions can be of varying depths, more or less visible, more or less lasting. They depend greatly on qualities of the surfaces on which they fall. Do these surfaces accept and preserve footprints? Few surfaces are as pristine as fresh sand at the beach; most real surfaces are scuffed and trammeled. Real programs rarely leave marks as distinguishing or as lasting as a fossil footprint.

Nevertheless, the metaphor of footprints is helpful in understanding the complexity of assessing program impact. What are the footprints left by calculus? They include cognitive and attitudinal changes in students enrolled in the class, but also impressions and reputations passed on to roommates, friends, and parents. They also include changes in faculty attitudes about student learning and in the attitudes of client disciplines towards mathematics requirements [20]. But how much of the calculus footprint is still visible two or three years later when a student enrolls in an economics or business course? How much, if any, of a student's analytic ability on the law boards can be traced to his or her calculus experience? How do students' experiences in calculus affect the interests or enthusiasm of younger students who are a year or two behind? The search for calculus footprints can range far and wide, and need not be limited to course grades or final exams.

In education as in industry, assessment is an essential tool for improving quality. The lesson learned by assessment pioneers and reflected in the activities described in this volume is that assessment must be broad, flexible, diverse, and suited to the task. Those responsible for assessment (faculty, department chairs, deans, and provosts) need to constantly keep several questions in the forefront of their analysis:

- Are the goals clear and is the assessment focused on these goals?
- Who has a voice in setting goals and in determining the nature of the assessment?

- Do the faculty ground assessment in relevant research from the professional literature?
- Have all outcomes been identified — including those that are indirect?
- Are the means of assessment likely to identify unintended outcomes?
- Is the mathematics assessed important for the students in the program?
- In what contexts and for which students is the program particularly effective?
- Does the assessment program support development of faculty leadership?
- How are the results of the assessment used for improving education?

Readers of this volume will find within its pages dozens of examples of assessment activities that work for particular purposes and in particular contexts. These examples can enrich the process of thoughtful, goal-oriented planning that is so important for effective assessment. No single system can fit all circumstances; each must be constructed to fit the unique goals and needs of particular programs. But all can be judged by the same criteria: an open process, beginning with goals, that measures and enhances students' mathematical performance; that draws valid inferences from multiple instruments; and that is used to improve instruction for all students.

References

[1]   American Mathematical Association of Two-Year Colleges. *Crossroads in Mathematics: Standards for Introductory College Mathematics Before Calculus.* American Mathematical Association of Two-Year Colleges, Memphis, TN, 1995.

[2]   Committee on the Undergraduate Program in Mathematics (CUPM). "Assessment of Student Learning for Improving the Undergraduate Major in Mathematics," *Focus: The Newsletter of the Mathematical Association of America*, 15 (3), June 1995, pp. 24–28, reprinted in this volume, pp. 279–284.

[3]   Douglas, R.G., ed. *Toward a Lean and Lively Calculus,* Mathematical Association of America, Washington DC, 1986.

[4]   Frechtling, J.A. *Footprints: Strategies for Non-Traditional Program Evaluation.*, National Science Foundation, Washington, DC, 1995.

[5]   Glassick, C.E., et. al. *Scholarship Assessed: Evaluation of the Professoriate*, Carnegie Foundation for the Advancement of Teaching, Jossey-Bass, San Francisco, CA, 1997.

[6]   Hoaglin, D.C. and Moore, D.S., eds. *Perspectives on Contemporary Statistics*, Mathematical Association of America, Washington, DC, 1992.

[7]   Joint Policy Board for Mathematics. *Recognition and Rewards in the Mathematical Sciences*, American Mathematical Society, Providence, RI, 1994.

[8]   Loftsgaarden, D.O., Rung, D.C., and Watkins, A.E. *Statistical Abstract of Undergraduate Programs in the Mathematical Sciences in the United States: Fall 1995 CBMS Survey*, Mathematical Association of America, Washington, DC, 1997.

[9]   Madison, B. "Assessment of Undergraduate Mathematics," in L.A. Steen, ed., *Heeding the Call for Change: Suggestions for Curricular Action*, Mathematical Association of America, Washington, DC, 1992, pp. 137-149.

[10] Mathematical Sciences Education Board. *Moving Beyond Myths: Revitalizing Undergraduate Mathematics,* National Research Council, Washington, DC, 1991.

[11] Mathematical Sciences Education Board. *Measuring What Counts: A Conceptual Guide for Mathematics Assessment,* National Research Council, Washington, DC, 1993.

[12] National Council of Teachers of Mathematics. *Assessment Standards for School Mathematics*, National Council of Teachers of Mathematics, Reston, VA. 1995.

[13] National Council of Teachers of Mathematics. *Curriculum and Evaluation Standards for School Mathematics*, National Council of Teachers of Mathematics, Reston, VA, 1989.

[14] Roberts, A.W., ed. *Calculus: The Dynamics of Change*, Mathematical Association of America, Washington, DC, 1996.

[15] Schoenfeld, A. *Student Assessment in Calculus*, Mathematical Association of America, Washington, DC, 1997.

[16] Steen, L.A., ed. *Calculus for a New Century: A Pump, Not a Filter*, Mathematical Association of America, Washington, DC, 1988.

[17] Steen, L.A., ed. *Reshaping College Mathematics*, Mathematical Association of America, Washington, DC, 1989.

[18] Stenmark, J.K., ed. *Mathematics Assessment: Myths, Models, Good Questions, and Practical Suggestions*, National Council of Teachers of Mathematics, Reston, VA, 1991.

[19] Stevens, F., et al. *User-Friendly Handbook for Project Evaluation*, National Science Foundation, Washington, DC, 1993.

[20] Tucker, A.C. and Leitzel, J.R.C. *Assessing Calculus Reform Efforts*, Mathematical Association of America, Washington, DC, 1995.

[21] Wiggins, G. "A True Test: Toward More Authentic and Equitable Assessment," *Phi Delta Kappan*, May 1989, pp. 703–713.

[22] Wiggins, G. "The Truth May Make You Free, but the Test May Keep You Imprisoned: Toward Assessment Worthy of the Liberal Arts," *The AAHE Assessment Forum*, 1990, pp. 17–31. (Reprinted in Steen, L.A., ed., *Heeding the Call for Change: Suggestions for Curricular Action*, Mathematical Association of America, Washington, DC, 1992, pp. 150–162.

Assessment and the MAA

Bernard L. Madison
University of Arkansas, Fayetteville

In August 1990 at the Joint Mathematics Meetings in Columbus, Ohio, the Subcommittee on Assessment of the Mathematical Association of America's (MAA) Committee on the Undergraduate Program in Mathematics (CUPM) held its organizational meeting. I was subcommittee chair and, like my colleagues who were members, knew little about the topic we were to address. The impetus for assessment of student learning was from outside our discipline, from accrediting agencies and governing boards, and even the vocabulary was alien to most of our mathematics community.

We considered ourselves challenging as teachers and as student evaluators. We used high standards and rigorous tests in our courses. What else could assessment be? We were suspicious of evaluating student learning through group work, student-compiled portfolios, and opinion surveys. And we were uncertain about evaluating programs and curricula using data about student learning gathered in these unfamiliar ways. Although many of us believed that testing stimulated learning, we were not prepared to integrate more complex assessment schemes into our courses, curricula, and other departmental activities.

We began learning about assessment. In its narrowest sense, our charge was to advise the MAA membership on the assessment of student learning in the mathematics major for the purpose of improving programs. Sorting out distinctions in the meaning of testing, student evaluations, program evaluations, assessment, and other recurring words and phrases was challenging, though easily agreed to by committee members accustomed to precision in meaning. We were to discover that assessment of student learning in a multi-course program was familiar to us, but not part of most of our departments' practices in undergraduate programs. In fact, departments and faculties had confronted a similar scenario in implementing a placement scheme for entering freshman students. The most comprehensive placement schemes used a variety of data about students capabilities in pre-college mathematics to place them in the appropriate college course. However, many placement schemes were influenced by the practices of traditional testing in undergraduate courses and relied on a single measurement tool, a test, often multiple-choice.

Another place where we had used a multi-faceted scheme for assessment of student learning was in our graduate programs, particularly the doctoral programs. Individual course grades are much less critical and meaningful in a doctoral program, where assessment of learning relies heavily on comprehensive examinations, interviews, presentations, and an unarticulated portfolio of interaction between the student and the graduate faculty. And, finally, there is the major capstone experience, the dissertation.

The subcommittee decided to draft a document that would outline what a program of assessment of student learning should be, namely a cycle of setting learning goals, designing instructional strategies, determining assessment methods, gathering data, and using the results to improve the major. Because of a lack of research-based information about how students learn mathematics, we decided not to try to address what specific tools measure what aspects of learning.

By 1993 we had a draft document circulating in the mathematics community asking for feedback. The draft presented a rather simple cyclical process with lists of options for learning goals, instructional strategies, and assessment tools. Some were disappointed that the draft did not address the more complex learning issues, while others failed to find the off-the-shelf scheme they thought they wanted. This search for an off-the-shelf product was similar to the circumstances in the 70s and 80s with placement schemes and, as then, reflected the lack of interest in ownership of these processes by many mathematics faculty members.

In spite of the suspicions and disinterest of many in our mathematics community, the discipline of mathematics was further along than most collegiate disciplines in addressing assessment of student learning. Through attendance at and participation in national conferences on assessment, we soon learned that we were not alone in our confusion and that the rhetoric about assessment was fuzzy, unusually redundant, and overly complicated.

The draft document, in general, received positive reviews, while eliciting minimal suggestions for improvement. Departmental faculties faced with a mandate of implementing an assessment program were generally pleased to have a simple skeletal blueprint. Feedback did improve the document which was published by the MAA in the June 1995 issue of *Focus*. [1]

After publication of the report to the MAA membership in 1995, the subcommittee had one unfinished piece of business, compiling and distributing descriptions of specific assessment program experiences by departmental faculties. We had held one contributed paper session at the 1994 Joint Mathematics Meetings and another was scheduled in January 1996, to be organized by subcommittee members Barbara Faires and Bill Marion.

As a result of the contributed paper sessions, we concluded that the experience within departments was limited, but by 1996, there were encouraging reports. At the 1996 meetings Bill Marion, Bonnie Gold, and Sandra Keith agreed to undertake the compilation the subcommittee had planned and to address a broader range of assessment issues in a volume aimed for publication by the MAA. This volume is the result.

The articles on assessing the major in the volume make at least three points. First, the experiences related here show that attitudes have changed over the past seven years about various instructional and assessment strategies. Group work, comprehensive exams, focus groups, capstone courses, surveys, and portfolios are now widely discussed and used. This change has not been due to the assessment movement alone. Reform in teaching and learning, most notably in calculus over the past decade, has promoted these non-traditional teaching and learning methods, and, as a result, has called for new assessment strategies. The confluence of pressures for change has made these experiences more common and less alien.

Second, the articles show that our experience is still too limited to allow documentation of significant changes in learning. As the articles indicate, the symptoms are encouraging, but not yet conclusive.

Third, the articles make it clear that developing an assessment program requires intense faculty involvement at the local level and a strong commitment by individual faculty leaders. The variety of experiences described in the section on assessing the major spans several non-traditional methodologies: portfolios, capstone courses, comprehensive examinations, focus groups, and surveys of graduates and employers. The variety enriches the volume considerably.

Some who read this volume may be disappointed by again not being led to a recipe or an off-the-shelf program of assessment. Only after one investigates the complexity of establishing a promising assessment program will one fully appreciate the work of those who are relating their experiences here and those who have compiled and edited this volume. This volume contributes significantly to an ongoing process of improving teaching and learning in collegiate mathematics. No doubt, someday it will be valuable only as a historical document as our experiences grow and we understand better how students learn and how to measure that learning more effectively.

But, for now, this is a valuable volume recounting well thought out experiences involving teachers and students engaged in learning together. The volume is probably just in time as well: too late for the experiences to be mistaken for off-the-shelf recipes, but in time to share ideas to make assessment programs better.

Reference

[1] Committee on the Undergraduate Program in Mathematics (CUPM). "Assessment of Student Learning for Improving the Undergraduate Major in Mathematics," *Focus: The Newsletter of the Mathematical Association of America*, 15 (3), June 1995, pp. 24–28. Reprinted on pp. 313 of this volume.

How to Use This Book

Lynn Steen has set the stage for this book by describing, in the preface, the pressures that bring many mathematicians to seek help with assessment, as well as some of the resistance to assessment that many of us in mathematics may feel. "Assessment" to many is mere jargon; it's a code word for more administrative red tape. Some fear that it will be used to push us to lower our expectations of our students, to lower our "standards." We hope our book will help allay some of these fears, and will go beyond that to encourage you, the reader, to view assessment as a natural part of the process of creating a positive and exciting learning environment, rather than as a duty inflicted from outside.

Many mathematicians are already taking seriously the challenge of finding better methods of assessment. What we offer here is essentially an album of techniques, from a wide assortment of faculty across the nation. We, as editors of this collection, are amazed at the energy of the individuals and schools represented here, all of whom care deeply about learning. We delight in their successes, insights, inventiveness, and the sheer combined diversity of their efforts, and we appreciate the frankness with which they have been willing to share the down-side of what they've tried. Their schools are in the midst of experiencing change. These are contributors who are courageous in experimenting with assessment and generous in their willingness to share with you ideas which are still in the process of development. Yet this book is not premature: good assessment is a cyclic process, and is never "finished."

Take a look at the table of contents. This is a book for browsing, not to be approached as a novel; nor is it merely an encyclopedia for reference. Skim over the topics until you find one that appeals to you — and we guarantee you will find something here that intrigues. Perhaps it will be something, such as capstone courses, that you have wondered about, or an in-class writing assignment that you have tried

yourself. Who you are will affect how you approach this book. The chair of a department in a liberal arts college, faced with developing a departmental assessment plan, may turn first to the section on assessing the major, and be curious about the assessment of general education courses or placement programs. The chair of a department at a comprehensive university might view the assessment of a department's role on campus as most important. An individual faculty member, on the other hand, might find immediately useful the section on classroom assesssment. We hope, above all, to draw you into seeing that assessment is conducted in many different ways, and that it pertains to almost everything we do as teachers.

To make the book easy to use, each article follows the same basic format. At the top of each article, there is a brief description, to help you decide whether it is the article you were expecting from its title. The article begins with a "Background and Purpose" section, which gives the context (type of institution, etc.) of the assessment method being described, and some history of the author's use of the technique. "Method" describes the technique in sufficient detail, we hope, for you to begin picturing how you might adapt it to your situation. "Findings" and "Use of Findings" describe the results the author has experienced, and how these can be used to improve student learning. At the end of each article, "Success Factors" cautions the readers on potential pitfalls, or explains what makes the method tick.

Each author writes from his or her personal perspective. What our authors present, then, may not translate instantly from one institution to another. However, our hope is that with the sheer diversity of institutions represented (ranging from 2-year colleges, to research universities) you will find something that attracts your interest or inspires you. Make it a personal book. Because of the wide variety of perspectives and philosophies of our authors, some of the ideas presented here you will take delight in, and others you will dislike. As editors, we found ideas which we wanted to try immediately in our own institutions, as did the MAA Notes Committee as they reviewed the manuscript. However, after reading an article which interests you, ask yourself: "Since the author's students are stronger/ weaker/harder working/ lazier/ more advanced/ less sophisticated than mine: how can I make this work for *my* class? Exactly what questions should I ask? How will I collect and analyze the data? How will I use the results?" You may even decide to contact the author for help as questions arise; authors' addresses are at the end of the book. Then try the activity on yourself. What is your response? You may also want to try it, if appropriate, on a colleague, to get feedback on how the questions are heard by someone else. Be prepared, however, for very different responses from the students, as they bring their own level of sophistication to the process. And be sure to let students know that the purpose of the activity is to improve what they get out of the program, and that their input is essential to this process. Then, after using the technique once, revise it. Or look for another technique which better focuses on the question you're concerned with.

Keep in mind that assessment must be cyclic. The cycle includes deciding what the goals of the activity are, figuring out what methods will allow you to reach those goals, deciding on a method of assessment which will let you know how well you have met the goals, administering the assessment instrument, compiling the data, and reporting the results. Then you return to the beginning: deciding whether the goals need revision given your new understanding of what is happening, what revision of methods will be needed, etc. Throughout the process, students should be kept informed of your reasons for doing the assessment, and of the findings and how they will be used. Of course, unless it's a completely new activity being undertaken, you will probably start somewhere in mid-cycle—you're already teaching this course, or offering this major, and had some goals in mind when you started, though these may have receded into the mists of the past. However, keeping the assessment cycle in mind will improve progress toward the ultimate goal, improving student learning.

We have tried both to offer as wide a variety of assessment methods as possible, and to show their interrelation, so that it is easy to move from one techinque to another. We've done this in several ways. In this introductory material, there is a chart of which topics occur in which articles, so that if, for example, you're looking for assessment via portfolios, you can look down the chart to see which articles describe their use. In the introduction to each section of the book, we've tried to show how the various articles in the section are related to each other. Finally, at the end of the book is a list of other books you may want to look at, which aren't specific to mathematics but which offer ideas that can be adapted to mathematics programs.

We must warn you what our book cannot achieve for you. Although some individual pieces may give you the flavor of research in several directions, this is not a collection of research articles in assessment in undergraduate mathematics. Nor is this book a collection of recipes ready-made to use. There is no one overarching theoretical view driving these articles, although most do assume that students learn better when active rather than passive. But all authors are concerned with improving student learning.

As the mathematical community as a whole has only recently begun to think seriously about assessment, we recognize that this book is only a beginning. We hope that others will take up ideas offered here, and look into them in greater depth. We offer you not only a compendium of ideas, but a source of individuals with whom you might correspond to learn more or develop collaborations. We hope that in several years, there will be a greater variety of methods, many of which will have gone through several full assessment cycles. Once this has happened, a more definitive volume can be written. But the book we have brought together here could not afford to come any later; at this stage, information and inspiration are needed above all.

Bonnie Gold, Monmouth University
Sandra Z. Keith, St. Cloud State University
William A. Marion, Valparaiso University

ARTICLES ARRANGED BY TOPIC

Course Evaluation

Developmental Courses

Exams

Exit Interviews, Focus Groups, Surveys

General Studies Courses

Institution-Wide Changes

Large-Class Methods

Peer Visitation

TEACHING GOALS INVENTORY, SELF-SCORABLE VERSION

The Teaching Goals Inventory below, is reprinted from *Classroom Assessment Techniques,* by Angelo and Cross. We suggest that you, the reader, take this inventory (with a specific course in mind, as the authors suggest) in part for your own information, and also because it can help you locate your own teaching styles and priorities in the context of this book.

Similar inventories can be found at a variety of web-sites; some colleges seem to have goals inventories such as this for students as part of their guidance policies. In Angelo and Cross, compiled results from this inventory from community and four-year colleges are offered as well; these can be useful for study and comparison. This inventory can also be effective in committee meetings (as, for example, when faculty are restructuring a course or curriculum), or to initiate a workshop session. It can also be used at the beginning and end of a general education course, as a method for observing changes in the values and goals of students. Even better than taking the inventory, perhaps, would be to design a comparable inventory of your own.

Purpose: The Teaching Goals Inventory (TGI) is a self-assessment of instructional goals. Its purpose is threefold: (1) to help college teachers become more aware of what they want to accomplish in individual courses; (2) to help faculty locate Classroom Assessment Techniques they can adapt and use to assess how well they are achieving their teaching and learning goals; and (3) to provide a starting point for discussion of teaching and learning goals among colleagues.

Directions: Please select ONE course you are currently teaching. Respond to each item on the inventory in relation to that particular course. (Your response might be quite different if you were asked about your overall teaching and learning goals, for example, or the appropriate instructional goals for your discipline.)

Please print the title of the specific course you are focusing on:

_____

Please rate the importance of each of the fifty-two goals listed below to the specific course you have selected. Assess each goal's importance to what you deliberately aim to have your students accomplish, rather than the goal's general worthiness or overall importance to your institution's mission. There are no "right" or "wrong" answers; only personally more or less accurate ones.

For each goal, choose only one response on the 1-to-5 rating scale. You may want to read quickly through all fifty-two goals before rating their relative importance.

In relation to the course you are focusing on, indicate whether each goal you rate is:

(5)	Essential	a goal you always/nearly always try to achieve
(4)	Very Important	a goal you often try to achieve
(3)	Important	a goal you sometimes try to achieve
(2)	Unimportant	a goal you rarely try to achieve
(1)	Not applicable	a goal you never try to achieve

Rate the importance of each goal to what you aim to have students accomplish in your course.

1. Develop ability to apply principles and generalizations already learned
   to new problems and situations   5 4 3 2 1
2. Develop analytic skills   5 4 3 2 1
3. Develop problem-solving skills   5 4 3 2 1
4. Develop ability to draw reasonable inferences from observations   5 4 3 2 1
5. Develop ability to synthesize and integrate information and ideas   5 4 3 2 1
6. Develop ability to think holistically: to see the whole as well as the parts   5 4 3 2 1
7. Develop ability to think creatively   5 4 3 2 1
8. Develop ability to distinguish between fact and opinion   5 4 3 2 1

9. Improve skill at paying attention	5	4	3	2	1
10. Develop ability to concentrate	5	4	3	2	1
11. Improve memory skills	5	4	3	2	1
12. Improve listening skills	5	4	3	2	1
13. Improve speaking skills	5	4	3	2	1
14. Improve reading skills	5	4	3	2	1
15. Improve writing skills	5	4	3	2	1
16. Develop appropriate study skills, strategies, and habits	5	4	3	2	1
17. Improve mathematical skills	5	4	3	2	1
18. Learn terms and facts of this subject	5	4	3	2	1
19. Learn concepts and theories in this subject	5	4	3	2	1
20. Develop skill in using materials, tools, and/or technology central to this subject	5	4	3	2	1
21. Learn to understand perspectives and values of this subject	5	4	3	2	1
22. Prepare for transfer or graduate study	5	4	3	2	1
23. Learn techniques and methods used to gain new knowledge in this subject	5	4	3	2	1
24. Learn to evaluate methods and materials in this subject	5	4	3	2	1
25. Learn to appreciate important contributions to this subject	5	4	3	2	1
26. Develop an appreciation of the liberal arts and sciences	5	4	3	2	1
27. Develop an openness to new ideas	5	4	3	2	1
28. Develop an informed concern about contemporary social issues	5	4	3	2	1
29. Develop a commitment to exercise the rights and responsibilities of citizenship	5	4	3	2	1
30. Develop a lifelong love of learning	5	4	3	2	1
31. Develop aesthetic appreciation	5	4	3	2	1
32. Develop an informed historical perspective	5	4	3	2	1
33. Develop an informed understanding of the role of science and technology	5	4	3	2	1
34. Develop an informed appreciation of other cultures	5	4	3	2	1
35. Develop capacity to make informed ethical choices	5	4	3	2	1
36. Develop ability to work productively with others	5	4	3	2	1
37. Develop management skills	5	4	3	2	1
38. Develop leadership skills	5	4	3	2	1
39. Develop a commitment to accurate work	5	4	3	2	1
40. Improve ability to follow directions, instructions, and plans	5	4	3	2	1
41. Improve ability to organize and use time effectively	5	4	3	2	1
42. Develop a commitment to personal achievement	5	4	3	2	1
43. Develop ability to perform skillfully	5	4	3	2	1
44. Cultivate a sense of responsibility for one's own behavior	5	4	3	2	1
45. Improve self-esteem/self-confidence	5	4	3	2	1
46. Develop a commitment to one's own values	5	4	3	2	1
47. Develop respect for one's own values	5	4	3	2	1
48. Cultivate emotional health and well-being	5	4	3	2	1
49. Cultivate an active commitment to honesty	5	4	3	2	1
50. Develop capacity to think for one's self	5	4	3	2	1
51. Develop capacity to make wise decisions	5	4	3	2	1

52. In general, how do you see your primary role as a teacher? (Although more than one statement may apply, please choose only one.)

1 Teaching students facts and principles of the subject matter
2 Providing a role model for students
3 Helping students develop higher-order thinking skills
4 Preparing students for jobs/careers
5 Fostering student development and personal growth
6 Helping students develop basic learning skills

Source: Classroom Assessment Techniques, by Thomas A. Angelo and K. Patricia Cross. Copyright© 1993. Used by permission. Publisher, Jossey-Bass, San Francisco, California.

Part I
Assessing the Major

Introduction

William A. Marion
Valparaiso University

In his introduction to this volume, Bernie Madison has described the process by which the MAA in the early 1990s became involved in the issue of assessing the mathematics major. Out of that effort came the document "Assessment of Student Learning for Improving the Undergraduate Major in Mathematics." [1] (You will find it reprinted on pages 279–284.) At about the time this document was published by the MAA (1995), a relatively small number of undergraduate mathematics programs across the country were faced with developing an "assessment plan" and the hope was that some of them would find the MAA's work helpful as a map to guide them in these uncharted waters. By this time (1999), the reality is that almost all undergraduate programs will have to have some type of assessment plan in place within the next couple of years. The hope is that what you will read in the next few pages will be helpful to you as you face this brave new world.

We, the editors, have chosen a representative sample of assessment programs from among those college and university mathematical sciences departments which have gotten into the game early. The sample contains articles from mathematics faculty at small schools, medium-size schools and large schools; at liberal arts colleges, regional comprehensive universities and major research institutions.

These departments are using a variety of methods to assess their programs, e.g., capstone courses, comprehensive exams, diagnostic projects, focus groups, student portfolios and surveys of various types. However, these techniques are not described in isolation; they are to be understood within the context of the assessment cycle. As you read these articles, you will see that each of the authors has provided that context. What follows is a brief description of each author's paper.

In the first two articles **student portfolios** are the principal assessment methods. Laurie Hopkins describes the as-sessment process at Columbia College in South Carolina. In this program certain courses are designated as portfolio development courses, and in these courses the student is asked to reflect in writing on the connection between the material included in the portfolio and the departmental goals. Linda Sons describes assessment portfolios at Northern Illinois University, which contain students' work from various points in their mathematics program and are examined by a department committee after the students have graduated.

Capstone courses as assessment techniques are discussed in two articles, one by Charles Peltier at St. Mary's College in Indiana and the other by Deborah Frantz of Kutztown University in Pennsylvania. Peltier describes a full-year senior seminar in which independent study projects (written and oral presentations) are developed as part of the seminar. Frantz provides us with the details of a one-semester Senior Seminar in Mathematics in which oral and written presentations are just two among a number of assessment techniques used.

Bonnie Gold and Dan Callon present us with different models for using **comprehensive exams** as assessment instruments. Gold describes a two-part exam consisting of a written component in the mathematics major and an oral component over the liberal arts. Callon discusses a rather unique approach to giving a comprehensive: a one-week, joint, written exam given to seniors during their fall semester. John Emert and Charles Parish discuss a "less" comprehensive exam developed to assess the core courses taken by all undergraduate mathematics majors.

Next we are introduced to an entirely different approach to assessing the mathematics major: the use of **focus groups**. Marie Sheckels gives us some insight as to how focus groups with graduating seniors can be incorporated into an assessment process.

The next two articles describe **overall departmental assessment plans** involving a variety of techniques to assess the major. Dick Groeneveld and Robert Stephenson describe the assessment measures used in their statistics program, particularly the use of grades of graduates, surveys of graduates and surveys of employers of graduates. Janice Walker also discusses all of her mathematics department's assessment techniques including the use of exit interviews and the Educational Testing Service's Major Field Test in Mathematics.

Three of the articles deal with assessing the mathematics program either at the **midway point** of a student's four-year career or at the freshmen level. Mark Michael discusses a semester-long Diagnostic Project which is part of a required Discrete Mathematics course, usually taken by mathematics majors in their sophomore or junior year. Judith Palagallo and William Blue of Akron University in Ohio describe a sophomore-level Fundamentals of Advanced Mathematics course and what information it reveals about the major early on. Elias Toubassi introduces us to a major effort at a large university to reform the entry-level mathematics courses through assessment and its subsequent effect on the mathematics major.

Finally, Deborah Bergstrand discusses assessing the major from the point of view of students who are **potential graduate students** in mathematics. Three measures — an Honors Project, a Senior Colloquium and Summer Undergraduate Research project — are used.

This section has been organized so that you can look at every article or pick and choose those which are of most interest to you. In some sense each of our undergraduate mathematics programs is unique and yet there is enough commonality so that you might be able to adapt at least one of the assessment processes described here to fit your own situation.

Reference

[1] Committee on the Undergraduate Program in Mathematics (CUPM). "Assessment of Student Learning for Improving the Undergraduate Major in Mathematics," *Focus: The Newsletter of the Mathematical Association of America*, 15 (3), June 1995, pp. 24–28. Reprinted on pp. 279–284 of this volume.

Assessing a Major in Mathematics

Laurie Hopkins
Columbia College

At a small, private women's liberal arts college in the South student portfolios have become the principal means for assessing the major. Unique to this program, certain courses are designated as portfolio development courses and in these courses the student is asked to reflect in writing on the connection between the material included in the portfolio and the department goals.

Background and Purpose

Columbia College is a small private women's college in Columbia, South Carolina, founded by the United Methodist Church. Although the college has a small graduate program in education, by far the major focus of the institution is on excellence in teaching undergraduates. The college is a liberal arts institution, but we do offer business and education majors. The number of students who major in elementary education, early childhood education, and special education form a significant percentage of the total student body. Also, many of the more traditional liberal arts majors offer secondary certifications in addition to major requirements. In mathematics, the majority of our majors also choose to certify to teach. The large commitment to pre-service teachers reinforces the institutional commitment to excellence in the classroom.

Our mathematics department has been experimenting with assessment of the major for improvement of student learning for about five years. Initially, the primary impetus for assessment was pleasing our accrediting body. As the college has embraced the strategic planning process, more and more departments are using planning and assessment for improvement purposes. Our first attempts centered around efforts to locate a standardized instrument. This search was unsuccessful for several reasons. First, the particular set of courses which our majors take was not represented on any test we examined. Moreover, without the ex-

pense and inconvenience of a pre-test and post-test administration, we felt that the information from such a measure would tell us more about ability than learning. Finally, if we did the dual testing we would only obtain answers to questions about content knowledge in various mathematical areas. Although there would clearly be value in this data, many unanswered questions would remain. In fact, many of the kinds of classroom experiences which we think are most valuable are not easily evaluated. Consequently, we chose student portfolios as the best type of assessment for our major. It seemed to us that this particular technique had value in appraising outcomes in mathematical communication, technological competence, and collaborative skills that are not measurable through standardized tests.

To assess the major, we first split our departmental goals into two sections, a general section for the department and a specific section for majors. We articulated one goal for math majors, with nine objectives. For each of the objectives we looked for indicators that these objectives had been met by asking the question "How would we recognize success in this objective?" Then we used these indicators as the assessment criteria. Some of our objectives are quite generic and would be appropriate for many mathematics departments, while others are unique to our program.

The goal, objectives, and assessment criteria follow.

Goal: The Mathematics Department of Columbia College is committed to providing mathematics majors the coursework and experiences which will enable them to attend graduate school, teach at the secondary level or enter a mathematics related career in business, industry or government. It is expected that each mathematics major will be able to:

1. Demonstrate the ability to write mathematical proofs.
   Assessment: Each student portfolio will contain at least four examples of proofs she has written in different classes.
2. Demonstrate the ability to read and evaluate a statistical study.
   Assessment: Each student portfolio will contain an evaluation of a statistical study that she has completed.
3. Demonstrate knowledge of women in mathematics.
   Assessment: Each student portfolio will contain evidence of knowledge of the contributions of women to mathematics.
4. Demonstrate topics in calculus from various perspectives — numerical, graphical, analytical and verbal.
   Assessment: Each student portfolio will contain at least one topic from calculus demonstrated from the numerical. graphical, analytical and verbal perspectives.
5. Use mathematics as a problem solving tool in real life situations.
   Assessment: Each student portfolio will contain at least four examples of her use of mathematics in solving a real life problem.
6. Work collaboratively to solve problems.
   Assessment: (i) Each student portfolio will contain at least four examples of problems that the student has solved collaboratively and a reflective statement about the collaborative process.
   (ii) 75% of course syllabi in upper level courses will indicate learning activities requiring collaborative problem-solving and/or presentation of in-depth problem solving to the class.
   (iii) A team of students selected and trained by the mathematics faculty will participate in the nationwide COMAP contest.
7. Organize, connect and communicate mathematical ideas effectively.
   Assessment: Each student portfolio will contain four examples with evidence of mathematical communication, either in written or spoken form.
8. Use technology appropriately to explore mathematics.
   Assessment: Each student portfolio will contain at least four examples of mathematical exploration using technology.
9. Feel adequately prepared for their post-graduation experience.
   Assessment: 90% of math major graduates who respond to a graduate survey will indicate that they felt mathematically prepared for the job they hold.

Method

Our initial effort to require the portfolio for each student was an informal one. We relied on a thorough orientation to the process of creating a portfolio, complete with checklists, required entries and cover sheets for each entry on which the student would write reflectively about the choice of the entry. We included this orientation in the advisement process of each student and stressed the value of the final collection to the student and to the department. Faculty agreed to emphasize the importance of collecting products for the portfolio in classes and to alert students to products which might make good portfolio entries.

We now designate a series of courses required for all majors as portfolio development courses. In these courses products are recommended as potential candidates for inclusion in the collection. Partially completed portfolios are submitted and graded as part of the courses. Perhaps most importantly, some class time in these courses is devoted to the reflective writing about ways in which specific products reflect the achievement of specific goals. Completing this writing in the context of classroom experience has had a very positive impact on the depth of the reflective work.

Findings

Our early efforts were a complete failure. No portfolios were produced. We probably should have realized that the portfolios would not actually be created under the informal guidelines, but in our enthusiasm for the process we failed to acknowledge two basic realities of student life: if it is not required, it will not be done and, equally important, if it is not graded, it will not be done well. After two years during which no student portfolios were ever actually collected, we made submission of the completed portfolio a requirement for passing the mandatory senior capstone course. The first semester this rule was in place, we did indeed collect complete portfolios from all seniors. However, it was obvious that these portfolios were the result of last-minute scrounging rather than the culmination of four years of collecting and documenting mathematical learning.

The second major problem was related to the first. Although the department was unanimous in its endorsement of the portfolio in concept, the faculty found it difficult to remember to include discussions related to the portfolio in advisement sessions or in classes. Incorporating the portfolio into specific mathematics classes was an attempt to remedy both situations.

The most striking problem surfaced when we finally collected thoughtfully completed portfolios and began to try to read them to judge the success of our program. We had never articulated our goals for math majors. The department had spent several months and many meetings formulating departmental goals, but the references to student behaviors were too general to be useful. The two relevant goals were "Continue to provide a challenging and thorough curriculum for math majors" and "Encourage the development of student portfolios to document mathematical development." In our inexperience we assumed that weaknesses in the major would be apparent as we perused each student's collection of "best work." Unfortunately, the reality was quite different. In the absence of stated expectations for this work, no conclusions could be reached. A careful revision of the departmental goals for a mathematics major was crafted in response to this finding. The resulting document is included in the first section. Examination of the portfolios pointed to a more difficult potential problem. In our small department, each student is so well known that it is difficult to inject any objectivity into program evaluation based on particular student work. We were forced to acknowledge that an objective instrument of some sort would be necessary to balance the subjective findings in the portfolios.

Use of Findings

One of the most important measures of the usefulness of a departmental strategy for assessment is that information gained can cause changes in the major, the department or the process. The changes made in the process include a revision of the goals and objectives, the addition of portfolio development in several required classes and the use of a post-graduation survey to determine graduates' perspective on the adequacy of their mathematical preparation for their post-graduation experiences. However, with the modified process now in place we have been able to recognize other changes that need to be made.

Examination of statistical entries in the portfolio indicated that there was not a sufficient depth of student understanding in this area. As a result we have split the single course which our students took in probability and statistics into a required series of two courses in an effort to cover the related topics in greater depth and to spend more time on the applications.

One graduate expressed her lack of confidence in entering graduate school in mathematics on the questionnaire. We have consequently added a course in Advanced Calculus to our offerings in an effort to help the students in our program who are considering graduate school as an option. Comments by several graduates that they did not understand their career options well have caused us to add a special section of the freshman orientation section strictly for mathematics majors. This course incorporates new majors into the department as well as the school and has several sessions designed to help them understand the options that mathematics majors have at our college and in their careers.

Success Factors

The assessment process is not a static one. Not only should change come about as a result of specific assessment outcomes, the process itself is always subject to change. The first series of changes made in our plan were the result of deficiencies in the plan. The next changes will be attempts to gain more information and to continue to improve our practices. We have recently added objective instruments to the data we are collecting. The instruments are a test to measure each student's ability in formal operations and abstract logical thought and a test to determine the stage of intellectual development of each student. We have given these measures to each freshman math major for the past two years and plan to administer them again in the capstone course. After an initial evaluation of the first complete data set, we will probably add an objective about intellectual development to our plan with these instruments as the potential means for assessment. We have recently begun another search for a content test to add to the objective data gathered at the beginning and the end of a student's college experience. There are many more of these types of measures now available and we hope to be able to identify one that closely matches our curriculum. With the consideration of a standardized content exam as an assessment tool, there is a sense in which we have come full circle. However, the spiral imagery is a better analogy. As we continue to refine and improve the process, standardized tests are being examined from a different perspective. Rather than abdicating our assessment obligations to a standardized test, we have now reached the point where the information that can be provided by that kind of tool has a useful position in our assessment strategy.

Portfolio Assessment of the Major

Linda R. Sons
Northern Illinois University

A Midwestern, comprehensive university which has five different programs in the mathematical sciences — General, Applied Mathematics, Computational Mathematics, Probability and Statistics, and Mathematics Education — requires students to maintain assessment portfolios in courses which are common to all five of the emphases. The portfolios of those who have graduated during the year are examined by a department assessment committee shortly after the close of the spring semester.

Background and Purpose

Evaluating its degree programs is not a new activity for the Department of Mathematical Sciences at Northern Illinois University. The Illinois Board of Higher Education mandates periodic program reviews, while the Illinois Board of Education, NCATE, and the North Central Association also require periodic reviews. The University regularly surveys graduates concerning their employment, their satisfaction with the major program completed, and their view of the impact the University's general education program had on their learning. The Department also polls graduates concerning aspects of the undergraduate degree program. However, in 1992 in response to an upcoming accreditation review by the North Central Association the Department developed a new assessment scheme for its undergraduate major program—one which is longitudinal in character and focuses on how the program's learning goals are being met.

The B.S. in Mathematical Sciences at Northern Illinois University is one of 61 bachelor's degree programs the University offers. As a comprehensive public institution, the University serves nearly 24,000 students of which about 17,000 are undergraduates who come geographically mostly from the top third of the State of Illinois. Located 65 miles straight west of Chicago's loop, Northern Illinois University accepts many transfer students from area community colleges. For admission to NIU as a native student an applicant is expected to have three years of college preparatory high school mathematics/computer science. Annually about 50 students obtain the B.S. in Mathematical Sciences of which about half are native students. Each major completes at least one of five possible emphases: General, Applied Mathematics, Computational Mathematics, Probability and Statistics, and Mathematics Education. Ordinarily, about one-half of the majors are in Mathematics Education, about one-quarter are in Probability and Statistics, and the remaining quarter is divided pretty evenly among the other three emphases.

Regardless of the track a student major may choose, the Department's baccalaureate program seeks to develop at least five capabilities in the student. During the Spring of 1992 our weekly meeting of the College Teaching Seminar (a brown-bag luncheon discussion group) hammered out a formulation of these capabilities which were subsequently adopted by the Department's Undergraduate Studies Committee as the basis for the new assessment program. These capabilities are:

1. To engage effectively and efficiently in problem solving;
2. To reason rigorously in mathematical arguments;
3. To understand and appreciate connections among different areas of mathematics and with other disciplines;
4. To communicate mathematics clearly in ways appropriate to career goals;
5. To think creatively at a level commensurate with career goals.

A discussion of a mechanism which could be used to gauge student progress towards the acquisition of these goals led to the requirement of an Assessment Portfolio for each student. The Portfolio would contain work of the student from various points in the program and would be examined at the time of the student's graduation.

Method

Common to all of the emphases for the major are the lower division courses in the calculus sequence, a linear algebra course, and a computer programming course. At the junior level, each emphasis requires a course in probability and statistics and one in model building in applied mathematics, while at the senior level each student must take an advanced calculus course (which is an introduction to real analysis). Each emphasis requires additional courses involving the construction of rigorous mathematical proofs and at least one sequence of courses at the upper division intended to accomplish depth.

The choices made for material for the assessment portfolio were:

a) the student's final examination from the required lower division linear algebra course (for information related to goals 1 and 2);

b) the best two projects completed by the student in the model building course—the course requires 5 projects for each of which a report must be written up like a technical report (for information related to goals 1, 3, and 4)

c) two homework assignments (graded by the instructor), one from about midsemester and one from late in the semester from the advanced calculus course (for information related to goals 1, 2, 4, and 5)

d) the best two homework assignments from two other advanced courses determined according to the student's emphasis, for example, those in the probability and statistics emphasis are to present assignments from a theoretical statistical inference course and a methods course, while those in the general emphasis present work from the advanced calculus II course and from a second senior-level algebra course (for information related to all five goals);

e) a 250–300 word typed essay discussing the student's experience in the major emphasizing the connections of mathematics with other disciplines (for information related to goals 3 and 4).

Items a)–d) are all collected by instructors in the individual courses involved and placed in files held in a Department Office by office personnel. The student is responsible for the essay in e) and will not be cleared for graduation until the essay is written and collected by the Department's Director of Undergraduate Studies. Thus, all items to be in the portfolio are determined by the faculty, but the student's essay may offer explanation for the portfolio items. The student's graduation is contingent upon having the portfolio essay submitted, but otherwise the portfolio has no "control" in the graduation process. Students are told that the collection of the items for the portfolio is for the purpose of evaluating the degree program.

After the close of the spring semester each year, a committee of senior faculty examines the assessment portfolios of the students who have graduated in the just completed year and is expected to rate each capability as seen in the portfolio in one of three categories—strong, acceptable, and weak. After all portfolios have been examined, the committee may determine strengths and weaknesses as seen across the portfolios and make recommendations for future monitoring of aspects of the program, or for changes in the program, or for changes in the assessment procedures. These are carried to the Department's Undergraduate Studies Committee for consideration and action.

Findings

The implementation of this new assessment scheme has resulted in the third review of portfolios being conducted in May–June 1997. Given the time lag inherent in the collection of the portfolios, the initial review committees have not been able to work with complete portfolios. However, these year-end committees made some useful observations about the program and the process. They discovered:

1. that the linear algebra course appeared to be uneven as it was taught across the department — on the department-wide final examinations, student papers showed concentrated attention in inconsistent ways to certain aspects of the examination, rather than a broad response to the complete examination;

2. that some students who showed good capability in mathematical reasoning on the linear algebra final examination experienced a decline in performance on homework assignments completed in the upper division courses emphasizing rigorous mathematical reasoning;

3. that evaluating capabilities was difficult without having a description of those capabilities in terms of characteristics of student work;

4. that student essays were well written (a pleasant surprise!), but the understanding expressed concerning connections between mathematics and other disciplines was meager;

5. that there appeared to be a positive influence on course-work performance when courses were sequenced in certain orders (e.g., senior-level abstract algebra taken a semester before the advanced calculus).

Use of Findings

In response to the discoveries made by the year-end committees, the Department took several actions.

First, to address the concerns related to the linear algebra course, the coordinators for the course were asked to clarify for instructors that the course was intended not only to enable students to acquire some computational facility with concepts in elementary linear algebra, but also to provide students with a means to proceed with the gradual development of their capacity in rigorous mathematical reasoning. This meant that some faculty needed to teach the course with less concentration on involvement with numerical linear algebra aspects of the course and greater concentration on involvement with students' use of precise language and construction of "baby" proofs. Further, a lower division course on mathematical reasoning was introduced as an elective course. This elective course which enables students to study in mathematical settings logical statements, logical arguments, and strategies of proof can be recommended to those students who show a weak performance in the elementary linear algebra course or who have difficulty with the first upper division course they take which emphasizeds proof construction. It can, of course, also be taken by others who simply wished to take such a course.

A second area of response taken by the Department was to ease the execution of the assessment process by introducing a set of checklists. For each capability a checklist of characteristics of student work which that capability involved were defined. For instance, under problem solving is listed: student understands the problem (with minor errors? gross errors?), approach is reasonable or appropriate, approach is successful, varied strategies and decision making is evident, informal or intuitive knowledge is applied.

Regarding the need for more connections to disciplines outside mathematics, as yet no satisfactory plan has emerged. While students in the applied emphasis must choose an approved minor area of study in the University, there is reluctance to require all majors to have such a minor and a lack of appropriate "single" courses offered in the University which easily convey the mathematical connections. For now the agreed upon strategy is to rely on connections expressed in the current set of required courses (including the model building course) and in programs offered by the Math Club/ MAA Student Chapter.

The observation concerning the performance of students when courses were taken in particular sequential patterns was not new to those who were seasoned advisers in the Department. But it did trigger the reaction that new advisers should be especially alerted to this fact.

Success Factors

While the work of the first review committee of faculty was lengthy and cautious because of the newness of the process, the next review committees seemed to move right along with greater knowledge of what to do and how to do it. The portfolio material enables the growth of a student to be traced through the program, and the decision of what to include makes it possible to have complete portfolios for most transfer students as well as for the other students. Further even incomplete portfolios help in the discernment of patterns of student performance. However, it does take considerable time to read through the complete folder for an individual student.

So far the University has been willing to provide extra monetary compensation for the additional faculty time involved in evaluating the portfolios. Should this no longer be the case, an existing Department committee could examine a subset of portfolios each year, or reduce consideration of all the objectives to tracing one objective in one year and a different objective in the next.

The program has not been difficult to administer since most of the collection of materials is done through the individual course instructors. The Department's Director of Undergraduate Studies needs to be sure these faculty are aware of what needs to be gathered in the courses and to see that the appropriate duplication of materials is done for inclusion in the folders. In addition, he/she must collect the student essays.

The process has the added benefit that involvement of many faculty in collecting the materials makes more faculty think about the degree program in its entirety, rather than merely having a focus on the individual course taught. Finally, the mechanism of the portfolios being collected when they are, and evaluated AFTER the student graduates, along with the qualitative nature of the evaluation process, insures that the assessment IS of the program rather than of individual students, faculty, or courses.

An Assessment Program Built Around a Capstone Course

Charles Peltier
Saint Mary's College

A full-year senior capstone course has evolved at a small, private women's liberal arts college in the Midwest to become the principal tool for assessing the major. Within this two-semester seminar each student has to develop an independent study project, known as the comprehensive project. Preliminary work on the project begins in the first semester and oral and written presentations of the completed project are given in the second semester.

Background and Purpose

Saint Mary's College is a catholic, liberal arts college for women with about 1500 students. We typically have ten to a dozen mathematics majors graduating each year; all of the mathematics and computer science courses are taught within the department, which consists of eight full-time faculty and two or three adjuncts teaching specialized courses. Since the 1930s the college has required that each student pass a comprehensive examination in her major. The requirement was rewritten in the early 1970s to give each department the freedom to determine the most appropriate form for examining its students. At that time the mathematics department replaced the "test" format with a requirement for an independent study project, which has come to be known as the comprehensive project.

A capstone seminar was developed to provide a framework for student work on the comprehensive project and to foster student independence in learning and skill in discussing mathematics. The spur for developing a formal program for overall assessment of student learning was more bureaucratic. In 1996 the college had to present to North Central, as part of the reaccreditation process, a plan for assessing student learning. The mathematics department had responsibility for developing the plan for assessment in the major and built its plan around the existing senior capstone course and the project.

In developing the assessment program, we had to articulate our goals for the major in terms of student achievement rather than department action.

Goals for a Mathematics Major

As a result of study in mathematics, a graduate of Saint Mary's College with a major in Mathematics will have met the following goals.

1. The graduate will have developed learning skills and acquired a firm foundation of knowledge of fundamental mathematical concepts, methods, reasoning and language sufficient to support further academic work or a career in an area that requires mathematical understanding.
2. The graduate will be able to apply her mathematical learning skills and knowledge and also to utilize appropriate technology to develop models for solving problems and analyzing new situations, both in mathematics and in areas that use mathematics.
3. The graduate will be able to communicate her ideas and the results of her work, both orally and in writing, with clarity and precision.
4. The graduate will be prepared to use her knowledge and learning skills to undertake independent learning in areas beyond her formal study.
5. The graduate will be prepared to use her critical thinking skills and mathematical knowledge as a contributing member of a problem solving team.

6. The graduate will have examined and formed ethical principles which will guide her in making professional decisions.

The senior seminar and senior comprehensive project, as assessment tools, are intended to provide both development and information on goals 1, 3, and 4. Some of the work in the first semester of the seminar touches on goal 5, and individual projects may get into actual problem-solving of the sort envisioned in goal 2. In practice, the senior seminar, like all of our courses, was designed to foster development toward these goals, as well as to assess their achievement.

In addition to the assessment of student learning described here, the department evaluates its programs through a biennial program in which recent graduates are invited back to campus for a day-long discussion with current students, through contact with alumnae, through contacts with interviewers who come to campus, and by using the information in the college's surveys of recent graduates.

The assessment method described here consists of (1) a full-year course meeting twice a week, called the Senior Seminar, and (2) an independent study project culminating in a formal oral and written presentation and response to faculty questions, known as the Senior Comprehensive Project. Roughly speaking, the first semester seminar begins the experience of extended independent learning and presentation and provides a vehicle for organizing the students' search for a topic and advisor. The second semester seminar is built around students' presentation of preliminary findings while they are working on their individual projects.

Method

During the first semester of the Senior Seminar, all students work on a common topic and from a common text. They are required to learn and present material on the level of a junior-senior mathematics course. The way this usually works is that the seminar director selects a text involving material not already covered by the students in other courses and determines a course schedule based on one-week sections of the material. Each section of material is assigned to a team of two students which is responsible, with advice from the seminar director, for learning the material, presenting the material to the class and assigning and correcting written problems based on the material. One semester generally allows time for two rounds of presentations, with the teams changed between first and second rounds, and for several problem presentation days in which individual students are assigned to present solutions of some of the more difficult problems. In weeks in which they do not present, students are responsible for reading the material, asking questions of the presenters, critiquing the presentations, and working the problems. Each student is required to have an advisor and topic for her senior comprehensive project by mid-October.

The writing assignment for the fall seminar often requires an introduction to the topic of the comprehensive project. The grade in the first semester of the seminar is based on demonstrated knowledge of the material, the presentations, participation (including asking questions, completing the critiques, etc.), and the required paper.

During the second semester, each student is working on her own senior comprehensive project. She works with the advisor in learning the material and preparing a paper and a formal public presentation based on her work. In the seminar, she presents two fifty-minute preliminary talks on her topic and is expected to follow other student talks in enough detail to ask questions and to critique the presentations. The seminar talks are expected to contain sufficient information for her classmates to understand her formal presentation. The grade in the second semester of the seminar is based on demonstrated learning of the topic, presentation of the topic, participation, and work with the comprehensive advisor.

The comprehensive project must develop material that is new to the student. This may be an extension of an area previously studied or a completely new area for the student and may be in any area of mathematics, including applications and computer science. The comprehensive advisor and the student work out a project that is of sufficient depth and breadth and that can be completed in the time available. The progress is monitored by both the advisor, who meets regularly with the student, and the seminar director, who observes the preliminary talks. The student provides a formal written presentation of her work, with a summary of the material covered in the seminar talks and a more detailed treatment of the final material. Once the paper is written, it is submitted to a committee of three faculty members — the advisor, the seminar director and one other — who read the paper and prepare questions related to the material. The formal presentation is made before an audience consisting of this committee, the seniors in the seminar, and any others who may wish to attend. It consists of a forty-minute presentation on the final material in her work and twenty minutes of response to questions from the faculty committee. The formal paper, the oral presentation, and the response to questions constitute the senior comprehensive examination required by the college. The paper for the senior comprehensive project also serves as the final submission for the student's advanced writing portfolio.

Grading of the senior comprehensive projects occurs in two stages. The committee determines whether the student passes or fails; there is usually some reworking of the paper needed, but except in extreme cases this not an impediment to success. After all the presentations have been completed, the department faculty meets to discuss all of the comprehensives and to determine which, if any, are to be awarded honors. The discussion is based on five criteria:

1) mastery of the subject,

2) the quality of the written paper,
3) the quality of the oral presentation,
4) the response to questions,
5) the independence and reliability shown in the work with the advisor.

This discussion also serves as an overview of the achievement of the students in the senior class, and the seminar director writes a report to the department based on the discussion.

Findings

Since the seminar and comprehensive project have been in place longer than our formal assessment program, most of the information has been gained and used in planning in an informal and implicit fashion. There are, however, a few results that can be stated directly.

The first is that taking courses does not prepare students to explain what they know to people who do not already know it — that is, to anyone other than teachers or other students who have seen the material. There are really two different but related issues here. The first is the identification of assumptions and background for a particular piece of work — understanding where it fits into a larger picture. The second is the need to move from the "show that I know it" mindset of passing a test to the "giving you an understanding of something I know and you don't" mindset. There is a stage of development here that does not occur automatically and it seems to be very closely related to learning to work independently and to critique the results. We do see growth here during the senior year.

A related fact is that students find it difficult to work with, and especially to combine, different presentations of the same idea. Searching for examples and dealing with varying notation in different sources is a major difficulty for many. Seeing old ideas in a new form and a new setting is often a challenge, especially to a student who is working at the limit of her experience. The failure to recognize ideas from previous courses, and a consequent inability to use them in new settings, is often another form of this problem. The seminar also provides a sometimes discouraging reminder that there are always topics from previous study that have been forgotten or were never really understood.

Discussions with alumnae and interviewers have strongly supported our belief that several aspects of the seminar and comprehensive project are very important for those graduates — a majority, in our case — who do not go into teaching or into explicitly mathematics-related fields. The important aspects seem to be the independent work on a long-term project and the organization and presentation of results. The practice in independent learning and interpretation has been equally important to those who have gone on to graduate study in mathematics or mathematics-related fields.

Use of Findings

The department's experience with the seminar and the comprehensive project has affected all planning and discussion for the last twenty years. One major result occurred at the level of the college curriculum when the department strongly supported the introduction of the "advanced writing" requirement, to be fulfilled in the student's major. Our experience in asking students to explain mathematics made it clear that we needed to develop this skill, and we had already begun introducing writing requirements within courses when this college-wide framework came under discussion. Currently the second year writing requirement in mathematics focuses on exposition and the third year adds technical writing, but we may find that some other approach works better.

The fall semester of the seminar has changed often, usually in small ways, from experience with the second semester and the comprehensive project. In order to make students more conscious of the decisions made in preparing a presentation, we introduced a student critique form, which must be filled out by each student for each speaker. We are considering having each student fill out such a form for herself, as well, to foster more reflection on the process. When the seminar was put in place, students worked on separate topics in both semesters — shorter topics in the fall, longer in the spring. We have found that the process of explanation takes more effort and learning, and that working on a common topic in the fall allows for more mutual support and cooperation. By encouraging more student questions of the presenters it also improves the feedback on the presentations and student awareness of the process of presentation. The use of teams of students arose for the same reason. The use of written exercises puts more pressure on the "audience" to really work at learning the material, and somewhat reduces the tendency to let the presenters get by with "good enough."

Our experience with the seminar and comprehensive has also contributed to an increase in assignments requiring longer explanations in earlier mathematics courses and an increase in requirements for in-class presentations by students in earlier years.

Success Factors

Alumnae who have gone to graduate programs and those who have gone to work report that the experience has been valuable in preparing them to deal with the work they have to do. A part of this is certainly the confidence they develop from having already worked on extended projects and reported on their work. We have also gained information on some things that work and some that do not work in developing the skills.

A major advantage to this program as an assessment tool is that it is a required part of the educational program of the department, not an added testing requirement. The seminar involves graded academic credit and the whole program is approached as a learning tool, so that problems of student participation are minimized. It is a great help that seniors in other majors are also involved in meeting the comprehensive requirement in ways tailored to these majors, even though the mathematics comprehensive is generally recognized as one of the most demanding.

There are two major drawbacks to this program — use of student time and use of faculty time. Since the seminar involves two semesters, the number of other courses taken by a student is reduced, though the fall seminar does, in fact, deal with standard content in a fairly standard way. The comprehensive project involves a great deal of time, and a student who has thought in terms of "a two-hour course" for the spring semester may find herself overextended. The program also uses considerably more faculty time than would appear on official records. The seminar director is credited with two semester hours of teaching in each semester, but the faculty who serve as project advisors receive no official teaching load credit. During the first semester, the seminar director spends a great deal of time in one-on-one or one-on-two discussion with the students as they first learn the material and then begin thinking about how to present it; this takes far more time than preparing lectures and exercises on the material. In addition, the seminar director provides written critiques to the students, incorporating the information from the student critiques. In the senior comprehensive phase, each faculty member is advising one or two students on projects which extend over several months and require planning of presentations and writing. Naturally, weaker students require a great deal of direction and advice, but faculty find that they are more likely to plan more ambitious projects with stronger students, so that these also require a lot of time. There is a positive side to this effort in the opportunity for faculty to explore new areas in their fields or in related fields. The seminar director has the unique opportunity of reading all the projects and seeing new areas in many fields of mathematics.

In working with a program such as this, it is essential to have requirements explicit and clear from the beginning. There is always time pressure on students, and for many the long-term nature of the project is a new experience. Without deadlines and dates the weaker students, especially, may get so far behind that they can catch up only with heroic effort — and at the cost of work in other courses. Grading criteria need to be as explicit as possible, because there are many opportunities for misunderstandings and incorrect expectations in a mathematics course that is very different from students' previous experience.

The involvement of all faculty in the senior comprehensive projects means that the program cannot work without full faculty support. Not only is there a serious time commitment, but it is necessary that all faculty members keep track of deadlines, formats for papers, etc., and be willing to adjust schedules to serve on the committees for other students. Working out the inevitable differences of opinion about interpretation of requirements before they become critical falls mainly to the seminar director, but would not be possible if other faculty were not interested in the program as a whole.

Using a Capstone Course to Assess a Variety of Skills

Deborah A. Frantz
Kutztown University

At a mid-sized, regional university in the East each student must complete a one-semester senior seminar in which a variety of assessment methods are used to assess student learning in the major. These methods include: a traditional final exam, a course project, an expository paper, reading and writing assignments, a journal and a portfolio — all of which are designed to assess a variety of skills acquired throughout a student's four years.

Background and Purpose

Kutztown University is one of fourteen universities in the State System of Higher Education in Pennsylvania. Of the approximately 7,000 students, four to six mathematics majors graduate each year. (This excludes those who plan teaching as a career.) As in many undergraduate institutions, a student with a mathematics major at Kutztown University is required to take a wide variety of courses that include a calculus sequence, several proof-based courses like abstract algebra and advanced calculus, and several application courses like linear algebra, and probability and statistics. It is typically the case, too, that a student's grade for each of these courses depends heavily (and sometimes entirely) on objective, computational exams. Because of options given to the students, rarely (if ever) do two students who finish our cafeteria menu of mathematics courses experience identical mathematics programs. How, then, does one measure the quality of the major? How can we ensure that each major is equipped to face a future in graduate school or in the work force?

One requirement for all of our mathematics majors is the successful completion of the Senior Seminar in Mathematics. It is designed as a culminating experience. It used to be taught using the lecture method on a mathematical topic of the instructor's choice, and was assessed using only traditional examinations. I was first assigned to teach the course in 1992.

Having already begun to incorporate oral presentations in some of my other classes, I decided to experiment with nontraditional teaching methods and alternative assessment techniques. The course has evolved to the point that no lectures are given. Students now study a variety of mathematical topics, keep a journal, participate in class discussions, write summaries of readings, write a resume, do group problem solving, give oral presentations, and complete a project. (The project consists of writing an expository paper and of delivering an oral presentation, both of which are results of an exploration of a mathematical topic of each student's choice.) Such activities enable students to more successfully achieve these course objectives:

1. increase his/her competence in the independent reading of mathematical materials
2. review, structure and apply mathematical knowledge
3. develop skills needed in presenting a rigorous argument
4. organize and deliver a mathematical presentation
5. pursue topics in mathematics not met in previous mathematics courses
6. explore further several topics that have been studied earlier
7. develop group problem solving skills
8. improve skills in utilizing resources
9. develop a global perspective of the role of mathematics in society.

Method

Due to the small class size (about 10), considerable individual attention can be paid to each student. Moreover, assignments can be adapted to meet the needs of a particular group of students. The course grade is determined by a collective set of assessment techniques that include holistic grading, traditionally graded problems, instructor evaluations based on preselected criteria, faculty-student consultations, and portfolios. In the paragraphs that follow, assignments are described along with the assessment techniques used for them. Objectives that are addressed by the assignments are identified by number at the end of each paragraph.

Early in the course, each student is given a different mathematical article to read and is asked to organize and deliver a five minute talk about it. The content is elementary in nature in order that the student focuses on the mechanics of communication and is not intimidated by the content. While each classmate identifies (in writing) at least one positive and one negative attribute, the instructor uses a set of criteria known to the students. Both forms of evaluation provide immediate feedback to the speaker. Using the same mathematical article, students give a second talk, implementing the assessment input from the first talk. This pair of talks serves as preparation for the 25-minute oral presentation later as part of the course project. This longer talk is assessed two ways: the instructor uses written evaluations based on preset criteria and a team of three members of the department faculty uses a pass/fail evaluation based on the student's apparent level of understanding, and on the cohesion of the talk. (Objectives 1, 2, 3, 4, 5, 6)

To indicate how the major program is complemented by the general education curriculum, several class discussions of preassigned readings are held. Topics for discussion include historical, ethical, philosophical, and diversity concerns and how they are related to the mathematics profession. Assessing performance in a class discussion is based on two criteria: Did the student understand the main idea of the assigned readings? Did the student convey individual interpretations of the issues at hand? (Objectives 1, 3, 9)

There are a variety of writing assignments. These include journal writings, summaries of readings from books and professional journals, the writing of a resume, and the composition of a 20-page expository paper. Students keep a journal that reflects their thoughts on mathematics-related topics. Three to five entries are made each week. The journals are holistically graded three times during the semester. In addition, each student uses the journal to provide a partial self-assessment on the final examination for the course. (Objective 9)

Students read articles from books and journals and then write a two-page summary of each one. These are graded holistically, as they are used primarily for diagnostic purposes. Students are also required to write their resume

for this course. They are urged to obtain assistance from the Office of Career Services. (Objectives 1, 3, 8, 9)

As part of the course project, each student writes a 20-page expository paper on a mathematical topic of his/her choice. The student is guided through several steps of the writing process throughout the semester. A draft of the paper is highly scrutinized based on preselected criteria (knowledge of the subject, breadth of research, clarity of ideas, overall flow of the paper, etc.). No grade is assigned to the draft. However, the same criteria are used in the final assessment of the paper. (Objectives 1, 2, 3, 4, 5, 6, 8)

Every college graduate should possess resource skills: library skills; internet skills; professional networking skills; and skills needed to find career-related information. To develop skills, students are given sets of specific questions, often in the form of a scavenger hunt. For example, they may be asked to find the number of articles published by Doris Schattschneider in a given year and then conjecture about her field of expertise, or they may be asked to find out as much as possible about Roger Penrose from the World Wide Web. These assignments are graded on correctness of answer. (Objectives 1, 2, 5, 6, 7, 8, 9)

The content portion of the course includes a wide variety of mathematical topics that are typically not covered in other courses required of our majors. These include graph theory, complex variables, continued fractions, fractals, non-Euclidean geometry, as well as topics in more classical areas of mathematics. (I have used books by William Dunham, *The Mathematical Universe* and *Journey Through Genius: Great Theorems of Mathematics*, as stepping stones into a vast range of topics.) This portion of the course is designed so that students both learn and tackle mathematics problems in small groups. The grouping of students is done by the instructor and is changed with each new topic and set of problems. Students earn both individual grades for correct solutions, and class participation points for team interaction. Groups are changed frequently in an effort to develop some degree of leadership skill in each student, and to allow students to learn of each others' strengths and knowledge base. (Objectives 1, 2, 3, 5, 6, 7, 8)

The final examination for the course consists of two parts: a set of questions, and a portfolio. The set of questions are principally content-based problems, to measure the student's mathematical knowledge based on the problems that were worked in small groups. There are a few open-ended questions that ask students to compare and contrast perspectives found in the assigned readings, written summaries, and class discussions. Students also provide an assessment of their attitudes and study habits that is based on their own journal entries. (Objectives 2, 3, 5, 6, 9)

A portfolio is prepared by each student and is focused on the individual's perceived mathematical growth. Students are asked to describe and document the growth that they made during the past three months. Content of the portfolio need not be restricted to materials obtained from the Senior Seminar. They write an essay unifying the various items that they choose to include. Assessment of the portfolio is based upon the breadth of perceived learning and how it is unified. Preselected criteria are used in the assessment. (Objectives 3, 8, 9)

Findings

Results of assessments of the oral components of the course indicate that examinations are not always a true indicator of a student's level of understanding. Mathematical knowledge can often be conveyed more effectively by oral discourse than by written computations or expositions.

Assessment results from writing assignments indicate that our mathematics majors need more than basic composition skills prior to enrolling in the Senior Seminar in Mathematics.

Prior to the reading assignments, few (if any) connections were recognized between the knowledge gained through general education courses and the discipline of mathematics.

Although students are more technologically prepared as they enter college than those who entered four years ago, results from the resource skills assignments indicate that little demand is made on them to develop further their information-retrieval skills during their undergraduate years.

A significant amount of bonding occurs as a result of working in small groups. Consequently, classmates are often used as resource people while the course project is being prepared. Students are forced to work with a variety of personality combinations, which each of them will have to do in the future.

Use of Findings

The oral presentations illustrate that students often can better demonstrate understanding by presenting a topic orally. Other mathematics courses should (and have begun to) incorporate oral assignments to facilitate learning.

Our major programs already require that students complete a scientific writing course and it is strongly recommended that a student complete this course prior to enrolling in the Senior Seminar. More writing needs to be incorporated in other mathematics courses as well as remain in the capstone course. Ideally, *every* mathematics course should incorporate some form of nontechnical writing component; realistically, post-calculus applications-based courses such as differential equations, numerical analysis, probability and statistics, and operations research appear to be those most suited for such assignments.

Students with severe writing difficulties (grammar, punctuation, run on sentences, etc.) or with reading deficiencies are identified early in the semester and advised to get extra help. Such assistance can be obtained without cost to the student from the Writing Lab and from the Department of Developmental Studies.

Information gleaned from the journals is used to identify an individual's strengths and weaknesses and to identify questions and concerns that confront the student. The instructor is able to provide relevant, timely feedback and is better able to serve as a facilitator of learning.

As a result of producing a resume, students become aware of the skills that may be required for a career that uses mathematics. The academic advising of our majors should be improved so that such information is realized by students earlier in the program. Then appropriate coursework could be recommended to help develop these skills. We also need to demand that more technological and resource skills be used in lower-level mathematics courses.

Class discussions indicate the need to guide the students in making connections: those among the many aspects of the discipline of mathematics; and those between mathematics and the general education component of the university degree. This type of assignment should remain part of the Senior Seminar course.

Results from the content-based portion of the course indicate that the assessment of previous coursework is reliable inasmuch as it measures computational skills and elementary critical thinking skills. Existing performance standards in required mathematics courses appear to be consistent.

Information gleaned from student portfolios is sometimes enlightening. For example, one student wrote that she felt that the amount of writing done in this course far surpassed the total amount of writing in all of her other classes combined! This reinforces our conclusion that more writing needs to be incorporated in other courses, whether mathematics courses or not. The portfolio also provides us with an early warning system for potential problems. For example, one student complained that our mathematics program did not prepare him for entrance into the actuary profession. (He took and did not pass the first actuary exam three times.) This has alerted us to try to better explain the difference between an undergraduate degree in mathematics and an individual's ability to score well on standardized tests.

Success factors

By using a variety of assignments and assessment techniques, the Senior Seminar in Mathematics has become a place in which many skills can be developed. Assignments are all related to the successful pursuit of a career in mathematics, most of which require skills that lie outside the traditional scope of a mathematics content course. The course works because of its small enrollment, and the instructor's knowledge of a wide variety of writing techniques, mathematical fields, and assessment techniques, and the willingness of colleagues in the department to maintain quality programs. Faculty colleagues who have assessed an oral presentation very quickly gain a perspective of the strengths and weaknesses of our major programs. Continued support by them is an invaluable asset when changes in the programs are recommended.

However, it takes time and patience to build quality assignments and develop meaningful assessment devices, and this should be done gradually. The strengthening of the major that results is worth the effort.

The Use of Focus Groups Within a Cyclic Assessment Program

Marie P. Sheckels

Mary Washington College

An entirely different approach to assessing the mathematics major has been developed at a state-supported, coeducational, liberal arts college in the Midsouth. Graduating seniors participate in focus group sessions which are held two days prior to graduation. These are informal sessions with a serious intent: to assess student learning in the major.

Background and Purpose

Mary Washington College is a state-supported, coeducational, predominantly undergraduate residential college of the liberal arts and sciences. It is located in Fredericksburg, Virginia, a historic city, which is about halfway between Washington, D. C. and Richmond, Virginia. The College is rated as "highly selective" in its admission status, and enrolls approximately 3000 undergraduates. There is also a small graduate program. There are ten full-time faculty members in the Department of Mathematics. On the average, 20 students graduate each year with degrees in mathematics.

Mary Washington College began to develop a program of outcomes assessment in 1989. The College wisely decided that each department, or major program within a department, should be responsible for developing its own plan to assess how well it was preparing its majors. By 1991, both faculty and administrators had learned more about outcomes assessment, and decided that assessment should be conducted according to a four-year cycle. Each major program has one faculty member, the "Outcomes Assessment Coordinator," who is responsible for the assessment.

Method

During the first semester of Assessment Year 1, faculty examine and revise, if necessary, their list of "Outcomes Expected."

This list of goals and objectives details the essential knowledge, skills, and abilities that students who complete their major program should possess. Following this, faculty decide how they will determine the extent to which these outcomes are achieved by their major students. Faculty determine what methodologies and instruments will yield relevant data, and decide upon a timetable for these evaluation procedures. Over a four-year assessment cycle, each major program must collect data using at least one of each of the following: a direct measure (e.g. tests, capstone courses, portfolios), an indirect measure (e.g. focus groups, exit interviews), and a survey of program alumni. During the second semester of Assessment Year 1, and during Assessment Years 2 and 3, outcome assessment coordinators collect and analyze the data gathered through the various forms of assessment. Generally, it is expected that data will be collected during one semester and then analyzed and interpreted during the next semester.

During Assessment Year 4, assessment coordinators, along with other faculty, compile, analyze and interpret all of the findings of the past three years regarding their major program. Although changes in the major program which are based on assessment results can be proposed at any time during the assessment cycle, it is often helpful to wait until Year 4 when changes can be based on the cumulative results of various forms of assessment.

The 1996–97 academic year was Year 1 in the Mary Washington College assessment cycle. Departments have adjusted their schedules so that all are now on the same four-year cycle. This was the start of the mathematics departments' third cycle (although one cycle was three years rather than four). During the fall 1996 semester the Department of Mathematics extended and refined the outcomes expected of its majors. They are listed below.

Outcomes Expected

Interpretation of Mathematical Ideas
- Students will read and interpret mathematical literature.
- Students will read and interpret graphical and numerical data.

Expression of Mathematical Ideas
- Students will use mathematical symbols correctly and precisely in expressing mathematical information.
- Students will represent quantitative information by means of appropriate graphing techniques.
- Students will present well-structured and valid mathematical arguments.

Critical Thinking
- Students will employ critical thinking skills in their comprehension and application of mathematics.
- Students will analyze and construct logical arguments.

Discovery
- Students will discover mathematical patterns and formulate conjectures by exploration and experimentation.

Applications
- Students will express problems in mathematical terms.
- Students will identify areas of mathematics that are most useful in solving practical problems.
- Students will use technology appropriately in solving problems.

Appreciation
- Students will give examples of the beauty and power of mathematics.

During the past eight years the mathematics department has employed the following measures to assess its major program: mathematics tests, both alumni and faculty surveys, and focus groups with graduating seniors. During 1989, 1990 and 1995 we administered in-house assessment tests which asked the students to read, write and interpret mathematics. On each occasion, students enrolled in Calculus II (beginning-level "serious" mathematics students), and students enrolled in Real Analysis (senior level mathematics majors) took the test. Faculty compared and analyzed the answers given by the two groups of students.

In 1993, we conducted an alumni survey of recent mathematics graduates. The surveys included multiple-choice questions in which the alumni rated different aspects of their mathematics education as well as free response questions where students wrote suggestions and advice for improving the program. In 1994 we circulated a survey asking mathematics faculty for their comments on various aspects of the program. The outcomes assessment coordinator held individual comments in confidence and reported a summary to the department.

In 1992 and 1996, we conducted focus group meetings with graduating seniors. These focus groups were designed to provide information on how our majors perceived different aspects of the mathematics program and to elicit their opinions on ways the program might be improved. While all of the data have been useful, the focus groups of graduating seniors have been the most enlightening. On both occasions, we conducted the focus groups on the two days preceding graduation, after final grades had been submitted. Seniors knew that they could give their honest impressions of the program at this time without fear of negative repercussions. We invited all of our graduating seniors to attend one of two one-hour sessions. Nearly all of the seniors attended. We served drinks and snacks and there was a relaxed, congenial atmosphere; it was one last time for majors to get together. Two faculty members conducted the sessions. They took turns where one asked the questions and the other recorded. Neither faculty had taught these seniors in their upper-level courses so the students were free to make comments about the instruction in these classes. The faculty informed the students that the purpose of the session was to record the students' perceptions of their education in the mathematics program and asked that they be open and honest in their responses. Faculty distributed the list of questions to the students and then one of the faculty read aloud each question in turn and encouraged discussion. Examples of questions we asked included:

- What are some of your initial feelings, opinions or comments about your experience as a mathematics major?
- Do you have any strong feelings about any specific courses? What seemed to you to be the main concern or characteristic of the math department? Why do you say this? Can you give examples?
- In what areas do you feel most prepared?
- If you could make only one change in the mathematics department, what would that be? Why?

The faculty encouraged students to extend, agree or disagree with each other's remarks by asking questions such as "Do you all feel the same way?" or "Would anyone like to elaborate on that?" Everyone was encouraged to express their opinion. The students were very forthright, open and honest in their responses.

Findings

It was interesting to note that the data we obtained through the various types of assessment were remarkably consistent. The results of our assessments were very positive. However, we did find a few areas where our program could be strengthened. We were very gratified to hear our majors say that one of the strongest aspects of the program was that it had taught them to think and that they were confident in their abilities to solve problems. These, of course, are major goals of the program. The students were thankful for the small classes and personal attention. Seniors in the focus groups and alumni who were surveyed were very pleased with their preparation in mathematics in general. However, both groups stated that they would like to see more emphasis placed on the applications of mathematics.

Another recommendation of the focus groups was to further integrate technology into the program and encourage students to become familiar with computers. These students also recommended that we improve the department's career advising, and suggested that we integrate some group projects in our courses, since prospective employers seemed to value this kind of experience.

In addition to learning about our major program, we have also learned about the process of assessment. The cyclic approach to outcomes assessment seems to work very well. Within certain guidelines, major programs can choose what forms of assessment best meet their needs. Using a variety of assessment techniques helps departments see the broad picture and determine which findings are consistent across measures. Since assessment activities are spread over several years, they are not overwhelming.

The focus groups, in particular, seem to yield a large quantity of valuable information for the time and effort spent. Conducting the focus groups is quite enjoyable and rewarding for the faculty — assuming that you have a good program. The students enjoyed this chance to share their opinions with the faculty and in so doing to help maintain a strong mathematics program at their soon-to-be alma mater. Moreover, they were gratified that the faculty respected them enough to care about their opinions.

Use of Findings

The results from the focus group, along with the results from our other assessment activities, helped the mathematics department determine how well we were meeting our goals and plan actions to maintain a strong major program. Partly in response to the recommendation to place more emphasis on mathematical applications, we hired two new faculty whose areas are in applied mathematics and we strengthened the program offerings in these areas. These faculty have allowed us to offer more sections of our "applied" courses and they have also developed and taught new upper-level courses in Chaos, and Linear Models. They also have developed a new sophomore-level course entitled "Mathematical Modeling," an interdisciplinary course in basic mathematical modeling investigating various scientific

models with an environmental theme. These faculty have both worked with students on independent studies in various applied fields, and have sponsored several students in internships. In addition to hiring these new faculty, we urged all faculty members to integrate more applications of mathematics into their courses wherever possible. These modifications should help students meet the outcomes expected in the application area.

We have further integrated technology into our program by using graphing calculators in all of the Precalculus and Calculus courses and by using computers in the Statistics, Differential Equations and Linear Algebra courses, as well as in some Special Topics courses. Also, we now strongly encourage students to take computer science courses to complement their program in mathematics.

In response to improving career advising, we have hosted professionals from the area to speak to students and we now have a "Careers Information Link" on the mathematics department's home page where students can find general information on careers for mathematics majors, internship possibilities, and specific information about organizations which are currently hiring.

Also in response to focus group recommendations, we have improved placement in freshmen courses, fought to keep class sizes as small as possible to ensure quality instruction, and have integrated group work and projects into existing coursework.

The results from the senior focus groups were so informative that the mathematics department is currently planning to conduct alumni focus groups. We will invite recent alumni in this area to meet with us to seek their perceptions of how well our mathematics program prepared them. We trust that we will get the same in-depth, thoughtful responses that we did from our senior majors. We then plan to use this information to help us write an alumni survey for those mathematics graduates who could not attend the focus groups.

Success Factors

We have learned much from the results of our various forms of outcomes assessment. We understand, however, that we must be cautious in interpreting the results of assessment. For example, the responses in both the focus groups and on alumni surveys were their *perceptions* of the program and we viewed them as such. A few of these perceptions were factually incorrect, such as on issues related to course scheduling. Student and faculty opinions did not always agree, such as on the value of certain courses and major requirements. However, this does not negate the importance of students' perceptions and opinions.

We believe that our focus groups have been successful due, in part, to precautions that we took in planning and conducting the focus groups. We scheduled the group sessions for a convenient time after grades had been turned in. Students signed up during their classes for one of the sessions. The day before their session, the department secretary called with a reminder. We selected faculty with whom the students could be open and honest and chose questions which were open-ended and invited discussion. While conducting the sessions, faculty waited patiently for responses and accepted the students' opinions without being defensive toward negative comments. Faculty encouraged the students to respond to and extend each others' responses. This was the students' turn to talk and the faculty's turn to listen.

Assessing the Major Via a Comprehensive Senior Examination

Bonnie Gold

Monmouth University (formerly at Wabash College)

At a small, private men's liberal arts college in the Midwest, a comprehensive examination, known as comps, has been a tradition for seventy years. It has evolved into the assessment technique which the department uses to assess student learning. Comps are taken by seniors over a two-day period just prior to the start of the spring semester. The exam consists of two parts: a written component in the mathematics major and an oral component over the liberal arts.

Background and Purpose

Wabash College is a small (850 student), non-religious private college for men, founded in 1832 in a small town in Indiana. The department of mathematics and computer science, with 7 full-time faculty members, offers a major and a minor in mathematics, and a minor in computer science. We graduate between 2 and 18 mathematics majors per year, many of them double majors with the other major in a field which uses mathematics heavily. In order to graduate, every student must pass a comprehensive examination (often called "comps"). This requirement has been part of the college curriculum since 1928. This examination consists of two parts: (1) a written component in the major, two days, three hours per day, and (2) an oral component over the liberal arts. The purposes of the written part of the examination include (1) having students review their courses in their major so that they will become aware of connections which may have been missed in individual courses and (2) satisfying the faculty that the graduating seniors have an acceptable level of competence in their major. The content of the written component is left to each department to put together as it pleases. The purpose of the oral component is to cause students to reflect on their whole liberal arts experience: how they have examined, and perhaps changed, their values and beliefs during their time here; what were the positive and negative aspects of their Wabash education; how they may

begin to fit into the world beyond Wabash; what directions they still need to grow in and how they can continue to grow after they leave. The oral is a 50-minute exam with each student examined by three faculty members, one from the major, one from the minor, and one at-large. Thus, the major is also examined in this component. It provides for the department very different information from that provided by the written part, as will be discussed below. Because we have been giving these examinations for many years, there has been a lot of opportunity to reflect on and change the curriculum in response to weaknesses we find in our seniors on comps.

Method

The written part of the comprehensive examination is given two days before the start of the spring semester. The mathematics exam is always given in the afternoons, while most other disciplines give theirs in the mornings. Thus, mathematics majors who are double majors (many are) can take two sets of exams during the two days. In November, the department discusses exactly what format that year's comps will take. There is always an essay or two (recently over post-calculus core courses, e.g., abstract algebra and real variables), as well as problems over the calculus sequence, on the first day, and advanced topics on the second day. However, the number of essays, how the topics are selected, exactly where the dividing line between the calculus sequence and advanced topics will be, and whether students are allowed aids such as calculators or textbooks, changes from year to year based on changes we have been making in the courses students take. For example, recently we decided that we were interested in how well students understand calculus concepts, not in how well they remembered formulas they had learned three years earlier. So we allowed them to bring texts or a sheet of formulas to the first day's exam. We recently changed from requiring all students to take both real variables and abstract algebra for the major to requiring only one of these courses. Therefore this year's exam only had one essay question, a choice between a topic in algebra and a topic in analysis. However, beginning with next year's class, all students will have taken a full semester of linear algebra as sophomores, and so we will then be able to require everyone to write an essay on a topic from linear algebra and one from either real variables or abstract algebra.

The essay topics are rather broad; from algebra: algebraic substructures or homomorphisms of algebraic structures; from real variables: differentiability in R and R^n, or normed vector spaces. The instructions say: write an essay which includes definitions of appropriate terms, relevant examples and counter-examples, statements of several important theorems, and the proof in detail (including proofs of lemmas used) of at least one of these theorems. Students are given a short list of potential topics in advance, and are told that the topics they will have to write on will be chosen from among these. They are encouraged to prepare these essays in advance, and even speak with members of the department about what they're planning to include. However, they write the essay without notes on the day of the exam. We're looking for their ability to choose appropriate examples and theorems, to explain them, and to put this all together into a coherent whole.

The non-essay portions of the exam are typical course test problems. Whoever has taught a given course within the last two years writes three problems over that course, and gives them to the exam committee (which consists of two people). The committee then takes all the submitted questions and chooses which to use. They try to have a good balance of easy and difficult problems. Since over the course of their four years the students have had a wide choice of elective courses, the exam committee must make sure that the second day's examination on advanced topics has a sufficient balance of problems to enable each student to choose problems from courses he has studied. The problems are not supposed to require memorization of too much information or otherwise be too obscure; rather, they should cover the main points of the courses. A few sample questions:

From calculus: Find a cubic polynomial $g(x) = ax^3 + bx^2 + cx + d$ that has a local maximun at $(0, 2)$ and a local minimum at $(5, 0)$.

From number theory: Find all solutions in positive integers for $123x + 360y = 99$.

Students are given identification numbers (the correspondence between names and ID numbers is sealed until the written exams have been graded and grades decided on), and all questions are double-graded, with the department sharing equally the job of grading. When there is a substantial disagreement between the two graders of a given question, they look over all the papers which answer that question and regrade that question together. The college requires that the oral component comprise 1/4 of the grade, but that, to pass the comprehensive examination, students must pass both the oral and written parts separately.

Each student has a different oral examination committee. All examiners are to examine the student over the liberal arts generally, but usually each one has about 15 minutes during which (s)he has principal responsibility for the questions (but the others are encouraged to pursue the student for more details if they feel the response is inadequate). Generally, the questions mathematicians ask of students who are mathematics majors are questions which would not have been asked in the written portion. They involve philosophical questions ("is mathematics a science or an art"), or test the student's ability to speak about mathematics to non-mathematicians ("please explain to Professor Day, who teaches classics, the idea of the derivative"). While we have several other methods of assessing the major, the oral exam is the only assessment tool for the minor (other than examinations in individual courses), and students with mathematics minors are often asked questions about the relation between mathematics and their chosen major (usually a physical or social science).

Findings

When I first came to Wabash, the comprehensive exams were often a source of considerable discomfort to the department. When asked to explain what the derivative was, mathematics majors would often start (and end!) by explaining that the derivative of x^n was nx^{n-1}. Students' grades, even those of good students, on the written comprehensives were often poor. Many majors have advisors outside of mathematics, and in the past, some accumulated a number of D's in mathematics courses. They then came to the comprehensive examination unable to pass. It's too late, in the second semester of the senior year, to tell a student he'd better get serious or find another major. We have made numerous changes in the program due to these results, and have also started failing students when they do sufficiently poorly (below 60% overall). When this happens, the department must write a new examination and the student must take the comps again in April in order to be able to graduate with his class. If he fails again, then he has to wait until the following year's exams. This has made most students take the exams fairly seriously. (About 10 years ago, 4 students failed on the first try, none of whom had done any serious studying. All passed on the second attempt. In the last 5 years, we've only had to fail one student.) At the moment, when students have studied for the exams, the results largely are consistent with the department's view of the student: good students (usually 1 or 2 each year) — heading to graduate school or promising careers — write exams which merit distinctions (90% or better, approximately), mediocre students (roughly 1/3) barely pass, and middle level students — the remainder — either do a good pass or a "high pass" (80% to 90%). Mathematics majors are now graduating with a better understanding of modern mathematics than in the past. The performance of minors and other students on the oral portion, however, is still causing some changes in the lower level curriculum.

Use of Findings

Over the years, having obtained unsatisfactory results in comprehensive exams, we have made many changes in departmental offerings and requirements. We started requiring abstract algebra and real variables when it became clear (from oral exam answers like the one mentioned above) that students were not getting a sufficiently rigorous theoretical background in mathematics. We started trying calculus reform when we saw how little students remembered from their calculus experience, and the minimal understanding demonstrated on the oral exam by minors and others who had not gone further. These changes in turn led to other changes, such as the introduction of a required, theorem-based linear algebra course in the sophomore year. On the other hand, because we want to allow a variety of electives, having added the linear algebra course, we decided to allow students to choose between real variables and abstract algebra, while encouraging them to take both. We have also moved to more active learning in the classroom, to help students become better able to handle the material. Students are learning more: we are graduating fewer students that we are embarrassed to acknowledge as mathematics majors than we had in the past.

To deal with the problem of students having advisors outside mathematics and surprising us by taking and failing comps, we adapted St. Olaf College's contract major. Each student signs a contract with the department describing what direction he plans his major to take (towards actuarial work, applied mathematics, pure mathematics, etc.). The contract is signed before registration for the second semester of the sophomore year, when students have their first choice of mathematics courses (between multivariable calculus and differential equations, though both are recommended). Each contract has some flexibility within it, and by petitioning the department, students can switch contracts if they find that their interests change. But by requiring them to sign a contract as sophomores, students begin working with the department much earlier in their student careers, when it is still possible to prod the student to the required level of effort.

Success Factors

Our capstone course, which involves readings in history of mathematics and preparation of a senior expository paper, is taken in the semester just before the comprehensive examination. The combination of a course spent reflecting on the history of mathematics together with the comprehensive examination for which the students review the courses they have taken helps students place the individual courses in perspective. Toward the end of the capstone course students are given a copy of the previous year's examinaton and there is some opportunity to discuss the upcoming comps.

The faculty meets weekly over lunch to discuss departmental business. Comps help focus our discussions, both before writing the test in the fall and after finishing grading them in the spring. In the fall we think about the progress our seniors have made, and what their focuses are. In the spring, we think about what changes we need to make in the program due to weaknesses which show up on the students' examinations.

The timing of the written comps has changed over the years. A long time ago, they were in April. However, if a student failed the examination he would have to wait until the following year to retake it and graduate. So they were moved to late January, during classes. Then, because seniors not only missed two days of classes, but often a whole week (a few days before to study, and a day or so after to celebrate), the written part of the comps was moved to the first two days of the spring semester. This way, grades are turned in by spring break, and any students who fail can retake them once in April and thus have a chance to graduate with their class.

We have had to fail students every several years on the exams; otherwise, some students do not take them seriously. However, since we switched to the contract major, students who have failed the written part have admitted that they didn't put much effort into studying for the exam. Nonetheless, both because of the additional trouble it causes the department, and because sometimes we're not convinced that a given student is capable of performing better on a second attempt, there is a contingent of the department which opposes failing anyone. It is always painful to fail our students, both because it's not clear whether we failed them or they failed us, and because of the additional time and suffering it causes us all. However, having established a history of doing this, our students in the last few years have taken the examinations fairly seriously and the results seem to reflect well what they have learned.

Alumni remember their comprehensive exams, especially the oral portion, vividly, particularly the composition of the committee and the questions they struggled with. They view the process as an important rite of passage, and one of the hallowed traditions of the college.

A Joint Written Comprehensive Examination to Assess Mathematical Processes and Lifetime Metaskills

G. Daniel Callon

Franklin College

A rather unique approach to giving a comprehensive exam to seniors is described in this article by a faculty member at a small, private co-ed college in the Midwest. The exam is taken by seniors in their fall semester and lasts one week. It is a written group exam which is taken by teams of three to five students. Currently, the exam is written and graded by a faculty colleague from outside the college.

Background and Purpose

Franklin College is a private baccalaureate liberal arts college with approximately 900 students. It is located in Franklin, Indiana, a town of 20,000 situated 20 miles south of Indianapolis. About 40% of our students are first-generation college students, and about 90% come from Indiana, many from small towns.

The Department of Mathematical Sciences offers majors in applied mathematics, "pure" mathematics, mathematics education, computer science, and information systems, and graduates between 8 and 15 students each year. The department has actively pursued innovative teaching strategies to improve student learning and has achieved regional and national recognition for its efforts, including multiple grants from the Lilly Endowment, Inc., and the National Science Foundation, and was named in 1991 by EDUCOM as one of the "101 Success Stories of Information Technology in Higher Education." The department seeks to promote active learning in the classroom through the implementation of cooperative learning and discovery learning techniques and the incorporation of technology. Members of the department have been unanimous in their support of the department's initiatives, and have participated on their own in a variety of programs at the national, regional, and local levels to improve teaching and learning. The department also maintains close ties with local public and private school systems, and has worked with some of those schools to aid in their faculty development efforts.

As part of a college-wide assessment program, the Department of Mathematical Sciences developed a departmental student learning plan, detailing the goals and objectives which students majoring in mathematics or computing should achieve by the time of graduation. For mathematics, there were three major goals. The first goal related to an understanding of fundamental concepts and algorithms and their relationships, applications, and historical development. The second centered on the process of development of new mathematical knowledge through experimentation, conjecture, and proof. The third focused on those skills which are necessary to adapt to new challenges and situations and to continue to learn throughout a lifetime. These skills, vital to mathematicians and non-mathematicians alike, include oral and written communication skills, the ability to work collaboratively, and facility with the use of technology and information resources.

The focus of efforts to assess students' attainment of these goals on a department-wide basis is a college-mandated senior comprehensive practicum, the format of which is left to the discretion of individual departments. In mathematics, the senior comprehensive is a component of the two credit hour senior seminar course, which is taught in the fall (to accommodate mathematics education majors who student teach in the spring). The seminar is designed as a capstone course, with its content slightly flavored by the interests of the faculty member who teaches it, but generally focusing on mathematics history and a revisiting of key concepts from a variety of courses and relying largely on student presentations. The latter provide some information for departmental assessment of student achievement in some of the areas of the three major goals. Other assessment methods, including departmental exit interviews and assessment efforts in individual classes, provide useful information but lack either the quantitative components or the global emphasis needed for a complete picture of student achievement.

Previously our senior comprehensive had consisted of a one-hour individual oral examination, with three professors asking questions designed to assess the student's mastery of concepts from his/her entire mathematics program. The emphasis of the oral exam was to draw together threads which wound through several different courses. Those professors evaluated the quality of the responses and each assigned a letter grade; those grades were then averaged to produce the student's grade on the examination. Students were given some practice questions beforehand, and the exam usually started with one or more of those questions and then branched into other, often related topics.

When additional money was budgeted for assessment efforts, a decision was made to add the Major Field Achievement Test (MFAT) from Educational Testing Service as an additional component of the senior comprehensive to complement the oral examination. However, that still left two of the three goals almost untouched, and we could find no assessment instruments available that came close to meeting our needs.

So we decided to come up with our own instrument. We wanted to address essentially the second and third goals from above, consisting of the experimentation-conjecture-proof process and the lifetime metaskills. We also wanted to include the modeling process, which is part of the first goal dealing with concepts and their application and development but is not covered in either the oral examination or the MFAT. We did not see any need to address the oral communication aspect of the lifetime metaskills in the third goal, which we felt was sufficiently covered in the oral examination. The result was the joint written comprehensive examination, which we have used since 1993.

Method

The joint written comprehensive examination is given around the fifth or sixth week of the senior seminar class. Students are informed at the beginning of the semester about the purpose and format of each of the components of the senior comprehensive practicum, but no specific preparation is provided for the joint written exam other than the general review included in the senior seminar course. Students in the class are divided into teams of 3–5. The class determines how the teams are to be selected if there are more than five students in the class. They are given a week to complete the test, and the class does not meet that week. Tests consist of four to five open-ended questions, and involve modeling, developing conjectures and writing proofs, and use of library and electronic information resources. Any available mathematical software is allowed. The team distributes the workload in any manner it deems appropriate, and is responsible for submitting one answer to each question.

Since the senior oral examination is evaluated by departmental faculty, we have a colleague from another college or university who is familiar with our environment write and grade the joint written comprehensive exam for a small stipend. The first two years we were fortunate to have a colleague who formerly taught at Franklin and now teaches at a slightly larger university in Indianapolis develop the exam, and her efforts helped smooth out many potential rough spots. The last two years we have employed colleagues at other similar institutions in Indiana, which has provided some variety in the questions and therefore allowed us to obtain more useful assessment information.

The following are a few excerpts from questions from the tests. Each of the four tests given thus far is represented with one question. (If taken in their entirety, the four questions together are a little more than the length of a typical exam.)

1. To facilitate a presentation for the annual Franklin College Math Day, it is necessary to construct a temporary computer communication link from the Computer Center to the Chapel. This link is to be strung by hanging cable from the tops of a series of poles. Given that poles can be no more than z feet apart, the heights of the poles are h feet, cable costs $\$c$ per foot and poles cost $\$p$ each, determine the minimum cost of the materials needed to complete the project. (Be sure to include a diagram and state all the assumptions that you make.)

2. Find and prove a simple formula for the sum

$$\frac{1^3}{1^4+4} - \frac{3^3}{3^4+4} + \frac{5^3}{5^4+4} + \cdots + \frac{(-1)^n(2n+1)^3}{(2n+1)^4+4}$$

3. Geographers and navigators use the latitude-longitude system to measure locations on the globe. Suppose that the earth is a sphere with center at the origin, that the positive z-axis passes through the North Pole, and that the positive x-axis passes through the prime meridian. By convention, longitudes are specified in degrees east or west of the prime meridian and latitudes in degrees north or south of the equator. Assume that the earth has a radius of 3960 miles. Here are some locations:

   St. Paul, Minnesota (longitude 93.1° W, latitude 45° N)

   Turin, Italy (longitude 7.4° E, latitude 45° N)

   Cape Town, South Africa (longitude 18.4° E, 33.9° S)

   Franklin College (longitude 86.1° W, latitude 39.4° N)

   Calcutta, India (longitude 88.2° E, latitude 22.3° N)

   ...

(b) Find the great-circle distance to the nearest mile from St. Paul, Minnesota, to Turin, Italy.

(c) What is the distance along the 45° parallel between St. Paul and Turin?

...

(f) Research the "traveling salesman problem." Write a well-prepared summary of this problem. In your writing indicate how such a problem might be modeled in order to find a solution.

(g) Starting and ending at Franklin College indicate a solution to the traveling salesman problem which minimizes the distance traveled. (Assume there is a spaceship which can fly directly between cities.)

4. One of the major ideas of all mathematics study is the concept of an Abstract Mathematical System (AMS for short). Consider the AMS which is defined by:
A nonempty set $S = \{a, b, c, \ldots\}$ with a binary operation $*$ on the elements of S satisfying the following **three** assumptions:

    A1: If a is in S and b is in S then $(a*b)$ is in S.
    A2: If $a*b = c$ then $b*c = a$.
    A3: There exists a special element e in S such that $a*e = a$ for each a in S.

Which of the following are theorems in the above abstract mathematical system? Prove (justify) your answers.

    (a) e is a left identity, that is, $e*a = a$ for all a in S.
    (b) $a*a = e$ for all a in S.

...

    (d) $*$ is a commutative operation.

...

Findings

For individual students, results from each of the components of the senior comprehensive practicum (the oral examination, the joint written examination, and the MFAT) are converted into a grade. These grades are assigned a numerical value and averaged on a weighted basis for a grade on the entire practicum, which appears on the transcript (with no impact on the GPA) and also is part of the grade for the senior seminar course. The oral examination is weighted slightly more than the other two due to its breadth and depth, and has remained unchanged since the tests complement each other so well. For the department, the results of all three components are evaluated to determine whether any modifications to departmental requirements or individual courses are indicated.

In analyzing the results of the joint written comprehensive exam, what has struck us first and most strongly has been the need for more focus on modeling. Students have had difficulty in approaching the problems as well as in communicating their assumptions and solutions. We have

also noted room for improvement in the use of information resources. Real strengths have been the use of technology, particularly in the experimentation-conjecture-proof process, and working collaboratively, although the latter seems stronger in the development of solutions than in putting results in writing.

The remainder of the senior comprehensive, the oral exam and the MFAT, have confirmed the impressions of our students which faculty members have developed by observing and interacting with them over four years, albeit with an occasional surprise. The difference in the tests' formats has allowed almost all students to showcase their abilities and accomplishments. In addition, our students have compiled a strong record of success both globally and individually on the MFAT, with consistently high departmental averages and very few students falling below the national median.

Use of Findings

Two major changes have resulted from these findings. First, a modeling requirement (either a course in simulation and modeling or a course in operations research, both part of our computing curriculum) has been added to the related field requirement for mathematics education and "pure" mathematics majors, solely based on the outcomes of the joint written comprehensive exam. We have also tried to be more conscious of the process of developing and analyzing models in the applied mathematics curriculum. Second, as a result of trying to answer the question of how to develop the competencies we are testing in the senior comprehensive exams, our department has also moved beyond curriculum reform of single courses to programmatic reform, in which we develop goals and objectives for individual courses and course sequences under the framework of the departmental goals and objectives. This has led to the identification of developmental strands, in which each objective (such as the development of oral communication skills) is specifically addressed in three or four courses with an emphasis on building on previous accomplishments rather than each course standing alone. These strands begin with the freshman calculus courses and continue through the entire four-year program. Sophomore-level courses in multivariable calculus and linear algebra have been particularly focused as a result, whereas in the past we would fluctuate in which goals were emphasized and to what degree.

A more subtle effect has been in faculty's awareness of what goes on in other courses. Not only are we more in touch with what content students have seen before they arrive in our classes, we also know (or can determine) how much exposure they have had to applications, written and oral communication skills, and other components of our goals and objectives. Although our department has had a long-standing tradition of effective working relationships and good communications, we have been surprised at how helpful this process has been.

The influences of the other two components of the senior comprehensive have been less pronounced. The oral exam has emphasized to faculty the importance of tying concepts together from course to course. The success of our students on the MFAT has been a helpful tool in recruiting and in the college's publicity efforts as well as in demonstrating the quality of the department's educational efforts to the college administration.

Success Factors

The joint written comprehensive examination fits well into Franklin's liberal arts philosophy since students are accustomed to being asked to draw together threads from a variety of courses and topics, both within and across disciplines. The fact that our department unanimously endorses the importance of all three components of the senior comprehensive results in an emphasis on similar themes in a variety of courses and encourages students to take the exams seriously. The willingness of our test authors to take on such a unique challenge has played a major role in the progress we've achieved. We have talked about the possibility of arranging for a consortium of similar colleges and universities to work together on a common joint written examination, with the result that the workload could be spread around and additional data and ideas generated.

Probably the biggest practical drawbacks to the implementation of a joint written comprehensive examination in the manner we have are logistical, including funding to pay the external reviewer and identifying willing colleagues. It would also be difficult for those departments which do not have a specific course or time frame in which such an instrument can be administered, since the investment of student time is quite large. It is also vital to get departmental consensus about what students should be able to do and how to use the information acquired, although that will probably be less of an issue as accrediting agencies move strongly toward requiring assessment plans to achieve accreditation.

Undergraduate Core Assessment in the Mathematical Sciences

John W. Emert and Charles R. Parish
Ball State University

In this article a "less" comprehensive exam to assess student learning in the core courses taken by all undergraduate mathematics majors at a regional, comprehensive university in the Midwest is discussed. We are guided through the process involved in developing the assessment instrument which is used in all four tracks of the mathematical sciences program: Actuarial Science, Mathematics, Mathematics Education and Statistics.

Background and Purpose

Ball State University is a comprehensive university with approximately 17,000 undergraduates. The University offers a strong undergraduate liberal and professional education and several graduate programs. The Department of Mathematical Sciences includes about 40 faculty, and graduates about 50 undergraduate majors each year.

Assessment activities for mathematics majors began at Ball State University in 1988 with data collection using the Educational Testing Services (ETS) Major Field Test in Mathematics. Coincidentally, our department initiated in the same year a new core curriculum to be completed by all of our undergraduate majors. These students pursue one of four majors: Actuarial Science, Mathematics, Mathematics Education, or Statistics. The core of courses common to these programs has grown to now include: Single- and Multi-variable Calculus, Linear Algebra, Discrete Systems, Statistics, Algebraic Structures (except Actuarial Science) and a capstone Senior Seminar.

Experience with the ETS examinations and knowledge of the mathematical abilities needed by successful graduates suggested that a specifically focused evaluation of the new core curriculum was both appropriate and desirable. It became apparent that there were several expectations relative to the core courses: the ability to link and integrate concepts, the ability to compare and contrast within a conceptual framework, the ability to analyze situations, the ability to conjecture and test hypotheses, formation of reasoning patterns, and the gaining of insights useful in problem solving. That is, the core courses (and thus the assessment) should maximize focus on student cognitive activity and growth attributable to concept exposure, independent of the course instructor.

Method

Development and construction of a pilot instrument (*i.e.,* a set of questions to assess the core curriculum) were initiated during 1991 and 1992. We independently formulated a combined pool of 114 items using various formats: true/false, multiple choice, and free response. Several items were refined and then we selected what to keep. Selection of items was based on two criteria: there should be a mix of problem formats, and approximately half of the questions should be nonstandard (not typically used on in-class examinations). Part I, containing 21 items covering functions and calculus concepts, and Part II, containing 18 items covering linear algebra, statistics, discrete mathematics, and introductory algebraic structures, were constructed. Each part was designed to take 90 minutes for administration.

The nonstandard questions asked students to find errors in sample work (*e.g.,* to locate an incorrect integral substitution), extend beyond typical cases (*e.g.,* to maximize the area of a field using flexible fencing), or tie together related ideas (*e.g.,* to give the function analogue of matrix inverse).

These questions often asked for justified examples or paragraph discussions.

Departmental faculty members other than the investigators were assigned to select subjects and administer the instruments in Fall 1991 (Part I) and Fall 1992 (Part II). Subjects were students who had completed appropriate core components, and were encouraged by the department to participate on a voluntary basis during an evening session. The students were not instructed to study or prepare for the activity, and there was no evidence to indicate that special preparation had occurred. While the student population for each part was relatively small (between 10 and 15), each population represented a good cross-section of the various departmental majors. Subjects were encouraged to give as much written information as possible as part of their responses, including the true/false and multiple choice items. Responses proved to be quite authentic and devoid of flippant, inappropriate comments. The subjects gave serious consideration to the instruments and gave fairly complete responses.

The responses were evaluated by the investigators independently, using a scoring rubric developed for this purpose. This rubric is presented in Table I. Differences in interpretation were resolved jointly through discussion and additional consideration of the responses. However, we did not formally address the issues of inter-rater reliability and statistical item analysis.

Table 1: Scoring Rubric

Score   *Criteria*

3   Conceptual understanding apparent; consistent notation, with only an occasional error; logical formulation; complete or near-complete solution/response.

2   Conceptual understanding only adequate; careless mathematical errors present (algebra, arithmetic, for example); some logical steps lacking; incomplete solution/response.

1   Conceptual understanding not adequate; procedural errors; logical or relational steps missing; poor response or no response to the question posed.

0   Does not attempt problem or conceptual understanding totally lacking.

Findings

Data collected suggested that subjects appear to be most comfortable with standard items from the first portion of the Calculus sequence—functions, limits, and derivatives. Subjects showed marginally better aptitude through nonstandard items than standard items in the subsequent topics of Calculus—integration, sequences, and series. In fact, by considering the frequencies associated with the Calculus items, a clearer picture of subject responses emerged. Of the items attempted by the subjects, slightly more than half of the responses were minimally acceptable. The remainder of the responses did not reflect adequate conceptual understanding. About two-thirds of those giving acceptable responses presented complete or near-complete responses. The data suggested that most inaccurate responses were not due to carelessness, but rather incomplete or inaccurate conceptualization.

For the non-Calculus items, subjects demonstrated great variability in response patterns. A few subjects returned blank or nearly-blank forms, while others attempted a few scattered items. When only a few items were attempted, subjects tended to respond to nonstandard items more often than to standard items. It appeared that subjects were more willing to attempt freshly-posed, nonstandard items than familiar but inaccessible standard items. A few subjects attempted most or all of the items. Of the non-Calculus items attempted by the subjects, about half of the responses were minimally acceptable.

Subsequently, revised (shortened) instruments were administered to a second set of approximately 15 subjects in Spring 1996. The resulting data suggested that the same basic information could be obtained using instruments with fewer items (12 to 15 questions, balanced between standard and nonstandard items) and a 60 minute examination period. It was observed from response patterns to the original and revised instruments that significant numbers of subjects do not explain their answers when specifically asked to do so. By the students' responses, it appears that attitudinal factors may have interfered with the assessment process. It was perplexing to the investigators that departmental majors appear to possess some of the same undesirable attitudes toward mathematics that cause learning interference in non-majors. We are in the midst of a pre/post administration of a Likert-type instrument developed to assess our majors' attitudes toward mathematics.

It should be noted that all core courses during the period of assessment were taught by standard methods. Because course syllabi did not require a technology component at the time, the degree of use of available technology such as graphing calculators or available computer software varied widely among instructors and courses.

Use of Findings

Our students need experience pushing beyond rote skills toward a more mature perspective of the subject. In order to help our students to develop better discussion and analysis skills, restructuring of the core courses was carried out beginning in 1994. We expect these changes will introduce students to more nonstandard material and to utilize nonstandard approaches to problem solving. These changes should lead to better problem-solving abilities in our students than is currently apparent.

Restructured versions of our Calculus, Discrete Mathematics, and Algebraic Structures courses came online in Fall

1996. Our Calculus sequence has been restructured and a new text selected so that the organization of the course topics and their interface with other courses such as Linear Algebra is more efficient and timely. More discretionary time has been set aside for the individual instructors to schedule collaborative laboratory investigations, principally using Matlab and Mathematica. The extant Discrete Mathematics course has been restructured as a Discrete Systems course and expanded to include logic, set theory, combinatorics, graph theory, and number systems development. The Algebraic Structures course now builds on themes introduced in the Discrete Systems course. A restructured Linear Algebra course with integrated applications using graphing calculators or computer software is in process. We anticipate that these curricular refinements will be evaluated during the 1998–99 academic year.

Success Factors

This project forced us to grapple with two questions: "What do we really expect of our graduates?" and "Are these expectations reasonable in light of current curricula and texts?" This concrete approach to these questions helped us to focus on our programmatic goals, present these expectations to our students, and gauge our effectiveness to this end. The project has guided us to redefine our curriculum in a way that will better serve our graduates' foreseeable needs. As their professional needs change, our curriculum will need continued evaluation and refinement.

Developing a custom instrument can be time intensive, and needs university support and commitment. Proper sampling procedures and a larger number of student subjects should be used if valid statistical analyses are desired.

Reference

[1] Emert, J.W. and Parish, C.R. "Assessing Concept Attainment in Undergraduate Core Courses in Mathematics" in Banta, T.W., Lund, J.P., Black, K.E.,and Oblander, F.W. eds., *Assessment in Practice: Putting Principles to Work on College Campuses,* Jossey-Bass Publishers, San Francisco, 1996, pp. 104–107.

Outcomes Assessment in the B.S. Statistics Program at Iowa State University

Richard A. Groeneveld and W. Robert Stephenson
Iowa State University

Although the statistics program at this large Ph.D.-granting university in the Midwest is not housed within a mathematical sciences department, the wide variety of measures used to assess the statistics program can serve as one model for assessing the mathematics major. All of the assessment measures used are described with particular emphasis on surveys of graduates and surveys of employers of graduates.

Background and Purpose

Iowa State University (ISU) is a large land-grant institution with a strong emphasis in the areas of science and technology. It was, in fact, designated the nation's first land-grant college when Iowa accepted the terms of the Morrill Act in 1864. It is located in Ames, Iowa, near the center of the state about 30 miles north of Des Moines. The current enrollment is about 24,000 students and about 20,000 of these are undergraduates, eighty-five percent from Iowa. The B.S. program in statistics began with the establishment of the Department of Statistics at ISU in 1947. It was one of the first universities in the U.S. or elsewhere to offer a curriculum leading to the B.S. degree in statistics. The first B.S. degree in statistics was awarded in 1949, and through 1997 over 425 individuals have received this degree. The enrollment in the major has remained reasonably stable over the period 1975–1997 in the range of 30–40 students. The current curriculum for the B.S. statistics degree includes:

1. An introductory course in statistics.

2. A three semester sequence in calculus and a course in matrix algebra or linear algebra.

3. A two semester sequence in probability and mathematical statistics (with the prerequisites of (2) above).

4. A two semester sequence in statistical methods including the analysis of variance, design of experiments, and regression analysis.

5. A two semester sequence in statistical computing.

6. A course in survey sampling design.

7. Two or more elective courses in statistics at the senior level or above.

In addition, the College of Liberal Arts and Sciences (LAS) has a variety of distribution requirements in the arts and humanities, foreign language, communication, the natural sciences and the social sciences.

The goals of the B.S. program in statistics had been implicitly considered as early as 1947 when the curriculum for the B.S. degree was established. However, the aims of the program had not been written in explicit form. A memorandum in September 1991, from the Office of the Provost, required each department and college to develop a plan for assessing the outcomes of undergraduate instructional programs. In response, in spring 1992 the Department submitted a document entitled "Student Outcomes Assessment for the B.S. Program in Statistics," to the Dean of the LAS College, which summarized the intended outcomes for our undergraduate major as follows:

> Students completing the undergraduate degree in statistics should have a broad understanding of the discipline of statistics. They should have a clear comprehension of the theoretical basis of statistical reasoning and should be proficient in the use of modern statistical methods and computing. Such graduates should have an ability to apply and convey statistical

concepts and knowledge in oral and written form. They should have the technical background and preparation to assume an entry level statistics position in commerce, government or industry. Academically talented and strongly motivated B.S. level graduates should have adequate background to pursue study towards an advanced degree in statistics.

With the idea of finding suitable measures of program performance, the 1992 plan also included the following abilities, knowledge and/or skills expected of B.S. graduates in statistics.

1. A knowledge of the mathematical and theoretical basis of statistical inference and reasoning. At a minimum this includes a knowledge of calculus, through multivariable calculus, of matrix theory and of probability and mathematical statistics.

2. A knowledge of statistical methods commonly used in practice. These include the analysis of variance, the design of experiments, the design of statistical surveys, and multiple regression.

3. Competency in the use of modern statistical computing. This includes facility with one or more statistical packages such as the Statistical Analysis System (SAS) or the Statistical Package for the Social Sciences (SPSS). Graduates should be capable of using modern graphical and display methods with real data.

4. An ability to summarize and present the results of a statistical study, orally or in writing to an educated, but not necessarily statistically expert, audience.

5. Proficiency in the use of statistics in a particular area of application or proficiency in the statistical analysis of a particular type of data.

Method

The following measures and procedures have been used over the period 1992 to date to assess the success in achieving the goals or knowledge/skills for students in the B.S. program in statistics at ISU. A brief description of these methods is presented here together with an indication of which goal(s) or knowledge/ skills(s) mentioned in Section 1 they are designed to measure or reflect.

1. A distribution of B.S. level statistics graduates grades (without names) has been maintained for the following categories of courses:

    a) Mathematical and theoretical courses

    b) Statistical methods courses

    c) Computer science and statistical computing courses

    d) Statistics courses

    e) All courses.

Grade data is summarized annually by the Director of Undergraduate Studies and is included in an annual Outcomes Assessment report to the LAS Dean. This information is also available to students interested in majoring in statistics.

These grade records provide information about the goals of B.S. graduates having a clear comprehension of the theoretical basis of statistics and being proficient in the use of modern statistical methods and computing. They refer directly to assessment of program and student success mentioned in the first three knowledge/skills categories in the first section.

2. A record of the performance on Actuarial Examinations 100 (General Mathematics) and 110 (Probability and Statistics) has been kept.

While taken by only a small number of our graduates these standardized examinations assess a B.S. student's knowledge of the mathematical and theoretical basis of statistics.

3. The Department keeps a record of summer undergraduate internships involving statistics in the Statistical Laboratory Annual Report, which receives wide distribution.

4. A record of the first positions or activities of our graduates is maintained. The names of the corporations, governmental bodies, or other institutions employing B.S. graduates together with the positions our students have obtained are listed. These positions or activities are grouped as commercial, governmental, manufacturing, graduate school and other.

The success of our graduates in obtaining internships and first positions in statistics reflects the goal of preparing our graduates to obtain entry level positions in commerce, government and industry. Additionally, the later hiring of statistical interns in permanent positions and multiple hiring by well established corporations or other institutions indicates the success of these graduates in meeting the general goals of our program.

5. A record of graduate degrees obtained by B.S. graduates (and the institution from which these degrees have been received) has been kept since 1974.

Most of our graduates who continue to graduate school do so in statistics. A record of the success of our graduates in obtaining an advanced degree (usually the M.S. in statistics) indicates the success of the program in providing an adequate background for academically talented students to pursue graduate study in statistics.

6. During the period July–October 1992 a survey of employers of two or more of our B.S. graduates in statistics since 1980 was conducted to obtain information about the opinions of supervisors of our graduates about their educational background. Respondents were encouraged to give their views on the strongest and weakest abilities of our graduates and make recommendations for improvement of our B.S. program.

This survey was valuable in providing the viewpoint of the employers of our graduates on all aspects of the B.S. program. Questions were asked relating to all the goals and knowledge/skills mentioned in Section 1.

7. A questionnaire was sent in the academic year 1992–93 to 111 B.S. graduates of the Department of Statistics receiving their undergraduate degrees in the period 1981–1991. Of these surveys 55 (50 percent) were returned. The survey was developed by Professor W. Robert Stephenson with assistance from the Survey Section of the Statistical Laboratory at ISU. The survey was aimed at determining the current employment and graduate educational experience of our B.S. graduates. Questions were also included to determine these graduates' evaluation of their educational experience at ISU. Additional questions were included to determine how well graduates of our program thought individual courses offered in the undergraduate statistics program prepared them for further education and employment. Open-ended questions about strong and weak points of the undergraduate program and an opportunity to make suggestions for improving the undergraduate statistics program were also included. This survey was also valuable in providing responses from our B.S. graduates concerning all aspects of the B.S. program in statistics at ISU.

Findings

1. *Distribution of B.S. level grade point averages by categories.*

For the 62 graduates of the program over the period from fall 1990 to spring 1997 the average GPAs were calculated for the categories defined in Section 2 and are presented below.

Category	Mean	Standard Deviation
(a) Mathematical and theoretical courses	2.84	0.69
(b) Statistical methods courses	3.30	0.56
(c) Computer science and statistical computing courses	3.03	0.77
(d) Statistics courses	3.25	0.56
(e) All courses	3.06	0.56

The information concerning grades reinforced our general perception that statistics majors have most difficulty with theory courses and least with statistical methods courses. No trends over this relatively short period of time in these GPAs have been observed.

2. *Actuarial Examinations 100 (General Mathematics) and 110 (Probability and Statistics).*

During the 1991–97 period three students took the 100 Examination, all during the current (1996–97) academic year. Two passed and one did not.

3. *Summer Internships.*

As an example, in the summer of 1996 four students held such positions. Companies employing these students were John Deere Health Care in Davenport, IA, the Mayo Clinic in Rochester, MN, the Motorola Company in Mount Pleasant, IA, and the Survey Sampling Section of the Statistical Laboratory at ISU. The first of these students was hired permanently by his internship company upon his graduation in spring 1997. The second was rehired by the Mayo Clinic this summer before he begins graduate school in statistics in fall 1997 at ISU. The third has an internship with the 3M Corporation in Minneapolis, MN this summer before she begins graduate school in statistics at the University of Minnesota in the fall. The last is working for the Statistical Laboratory again, prior to his senior year at ISU. It is generally true that it is possible to place only our better students in these rewarding positions.

4. *First Positions or Activities.*

Of the 62 graduates from fall 1990 to spring 1997, 15 have gone to graduate school, 27 to positions in commerce (in banking, health services, insurance, research and similar organizations), 2 to positions in manufacturing, 7 to governmental positions and 11 are in other situations.

5. *Graduate School in Statistics.*

Eight B.S. graduates since 1990 have received M.S. degrees in statistics, four are currently in M.S. degree programs in statistics, and one (in a Ph.D. program in statistics) has completed the M.S. degree requirements.

6. *Survey of Employers of B.S. Graduates.*

The purpose of this survey was to obtain an evaluation of the strengths and weaknesses of graduates of our program as seen by employers of these graduates and to make suggestions to improve our program. The average overall response (on a scale of Poor=1 to Excellent=5) of respondents to the questions concerning the overall quality of the education of our B.S. graduates was 4.21/5.00, received with some gratitude by the current authors. The strongest abilities of our graduates noted by their supervisors were in general statistical background and in an area best described as "having a good work ethic."

Employers expressed concern about the following areas of the background of our graduates:

a) Knowledge of "real world" applications of statistics.

b) Ability to communicate statistical ideas well orally and in writing.

c) Having substantive knowledge in a specific field of application.

7. *Survey of Graduates of the B.S. Program in Statistics at ISU.*

Of the 55 individuals returning a survey 30 (54.5 percent) had continued to graduate school with 26 of 30 (86.7 percent) either having completed a graduate degree or continuing in graduate school. This probably indicates that this group was more likely to respond to the survey, as other information indicates that 35–40 percent of our B.S. graduates continue to graduate education. Almost all of the respondents have been employed since receiving the B.S. degree, with 87 percent of the respondents having taken a position in statistics or a related field.

B.S. graduates did indicate some areas of concern with our program. They indicated, as did employers, that there is a clear need for strong oral and written communication skills. This was particularly evident in the responses of individuals who went directly to positions in commerce, government or industry. Secondly, B.S. graduates reported a need for improved background in two topics in statistical computing — computer simulation and graphical display of data. Finally, the course on survey sampling was the lowest rated of the core (required) courses in the student survey.

Readers who wish to obtain copies of the survey instruments and/or a more complete summary of the results of the survey of B.S. graduates may contact W. Robert Stephenson via email at: wrstephe@iastate.edu.

Use of Findings

Explicit action has been taken regarding the findings under the headings 1, 6 and 7 in the previous section. Firstly, noting that the courses in probability and mathematical statistics are the most difficult core courses for our students, additional efforts have been made to provide regular problem sessions and review lectures prior to examinations in these courses in probability and mathematical statistics.

The Department has recognized the need for better communications skills and has for many years required a Junior/Senior level writing course (Business Communications or Technical Writing) in addition to the one year Freshman English sequence. We have also required a speech course, typically Fundamentals of Speech Communication. Both of these courses were ranked highly by students in response to the question: "How valuable has this course been in relation to your current employment?" As a result of the continued emphasis placed on communications by our graduates and their employers, we now incorporate writing projects in the statistical methods sequence required of all majors. These cover a wide range of topics and styles. At the beginning of the sequence students are asked to write short summaries of statistical results they obtain on weekly assignments. Later, they are asked to write a short newspaper-like article describing a scientific study that explains statistical analyses and concepts. The students also participate in group data collection and analysis projects that result in five to ten page technical reports. The Department now has a requirement that a sample of the technical writing of each graduate be evaluated and placed in their departmental file prior to graduation.

Concerning computing, the Department received an National Science Foundation (NSF) supported Instructional Scientific Equipment Grant (1992–94) which has permitted the acquisition of 22 Digital Equipment Corporation (DEC 5000/25) workstations in a laboratory in Snedecor Hall where the Department of Statistics is located. This has permitted substantial improvement in the teaching of computer graphics and simulation in many of the required courses in the B.S. program. For example, in the multivariate analysis course, students can now look at several dimensions simultaneously. It is now possible to rotate the axes of these plots to visually explore interesting patterns in multidimensional data. In the statistical computing courses, students are not only able to simulate sampling from populations, but they are able to visualize sampling distributions and statistical concepts such as the central limit theorem and confidence intervals.

Recognizing that computing is becoming extremely important for practicing statisticians, the Department has received funding from NSF (1997–99) to upgrade the equipment available to our undergraduates. All of the workstations described above will be replaced by high performance DEC Alpha workstations. This equipment will also allow faculty members to develop instructional modules that will go beyond traditional statistical methods. The modules will be developed around real world problems and will implement new statistical methods (often computationally intensive) to solve those problems. It is our intent to expose our students to these new methods so that they will be better prepared as statisticians.

Responding to the suggestions of our B.S. graduates, the survey sampling course has been redesigned by two new faculty members. They have introduced hands-on experience in designing surveys and analyzing their results utilizing a simulation package on the work-stations. The recent (S95 and S96) student evaluations have shown substantial improvement in student reception of the redesigned survey sampling course.

The documentation of summer internships achieved, first positions obtained, and graduate degrees received by B.S. graduates under headings 3, 4, and 5 in the previous section give a method of monitoring the success in achieving the goals of preparing graduates for entry level positions in commerce, government and industry and of preparation for graduate school stated in Section 1.

Success Factors

To implement a program of outcomes assessment of this type requires substantial effort on the part of all faculty associated with the B.S. program. It is important that a program of Student Assessment have support at a high administrative level (i.e., the presidential level) and the support of the Faculty Senate (or equivalent group). There must also be departmental support to provide adequate financial and secretarial aid to maintain records and carry out surveys. In our opinion this will take time equivalent to at least one course per year for a faculty member. Such an individual should be assigned to maintain such records and carry out and analyze appropriate surveys. We feel that this duty should be assigned over a period of years (about four). Such a program needs administrative support and understanding that improvements in undergraduate education require effort, resources and time. We believe that improvements in our B.S. program have justified these expenditures.

Assessment of the Mathematics Major at Xavier: the First Steps

Janice B. Walker
Xavier University

In an evolving assessment program at a private, medium-sized, comprehensive university in the Midwest a variety of assessment techniques are being developed to assess student learning. How two of them — exit interviews and the Educational Testing Service's Major Field Test in Mathematics — are part of the fabric of the mathematics program's assessment cycle is described in this article.

Background and Purpose

Xavier University is a Catholic, Jesuit institution having approximately 2800 full-time undergraduates and about 1,100 part-time undergraduates. The Department of Mathematics and Computer Science is housed in the College of Arts and Sciences and has fourteen full-time members and six part-time members. All but three teach mathematics exclusively. There are about 45 mathematics majors with an average of ten graduates each year. The mathematics major at Xavier must complete 42 semester hours (thirteen courses) of mathematics, which include three semesters of calculus, differential equations, two semesters of linear algebra, abstract algebra, real analysis, and four upper division electives.

In the fall of 1994, each department received a request from the office of the academic vice president to devise plans to assess programs for its majors by March of 1995. Each plan was to cover the following topics:

- goals and objectives
- the relationship of department goals to university goals
- strategies to accomplish the goals for student achievement
- assessment techniques
- feedback mechanisms (for students, faculty, and the university)

Although we have a bachelor's program for computer science in the Department, this paper will only address assessing the program for mathematics majors at Xavier University.

According to the mission statement of the University, the primary mission of the University is to educate, and the essential activity at the University is the interaction of students and faculty in an educational experience characterized by critical thinking and articulate expression with special attention given to ethical issues and values. Moreover, Jesuit education is committed to providing students with a supportive learning environment, addressing personal needs, and developing career goals along with the academic curriculum.

Method

Our assessment plan records students' growth and maturity in mathematics from the time they enter the program until they graduate. The plan was submitted in the spring of 1995 and put into place during the school year 1995-96. We chose the following methods for assessing the major:

- Portfolio
- Course Grades
- Course Evaluations
- MFT (Major Field Test)
- Senior Presentation
- Exit Questionnaire and Exit Interview
- Alumni Survey and Alumni Questionnaire

During the first few weeks of the fall semester, freshman mathematics majors receive letters stating the expectations for majors for the coming years. In particular, they are informed that they must maintain a portfolio, make an oral presentation during their senior year, and meet our standards for satisfactory performance on the MFT (Major Field Test) in Mathematics.

Portfolio: Portfolios are maintained by faculty advisors. Each semester the student must submit final examinations and samples of their best work from each mathematics class. A copy of any research or expository paper in an upper level class should also be included. Hopefully, these works will provide evidence of how well the students are learning to synthesize and integrate knowledge from different courses in the program. Moreover, the samples should indicate that a student can use various approaches to solve problems. Samples of work will be collected for the portfolio by advisors during formal advising periods.

A number of additional documents are standard. These include an entry profile, an entry questionnaire, counseling forms, and advisor notes. The entry profile contains information on a student prior to enrolling at the University, such as the mathematics courses taken in high school, grade point average and rank in graduating class, SAT and/or ACT scores, and advanced placement (AP) credits. It also contains the placement scores from the examinations in mathematics and foreign language which are administered by the University. The entrance questionnaire, a short questionnaire given in the fall term, reveals the student's perception of mathematics and his expectations of the program and faculty. (See [1] in this volume for further information on student portfolios).

MFT: The Department has given some form of a comprehensive exam for many years. The Advanced Test in Mathematics of the GRE served as the last measure until the 1994-1995 academic year. It was replaced by the Major Field Test (MFT) in Mathematics, which is distributed and scored by the Educational Testing Service. The profiles of the students taking the MFT are a far better match with the profiles of our students than with those taking the GRE. Moreover, not only are individual and mean scores reported, but also subscores which help point out the strengths and weaknesses of the examinees in broad areas within mathematics.

The Department has set as successful performance a score at the 60% mark or higher. However, this mark in not rigidly enforced. The Department undertakes an extensive study of the individual case whenever this goal is not met. Then a recommendation is made whether to pass the student or ask the student to retake the exam. A student may request a departmental comprehensive examination after two unsuccessful performances on the MFT. If a student's score on this exam is also unsatisfactory, the student may be advised to seek a bachelor's degree in liberal arts. A degree in liberal arts would necessitate no additional requirements to be fulfilled for graduation.

We expect very few students, if any, to find themselves in the position of having failed the MFT twice. Thus far, no one has been unsuccessful on their second attempt at making a passing score. During the fall semester the Department provides review sessions for seniors preparing for the MFT. The whole experience of the review sessions and exam-taking serves as a mechanism for students to rethink and synthesize mathematical concepts as well as to perfect skills that were covered in their mathematics courses.

Senior Presentation: Prior to our assessment plan, students rarely participated in departmental colloquia, but will now assume an integral and vital part. Beginning in the 1998-99 academic year each senior will be required to make an oral presentation. The topic for the senior presentation will be selected by the student with the approval of the advisor for senior presentations. The advisor will provide each senior with a list of steps that are helpful in preparing a talk. After a student chooses a topic, she must submit a written draft for a thirty-minute lecture. The advisor will provide feedback on the draft and set up a practice session for the student.

Questionnaires: Questionnaires and surveys are critical elements of the assessment process. Exit questionnaires are sent to graduating seniors about three weeks prior to the end of the spring semester and exit interviews are scheduled. An alumni survey will be administered every five years. This will help us track the placement of students in various jobs and professional programs. Between the years scheduled for alumni surveys, alumni questionnaires are mailed to graduates one year after graduation. These forms are much shorter and do not cover topics in the breadth and depth as does the alumni survey.

Findings

We have completed two years with the new assessment plan. The first steps proceeded fairly well. The freshman advisors created portfolios for the freshmen, the chair passed out entry profiles, and advisors collected samples of work for the portfolios throughout the year. There were very few freshmen in the second year and thus the job of the creating portfolios was easier, but that of recognizing a class profile was more difficult.

Although our experience with portfolios is very limited, portfolios have been useful. Not only is there an evolving picture of mathematical growth and progress, but also evidence of the content, level of difficulty, and teaching philosophy in the mathematics courses for majors. They provide information on the manner in which a course is taught which may not be obtained from course evaluations. Thus, a more thorough understanding of the overall curriculum is likely as more students and courses are tracked.

In 1996, seniors took the MFT exam and each met the departmental goal. In 1977 two of ten seniors did not meet the goal. They subsequently retook the exam and passed.

Although seniors are not yet required to make formal presentations, there have been presentations by the juniors and seniors in the departmental colloquium series for the last two years. Several students who participated in the 1996-97 Mathematics Modeling Competition gave talks which were very well attended by mathematics and computer science majors. Freshmen have been strongly encouraged and sometimes required to attend the colloquia.

Almost all majors in the classes of 1996 and 1997 returned their exit questionnaires and had an exit interview. The data from both 1996 and 1997 were similar and consistent with the information gathered in our alumni survey, which was carried out in the fall of 1995. Although these students and alumni have overall favorable impressions of the program and faculty, they raised issues that must be addressed. In particular, some felt that there should be more electives and fewer required courses, while others suggested that specific courses be added to or removed from the curriculum. A few thought that there should be a capstone course. Comments about the discrete mathematics course were generally negative and the merits of particular electives repeatedly surfaced.

There was substantial feedback on instructors, the use of technology in courses, and the availability of resources (particularly computers) in the exit questionnaires. In particular it was remarkably clear whom students classified as the very good and the not-so-good instructors and the reasons for their choices. Experiences with MAPLE, the computer algebra system employed in the calculus and linear algebra sequences during the first two years, drew mixed reviews. Students in Calculus III liked using MAPLE; a substantial number in Calculus I and Linear Algebra did not. Moreover, with the increasing use of computer labs on campus, accessibility to computer equipment is an issue that must be continually addressed.

The questionnaires and alumni survey also indicated a need for more counseling about career options. Prior to these results, most advisors believed that students were aware of the career opportunities that are available to them. This has proved to be false. Some students professed little knowledge about career opportunities and some had little idea of what they would do after graduation.

Use of Findings

The Department is responding to the issues that have surfaced from assessing the mathematics major. There are three main areas of focus. The one that has prompted the most scrutiny is the curriculum for the mathematics major itself. Some alumni who are currently teaching secondary mathematics remarked on the value of the elective Survey of Geometry that had been offered, but was dropped several years ago when the list of electives was streamlined. Some alumni and seniors suggested that statistics should be a requirement for all mathematics majors. In addition, many students questioned the intent of discrete mathematics and its contribution to their overall development. As a result of these findings, the Department will first thoroughly review the content in the discrete mathematics and the computer science courses in the freshman year. Then we will examine the cycle of electives. We are open to revising any aspect of the curriculum.

The use of MAPLE is the second area of concern. When there was a very heavy emphasis on MAPLE in a course, there was a considerable number of negative comments. In such cases, seniors clearly expressed the need for more balance in the use of technology. Many felt that the intense use of MAPLE left too little time for a deep understanding of basic concepts. This message was communicated to all departmental members, especially those named in the exit interviews and questionnaires. Moreover, the expansion of the use of MAPLE outside the calculus sequence has created a few problems of accessibility of computers.

The area of counseling is the third area of concern. To help students make better decisions about courses and career options, we are giving students more information early. In particular, during preregistration in the summer, the "fact sheet" on the mathematics program is given to all incoming majors. This brochure is put together jointly by the Department and the Office of Admissions (and revised annually). It contains the goals of the program, a description of the mathematics major, a recommended sequence of courses for the mathematics major, and information about careers in mathematics. In the fall semester, a packet containing additional information about careers from other sources will be given to freshmen majors.

It is quite apparent from the exit questionnaires and interviews which faculty members students perceive to be the best professors in the Department and the reasons for the acclaim. It is just as apparent which courses are the most difficult. Such information provides the chair with valuable information when composing a schedule of classes. The chair will speak to the individual faculty members whenever there is feedback which merits special attention.

The exit questionnaires are available to all members of the Department. Information in summary form is also made available to the Department and discussed in a department meeting. Thus, the Department is kept apprised of students' reactions.

Success Factors

Trends, problems, and successes have been fairly evident using data acquired from the groups of students with similar experiences at approximately the same time. Unfortunately, we have no means of requiring students to participate in the interviews and questionnaires, which are key elements of assessing the program for majors. Of course, students must take the MFT in Mathematics and make a senior presentation. These are now printed in the current University catalog. We hope to impress upon students the importance of their participation in the entire assessment process.

Assessing the program for majors is an ongoing project that will demand a commitment of time and energy from many department members. It is not difficult to foresee how the enthusiasm of some may wane and how those upon whom most of the work falls may become disgruntled. Because there are many aspects of the plan to monitor, it would be easy to let one or more measures slide from time to time. How the next chair will respond to the tasks is also an unknown. Hopefully, the Department will continue to assess the mathematics major beyond the "first steps" with the interest and energy that currently exist.

References

[1] Sons, L. "Portfolio Assessment of the Major," in this volume, p. 24.

Assessing Essential Academic Skills from the Perspective of the Mathematics Major

Mark Michael
King's College

At a private, church-related liberal arts college in the East a crucial point for assessing student learning occurs midway through a student's four year program. A sophomore-junior diagnostic project which is part of the Discrete Mathematics course, taken by all majors, is the vehicle for the assessment. Each student in the course must complete a substantial expository paper which spans the entire semester on a subject related to the course.

Background and Purpose

King's College, a church-related, liberal arts college with about 1800 full-time undergraduate students, has had an active assessment program for more than a decade. The program is founded on two key principles: (1) Assessment should be embedded in courses. (2) Assessment should measure and foster academic growth throughout a student's four years in college. The entire program is described in [1] and [2], while [4] details one component as it relates to a liberal arts mathematics course. The remaining components are aimed at meeting goal (2).

Of these, the key is the Four-Year Competency Growth Plans, blueprints designed by each department to direct the total academic maturation of its majors. These plans outline expectations of students, criteria for judging student competence, and assessment strategies with regard to several basic academic skills, e.g., effective writing. Expectations of students increase from the freshman year to the senior year. More importantly, the continued growth and assessment of competencies are the responsibilities of a student's major department. In the freshman year, all departments have similar expectations of their majors. In subsequent years, competencies are nurtured and evaluated in discipline-specific ways; mathematics majors are expected to know the conventions for using mathematical symbols in prose, while computer science majors must be able to give a presentation using a computer.

In addition to continual assessment charted by the growth plans, there are two comprehensive, one-time assessment events which take a wide-angle view of a student's development from the perspective of the student's major: the Sophomore-Junior Diagnostic Project and the Senior Integrated Assessment. The former is the subject of this report. It is a graded component of a required major course taken in the sophomore or junior year. Coming roughly midway in a student's college career, the Sophomore-Junior Diagnostic Project is a "checkpoint" in that it identifies anyone whose command of fundamental academic skills indicates a need for remediation. However, the Project is not a "filter;" it is, in fact, a "pump" which impels all students toward greater facility in gathering and communicating information. The variety of forms the Project takes in various departments can be glimpsed in [5].

Method

For Mathematics majors and Computer Science majors, the Project's format does not appear particularly radical. It resides in the Discrete Mathematics course since that course is required in both programs, is primarily taken by sophomores, and is a plentiful source of interesting, accessible topics. The Project revolves around a substantial expository paper on an assigned subject related to the course. It counts as one-fourth of the course grade. But how the

Project is evaluated is a radical departure from the traditional term papers our faculty have long assigned in Geometry and Differential Equations courses. Previously, *content* was the overriding concern. Now the emphasis is on giving the student detailed feedback on *all* facets of his/her development as an educated person.

Another departure from past departmental practice is that the Project spans the entire semester. It begins with a memo from the instructor to the student indicating the topic chosen for him/her. After the student has begun researching the topic, he/she describes his/her progress in a memo to the instructor. (This sometimes provides advanced warning that a student has difficulty writing and needs to be referred to the King's Writing Center for help.) It is accompanied by a proposed outline of the report and a bibliography which is annotated to indicate the role each reference will play in the report. This assures that the student has gathered resources and examined them sufficiently to begin assimilating information. It also serves to check that the student has not bitten off too much or too little for a report that should be about 10 double-spaced pages.

Requiring that work be revised and resubmitted is common in composition courses, less so in mathematics courses. It is crucial in this exercise. The "initial formal draft" is supposed to meet the same specifications as the final draft, and students are expected to aim for a "finished" product. Nonetheless, a few students have misunderstandings about their subject matter, and all have much to learn about *communicating* the subject. To supplement the instructor's annotations on the first draft, each student is given detailed, personalized guidance in a conference at which the draft is discussed. This conference is the most powerful learning experience to which I have ever been a party; it is where "assessment as learning" truly comes alive. Recognizing that all students need a second chance, I count the final draft twice as much toward the project grade as the first draft.

Also contributing to the project grade is a formal oral presentation using an overhead projector. As with the written report, students make a first attempt, receive feedback to help them grow, and then give their final presentation. Unlike the first draft of the report, the practice presentation has no impact on a student's grade. In fact, each student chooses whether or not the instructor will be present at the trial run. Most students prefer to speak initially in front of only their classmates, who then make suggestions on how the presentation could be improved.

While we mathematics faculty are increasingly recognizing the importance of fostering and evaluating our students' *non*mathematical skills, there is still the problem of our having to make professional judgments about writing and speaking skills without ever having been trained to do so. The solution in our department was to create "crutches": an evaluation form for the written report and another for the oral presentation. Each form has sections on content, organization, and mechanics. Each section has a list of questions, intended not as a true-false check-list but as a way to help the reader or listener focus on certain aspects and then write comments in the space provided. Some of the questions are designed for this particular project ("Are the overhead projector and any additional audio-visual materials used effectively?") while others are not ("Does the speaker articulate clearly?").

Findings

All students — even the best — find writing the Project report to be more challenging than any writing they have done previously. Student Projects have revealed weaknesses not seen in students' proofs or lab reports, which tend to be of a more restricted, predefined structure. In explaining an entire topic, students have many more options and typically much difficulty in organizing material, both globally and at the paragraph level. They also have trouble saying what they really mean when sophisticated concepts are involved.

Peer evaluations of the oral presentations have proved to be surprisingly valuable in several ways. First, by being involved in the evaluation process, students are more conscious of what is expected of them. Second, their perspectives provide a wealth of second opinions; often students catch flaws in presentation mechanics that the instructor might overlook while busy taking notes on content and technical accuracy. Third, students' informal responses to their peers' practice presentations have improved the overall quality of final presentations — partly by putting students more at ease!

Perhaps the most surprising finding, however, is how widely sets of peer evaluations vary in their homogeneity; some presentations elicit similar responses from all the audience, while in other cases it is hard to believe that everyone witnessed the same presentation. Peer evaluations, therefore, present several challenges to the instructor. What do you tell a student when there is no consensus among the audience? How does one judge the "beauty" of a presentation when different beholders see it differently? What are the implications for one's own style of presentation?

Use of Findings

Since the institution of the Sophomore-Junior Diagnostic Project, there has been an increased mathematics faculty awareness of how well each student communicates in the major. As the Projects are the most demanding tests of those skills, they have alerted instructors in other classes to scrutinize student written work. Before, instructors had discussed with each other only their students' mathematical performance. Moreover, there has been an increased faculty appreciation for writing and the special nature of mathematical writing. While national movements have elevated the prominence of writing as a tool for learning mathematics, this activity promotes good *mathematical* writing, something my generation learned by osmosis, if at all.

Furthermore, there is an increased realization that a student's ability to obtain information from a variety of sources — some of which did not exist when I was a student — is an essential skill in a world where the curve of human knowledge is concave up. Previously, some, believing that isolated contemplation of a set of axioms was the only route to mathematical maturity, had said, "Our students shouldn't be going to the library!"

Collectively, student performances on Projects have motivated changes in the administration of the Project. One such change was the addition of the trial run for oral presentations. A more significant change was suggested by students: replace the generic evaluation form used in the general education speech class (a form I tried to use the first time through) with a form specifically designed for these presentations.

While students were always provided detailed guidelines for the projects, I have continually endeavored to refine those handouts in response to student performances. Other forms of aid have also evolved with time. Exemplary models of exposition relevant to the course can be found in MAA journals or publications. For abundant good advice on *mathematical* writing and numerous miniature case studies, [3] is an outstanding reference. For advice on the use of an overhead projector, the MAA's pamphlet for presenters is useful. In addition, I now make it a point to use an overhead projector in several lectures expressly to demonstrate "How (not) to."

Success Factors

The primary key to the success of the Sophomore-Junior Diagnostic Project is in giving *feedback* to students rather than a mere grade. When penalties are attached to various errors, two students may earn very similar scores for very different reasons; a traditional grade or score might not distinguish between them, and it will *help* neither of them.

Another contributor to the Project's success as a learning exercise is the *guidance* students are given throughout the semester. Students need to know what is expected of them, how to meet our expectations, and how they will be evaluated. The instructor needs to provide specifications, suggestions, examples, and demonstrations, as well as the evaluation forms to be used.

The third factor in the success of the Projects is the *follow-up*. Best efforts — both ours and theirs — notwithstanding, some students will not be able to remedy their weaknesses within the semester of the Project. In light of this (and contrary to its standard practice), King's allows a course grade of "Incomplete" to be given to a passing student whose Project performance reveals an inadequate mastery of writing skills; the instructor converts the grade to a letter grade the following semester when satisfied that the student, having worked with the College's Writing Center, has remedied his/her deficiencies. This use of the "Incomplete" grade provides the leverage needed to ensure that remediation beyond the duration of the course actually occurs.

The many benefits the Sophomore-Junior Diagnostic Projects bring to students, faculty, and the mathematics program come at a cost, however. The course in which the Project is administered is distinguished from other courses in the amount of time and energy both instructor and student must consume to bring each Project to a successful conclusion. This fact is acknowledged by the College: the course earns four credit-hours while meeting only three hours per week for lectures. (Extra meetings are scheduled for oral presentations.)

For the instructor, this translates into a modicum of overload pay which is not commensurate with the duties in excess of teaching an ordinary Discrete Mathematics course. The primary reward for the extra effort comes from being part of a uniquely intensive, important learning experience.

References

[1] Farmer, D.W. *Enhancing Student Learning: Emphasizing Essential Competencies in Academic Programs*, King's College Press, Wilkes-Barre, PA, 1988.

[2] Farmer, D.W. "Course-Embedded Assessment: A Teaching Strategy to Improve Student Learning," *Assessment Update*, 5 (1), 1993, pp. 8, 10–11.

[3] Gillman, L. *Writing Mathematics Well*, Mathematical Association of America, Washington, DC 1987.

[4] Michael, M. "Using Pre- and Post-Testing in a Liberal Arts Mathematics Course to Improve Teaching and Learning," in this volume, p. 195.

[5] O'Brien, J.P., Bressler, S.L., Ennis, J.F., and Michael, M. "The Sophomore-Junior Diagnostic Project," in Banta, T.W., et al., ed., *Assessment in Practice: Putting Principles to Work on College Campuses*, Jossey-Bass, San Francisco, 1996, pp. 89–99.

Department Goals and Assessment

Elias Toubassi
University of Arizona

At a large, research university in the West, a major effort over the past 10 to 15 years has been underway to reform the entry level mathematics courses which the department offers. Assessment has been at the heart of this process. Focusing on all first year courses through assessment has had a positive effect on the undergraduate mathematics major and the courses in that major. This article describes the process and the ongoing assessment of student learning.

Background and Purpose

During the 1970s the University of Arizona, like many other colleges and universities, experienced a large growth in enrollment in beginning mathematics courses. This growth was coupled by a loss of faculty positions over the same period. The resulting strain on resources forced the department to make some difficult decisions in order to meet its teaching obligations. Notable among these was the creation of a self-study algebra program and the teaching of finite mathematics and business calculus in lectures of size 300 to 600. The situation reached its lowest point in the early 1980s when the annual enrollment in the self study algebra program exceeded 5000 students and the attrition rates (failures and withdrawals) in some beginning courses reached 50%. Before we describe the turnaround that began in 1985 we give some data about the department.

Currently the Mathematics Department has 59 regular faculty, 8 to 12 visiting faculty from other universities, 20 to 25 adjunct faculty, 6 to 8 visiting faculty from high schools and community colleges, 2 research post-doctorate faculty, and 2 teaching post-doctorate faculty. Each semester the Department offers between 250 and 300 courses serving about 10,000 undergraduate and graduate students. (The enrollment of the University is around 35,000.) Sixty to seventy percent of mathematics students take freshmen level courses in classes of size 35. There are approximately 350 majors in mathematics, mathematics education and engineering mathematics. In the past

five years the Department has secured external funding at an average of 2.5 million per year.

The overarching goal of the Mathematics Department at the University of Arizona is to provide intellectual leadership in the mathematical sciences. It spans the areas of research, undergraduate and graduate teaching, outreach, and collaboration with other University units. The Department's goals include:

1. To provide flexible yet solid undergraduate and graduate programs which challenge the intellect of students.
2. To prepare an ethnically diverse spectrum of entering students.
3. To embrace the notion that change such as is manifested in computer technology and educational reform can enhance learning and enrich the intellectual environment.
4. To be a resource in the mathematical sciences for other disciplines whose own activities have an ever-increasing need for the power of mathematics.
5. To work closely with colleagues from the local schools and community colleges who share with us the responsibility of ensuring the flow of a mathematically literate generation of students.

Method

The assessment of the freshman mathematics program began in 1983 with a Provost appointed University committee consisting mostly of faculty from client departments with three representatives from mathematics. This was followed by an intensive departmental review in 1984. The review included an examination of the department by an Internal Review Committee and an External Review Committee of distinguished mathematicians from other universities. The committees reviewed reports on enrollment, pass/failure rates in lower division courses, interviewed faculty members individually or in small groups, met with graduate teaching assistants, mathematics majors, representatives of client departments and the University Administration.

In the 1992/93 academic year the Department went through its second review to check on its progress to meet its goals. This included departmental committees on Faculty and Research, Undergraduate Education, Graduate Program, Academic Outreach, Collaboration with other University Units, Space, the Computer Environment, Library Facilities, and the Departmental Administration. In addition, there was a University Review Committee consisting of faculty from hydrology, chemistry, philosophy and biology and an External Review Committee with mathematics faculty from MIT, Rutgers, Wisconsin, Nebraska, and Texas.

Other reviews of the program are conducted regularly. One in particular that is worthy of mention is done by the Mathematics Center in the form of exit surveys of mathematics majors. The surveys cover three main areas. The background section covers such topics as why the student became a mathematics major and whether the student worked while in college. The mathematics experience section covers such areas as the courses and instructors the student appreciated and whether the overall educational experience met the student's expectations. The future plans section inquires into the student's immediate plans such as graduate school, employment or other options. Exit interview data is reviewed by the Associate Head for Undergraduate Programs and the Department Head. Comments and suggestions by students are considered in making changes in the undergraduate program and in continuing to find ways to support our undergraduates.

Findings

These reports documented several important findings. The following are three quotations from the 1983 and 1984 reports indicating the dismal nature of things at the time. Provost Committee: "Although other factors need to be considered, it is difficult to avoid the conclusion that many of the problems in freshman math can be traced to inadequate allocation of resources by the University Administration." Internal Review Committee: "The most important course in the lower division is calculus and this course is not a success ...the success rate is less than 50% ..." External Review Committee: "One of the best ways to help the research effort

... would be to tame this monstrous beast (precalculus teaching) ... it seems essential for everyone's morale that the pass/fail ratio be increased, section size reduced, ..."

The situation was different in 1992. The findings by the review committees showed that substantial improvements were accomplished since 1984. The following are some quotations from these reports. University Review Committee: "The Department of Mathematics has made tremendous strides since its last review in 1984. Its faculty has grown and established a firm scholarly reputation in key areas of pure and applied mathematics. Undergraduate education has improved greatly due to a reduction in entry level class sizes, the introduction of desktop computers in the classroom, and the vibrant activities of the mathematics education group that has evolved into a national leader ..." External Review Committee: "Since the last external review (1984), the Department has addressed the problems of entry-level instruction responsibly, effectively, and efficiently . . ."

Use of Findings

These reports led to the appointment of two significant committees: an Ad Hoc Intercollegiate Committee on Calculus and a Departmental Entry Level Committee with the charge to draft a plan to address the problems in beginning mathematics courses. The result was a five year implementation plan together with an estimate of the resources needed in each year. The plan was based on five fundamental premises.

1. Students must be placed in courses commensurate with their abilities and mathematical background.
2. Students must be provided with a supportive learning environment and a caring instructor committed to undergraduate education.
3. Entry Level courses must be structured to meet current student needs.
4. The future success of the program relies on an effective outreach program to local schools.
5. Students need to be exposed to technology to enhance the learning experience.

The following are the key recommendations of the Entry Level Committee's action plan. First, reduce class size. Second, institute a mandatory math placement and advising program. Third, institute a new precalculus course. Fourth, introduce a calculus I course with five credit hours. Fifth, replace the self-study intermediate and college algebra courses with a two semester sequence in college algebra to be taught primarily in small classes. (Note: This sequence is currently being upgraded to a one semester four credit algebra course.)

The findings by the various review committees coupled with the action plan of the Entry Level Committee began a process of change in 1985 which resulted in a dramatic turnaround in the educational environment. Over the past twelve years the University has provided mathematics with a dozen new faculty positions and a supplementary budget to improve undergraduate education. (This is currently at $900,000 per year.) These funds, augmented by grant monies from NSF, have allowed the Department to dramatically improve the quality of the undergraduate mathematics experience particularly for beginning students. This was accomplished under the leadership of the Head, a dozen or so faculty, and a dedicated cadre of adjunct lecturers. In sharp contrast to the situation pre 1985, currently all freshman level courses are taught in classes of size 35. Student performance has improved greatly and the passing rate in these courses has increased 30 to 40 percent. We have excellent relations with other departments in the University, with the high schools, and with the local community college.

The number of mathematics majors has increased steadily over the past eight years. In the 1988-89 school year the University awarded 12 BS or BA degrees while in 1996-97 we expect to award 52 degrees. In this period the number of degrees in mathematics education have been steady at around 12 while the number of engineering mathematics degrees has declined from 12 to 4 due to the establishment of undergraduate degree programs in computer science and computer engineering.

The increase in the number of mathematics majors is due to several factors but in particular to the increased time and effort put in to improve their lot. We have created a new Mathematics Center—an office with library materials, a full-time advisor, lockers, and computer facilities. The Center provides special services to mathematics majors including access to an academic advisor during working hours, upper division tutoring, an undergraduate mathematics colloquium, and information about graduate schools, careers, and summer programs. The Center is under the direction of a new position of Associate Head for Undergraduate Programs.

In the last six years the Department has been involved in education reform. It is becoming a recognized national leader in calculus and differential equations reform and the use of technology in the classroom. Its leadership role is evident from the amount of external funding it has received for educational endeavors, for the number of visiting scholars who come to Arizona to learn about our classroom environment, and the participation of faculty members in national meetings and service on national committees.

Success Factors

The Mathematics Department has tackled the challenge of improving the educational experience for students in what we feel is a wholly unique manner. The additional resources allowed the Department to augment its faculty with adjunct lecturers who are dedicated to teaching. This meant several things. First, the number of students who had to take the self-study algebra program decreased in stages to its discontinuation in Fall 1997. Second, our faculty noticed an improvement in the teaching environment. Because of the mandatory mathematics placement test their classes contained students with better mathematical preparation. Furthermore, the addition of adjunct faculty gave the regular faculty opportunities to teach the classes that they preferred, generally in smaller sections. The result is an improvement in faculty morale in the Department. A number of faculty have been involved in curriculum projects to improve undergraduate education. Notable among these is the Consortium Calculus Project, the development of material for two differential equations courses, and a geometry and algebra course for prospective teachers.

We close with a list of some of the ingredients that go into a successful program.

1. Broad faculty involvement.
2. General departmental awareness and support.
3. University encouragement and support for faculty.
4. Regular communication with interested university departments.
5. A mandatory mathematics placement test.
6. Careful teaching assignments.
7. Augmentation of regular faculty with instructors dedicated to teaching.
8. Open dialogue and development of joint programs with local school districts and community colleges.
9. Assessment of the curriculum and definition of the goals for each course.
10. Development of an effective training program for graduate teaching assistants.
11. Reward for faculty who make significant contributions to the undergraduate program.
12. Adequate staff support.

Analyzing the Value of a Transitional Mathematics Course

Judith A. Palagallo* and William A. Blue

The University of Akron

At a large, regional university in the Midwest a specific course (Fundamentals of Advanced Mathematics) at the sophomore level provides a transition for student majors from the more computationally-based aspects of the first year courses to the more abstract upper-division courses. Surveys have been developed to measure the effects of this course on upper level courses in abstract algebra and advanced calculus. In addition, these surveys provide information about student learning in the major.

Background and Purpose

The University of Akron is the third largest public university in the state of Ohio. The departmental goals for the mathematics major state that each student should be able to read and write mathematics, to communicate mathematical ideas verbally, and to think critically. In addition, each student should acquire a body of fundamental mathematical ideas and the skill of constructing a rigorous argument; each student should achieve an understanding of general theory and gain experience in applying theory to solving problems. Several years ago the mathematics faculty became concerned about the difficulty experienced by some students in making the transition from the traditional calculus courses, where few written proofs are required, to more advanced work in mathematics. The abrupt transition to rigorous mathematics often affected even those who had done superior work in the calculus courses and was targeted as an important issue involved in a student's decision to leave the study of mathematics. To address the problem we designed the course Fundamentals of Advanced Mathematics (FOAM) to help bridge the gap of the students' understanding and preparation for continued mathematical training.

The prerequisite for FOAM is the second semester of calculus, so students can take the course as early as their sophomore year. FOAM is required for mathematics majors and for those preparing to be secondary mathematics teachers. Other students with diverse majors such as physics, engineering, computer science, and chemistry also enroll in the course (especially those who desire a minor in mathematics). The topics included in FOAM were designed to introduce the students to the goals we have for our majors. Elementary symbolic logic is the first topic of the course and sets the foundation for later work. A study of standard proof techniques, including the Principle of Mathematical Induction, requires the application of symbolic logic to the proof process. The proof techniques are then applied to topics such as sets (including power sets and arbitrary families of sets), relations, equivalence relations, and functions. Discussion of these major topics shows students the interconnections of ideas that explain and give meaning to mathematical procedures. Students can also see how mathematicians search for, recognize, and exploit relationships between seemingly unrelated topics. Throughout the semester students write many short proofs. For nearly all, this is a first experience at writing a paragraph of mathematical argument. The proofs are graded so that each student has plenty of feedback. The students also read proofs written by others and evaluate them for approach, technique and completeness. Students are required to make periodic oral presentations of their own proofs at the board.

* Research supported in part by OBR Research Challenge Grant, Educational Research and Development Grant #1992-15, and Faculty Research Grant #1263, The University of Akron.

Once FOAM had been taught for a couple of years, an assessment was necessary to see if it achieved the broad goal of "bridging the gap." We needed to know if the mathematical topics covered were suitable preparation for more advanced courses. In addition we wanted the course to alleviate some of the difficulty in making the transition to proof-oriented courses.

Method

We decided to assess the goals for the course by developing a survey in two parts. In one part we would assess the perceived value to the student of the course content in preparation for more advanced work. The second part would evaluate the confidence level of the students in doing mathematics in the later courses. Before constructing a survey instrument, we interviewed students who were currently enrolled in FOAM to hear their reactions to the course. We also reviewed the literature for other attitude surveys, especially those done in mathematics. (See, for example, [1] and [2].) From the student responses and the literature, we selected statements to include in an attitude survey.

The content of FOAM was analyzed as viewed from the attitudes of students who took the course and then continued on to take advanced calculus or abstract algebra. Over the past five years, surveys were distributed to students near the end of the semester in the advanced course. Because the FOAM course is required, we do not have a control group of students who did not take FOAM before these advanced courses. Students were asked to rate each major topic in FOAM with respect to its value in their current courses. The topics and a frequency count of their responses are listed in Table 1. (Not all students responded to every question.) The arbitrary coding — extremely helpful = 3, moderately helpful = 2, slightly helpful = 1, not helpful = 0 — is used, and a mean is calculated.

The second phase of the evaluation process was to determine if this initial experience in conceptual topics increased the confidence of students in advanced mathematics courses. Confidence in learning mathematics was found to be related to student achievement in mathematics and to students' decisions to continue or not continue taking mathematics courses. (See, for example, the attitude studies in [1,3].) To gauge this, the same students responded to a sequence of statements on an attitude survey. The statements, and the frequencies and means of the responses are shown in Table 2.

Findings

A quick inspection of the frequency counts in Table 1 suggests that students feel the studies of proof techniques, logic, induction and set theory were helpful. They were divided on the study of functions, perhaps because this is a

---

Table 1

Perceived Value of Course Content

Evaluate how helpful the following topics from FOAM have been in Advanced Calculus and/or Abstract Algebra. Use a scale of

EH=extremely helpful   MH=moderately helpful
SH=slightly helpful   NH=not helpful

	EH	MH	SH	NH	Mean
A. Advanced Calculus					
Proof techniques	24	22	8	1	2.25
Set Theory	18	18	10	9	1.82
Functions	9	24	14	8	1.62
Induction	24	21	6	4	2.18
Logic	20	21	13	1	2.09
B. Abstract Algebra					
Proof techniques	24	34	10	5	2.05
Set Theory	20	33	14	7	1.89
Functions	16	27	20	7	1.74
Induction	23	25	11	12	1.83
Logic	22	29	11	10	1.88

Note: EH=3, MH=2, SH=1, NH=0.

Table 2

Attitude Survey Results

Respond to these statements using the given code

SA=strongly agree   A=agree   N=neutral/no opinion
D=disagree   SD=strongly disagree

	SA	A	N	D	SD	Mean
1. The experience of taking FOAM increased my confidence to take later mathematics courses.	21	37	20	15	8	2.48
2. The experience of taking FOAM increased my interest in mathematics.	13	31	27	24	6	2.21
3. I understood how to write mathematical proofs before taking FOAM.	0	16	13	50	22	1.23
4. After taking FOAM I am better at reading proofs.	37	46	8	9	1	3.08
5. After taking FOAM I can write better proofs.	36	43	12	9	1	3.03
6. I am still confused about writing proofs.	4	27	20	43	7	1.60
7. I like proving mathematical statements.	6	35	24	27	9	2.02
8. Taking FOAM has increased my general reasoning and logical skills.	21	52	12	13	2	2.77
9. If I had known about the abstract nature of mathematics, I would have chosen a different major.	2	8	21	46	24	1.19

Note: SA=4, A=3, N=2, D=1, SD=0.

final topic and may not have been covered completely in a given term.

From Table 2 we note that the students reported an increase in confidence and in general reasoning skills. Most noticeably, they feel they are better able to read and write mathematical proofs after taking FOAM. They were divided on whether their mathematical interest had increased. It is encouraging to see that most are still interested in continuing a mathematical career!

Use of findings

We can conclude that we are teaching correct and relevant topics to help prepare the students for later work. Students reported that their confidence and skills in mathematics increased with their experience in FOAM. The boost in confidence should help retain students in the field.

The surveys conducted also indicate that in certain areas the attitudes of mathematics education majors (prospective secondary mathematics teachers) differ from mathematics majors. The graphs in Table 3 report the percentages of strongly agree/agree responses on selected questions by the two groups of students. (The question numbers refer to the questions on the attitude survey in Table 2.) Both groups feel they are better able to read and write mathematical proofs after taking FOAM. However, only about half (49%) of the mathematics education majors reported an increase in confidence for later courses and fewer than a third reported an increase in interest (29%). From conversation with these students we infer that they do not recognize where they will use this type of material in their future classrooms and, at times, resent the requirement of the courses in their degree program. We are trying to address these issues now in FOAM, and also in the advanced courses, by introducing examples of the actual material in the secondary school setting.

This project has led to additional questions and studies. Using multiple linear regression, we built a model to predict a student's grade in FOAM based on grades in the first two semesters of calculus. The model predicts that a student's grade will drop about 0.5 (on a 4-point scale) in FOAM. However, for a certain collection of students, grades drop

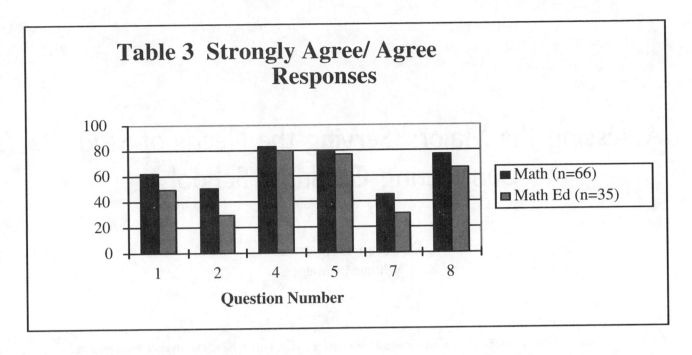

Table 3 Strongly Agree/ Agree Responses

dramatically when going from the calculus sequence to FOAM. These students are the subjects of a separate survey to determine where they attribute the cause of this decline in their performance. Finally, the original study concentrated on students who complete FOAM and continue into later courses. We have collected data on the students who do not complete the course, and thus decide to quit the study of mathematics at this point. Analyzing the perceptions of these students will be important to the extended project.

Success factors

A reliable assessment study is a slow process. The sample is small since there are not many mathematics/ mathematics education majors, even at a large university. Furthermore, individual classes may vary in size (in our case, from 12-30). Also student attitudes and reactions to a course are undoubtedly dependent on the instructor. Thus the data is best collected and evaluated over an extended time period.

Note: In [4], Moore reports on a study of cognitive difficulties experienced by students in a course similar to FOAM. He discusses experiences with individual students in learning proof techniques and includes a related bibliography.

References

[1] Crosswhite, F.J. "Correlates of Attitudes toward Mathematics," *National Longitudinal Study of Mathematical Abilities*, Report No. 20, Stanford University Press, 1972.

[2] Fennema, E. and Sherman, J. "Fennema-Sherman mathematics attitudes scales: Instruments designed to measure attitudes toward the learning of mathematics by females and males,"*JSAS Catalog of Selected Documents in Psychology* 6 (Ms. No. 1225), 1976, p. 31.

[3] Kloosterman, P. "Self-Confidence and Motivation in Mathematics," *Journal of Educational Psychology* 80, 1988, pp. 345–351.

[4] Moore, R.C. "Making the transition to formal proof,"*Educational Studies in Mathematics* 27, 1994, pp. 249–266.

Assessing the Major: Serving the Needs of Students Considering Graduate School

Deborah Bergstrand
Williams College

The focus of this article is on assessing student learning for a segment of the undergraduate mathematics majors: those talented students who are prospective mathematics graduate students. At this private, liberal arts college in the Northeast there are three aspects of the undergraduate mathematics experience outside the standard curriculum which are described in this paper: a senior seminar (required of all majors), an Honors thesis and a summer undergraduate research experience. All three combined are used to assess how well the department is preparing students for graduate school.

Background and Purpose

Williams College is a small liberal arts college in rural Massachusetts with an enrollment of 2000 students. The Mathematics Department has 13 faculty members covering about 9.5 FTE's. Williams students are bright and very well prepared: the median mathematics SAT score is 700. For the past 7 years, annual mathematics enrollments have averaged about 1050, about 30 students have majored in mathematics from each class of 500, an average of 3 or 4 students have gone on to graduate school in mathematics or statistics, and about half to two-thirds of those have completed or are making progress toward the PhD.

Three aspects of the mathematics experience outside the standard curriculum at Williams are particularly relevant to students considering graduate school: the Senior Colloquium, the honors thesis, and the SMALL Undergraduate Summer Research Project. The colloquium and honors thesis lie within the formal structure of the major; both have existed for decades. The SMALL project is a separate program which began in 1988. These three experiences help cultivate in students some of the qualities necessary for success in graduate school such as independent learning, creative thinking on open problems, and the ability to present mathematics well orally and in writing. How well our students respond to these undertakings in part measures how well the rest of the major prepares them for graduate work. Which students choose to do an honors theses or apply to SMALL is another measure of how appropriately we inspire our students to consider graduate school.

The Senior Colloquium and Honors Program have existed at Williams for many years. The former is required of all majors, a shared experience with many challenges and benefits. The latter is a special opportunity appropriate only for the most able and ambitious students. Each is a barometer in some obvious ways of how well we prepare students for graduate school. The ability of students to prepare clear and engaging talks on demanding topics they research with little faculty assistance is a sign of good preparation, not only for graduate school but for many other undertakings. Students who produce interesting results in their theses, written and presented orally in mature mathematical style, further demonstrate that we are helping their development as mathematicians. Likewise the performance of students in the SMALL project is an obvious preparation indicator.

Method

The Senior Colloquium is a colloquium series that meets twice a week throughout the school year. It is advertised and open to the entire community. The talks are given mainly by seniors, with occasional outside speakers, to an audience of mostly mathematics majors and faculty. Presentations last 30–40 minutes, followed by soda, cookies, and informal discussion. Senior majors are required to give one talk each, prepared under the guidance of a faculty member on a subject they have not studied in a class. After a few minutes of mingling with students, faculty members evaluate the talk together on a pass/fail basis. This private discussion lasts only a few minutes. The advisor verbally conveys the result to the student immediately, along with comments on the presentation. In an average year at most one or two failures occur. Seniors who fail the talk are usually required to give a second talk on a new topic. Seniors must also meet an attendance requirement to pass the Colloquium. Students do not receive formal credit for the Senior Colloquium, but it is a requirement for the major.

Evaluation of the Senior Colloquium from the student point of view takes place in two ways: the Mathematics Student Advisory Board and the senior exit interview. The Mathematics Student Advisory Board (MSAB) is a set of six students (three seniors, two juniors, and one sophomore) who meet with job candidates, help plan department social events, and serve as a sounding board for student concerns throughout the year. The senior and junior members are mathematics majors elected by their peers. The sophomore is appointed by the department. As with many such matters, the MSAB is routinely consulted about the Senior Colloquium, especially when a particular concern arises. Though seniors, both through the MSAB and through their exit interviews, are almost always the source of specific concerns or suggestions about the colloquium, the resulting issues are nearly always discussed with the MSAB as a whole. The input of underclass students is often very helpful.

Very rarely seniors will approach their colloquium so irresponsibly that their talks need to be canceled (the student, the advisor, or both may come to this decision). In such cases the student is usually required to give an extra colloquium talk. In general, we tend to deal with recalcitrant seniors on an ad hoc basis. Fortunately they are rare, and their circumstances often convoluted enough, that this approach is most effective and appropriate.

The Senior Colloquium can be a particularly valuable experience for students considering graduate school. The independent work required is good practice for more demanding graduate-level courses. The effort required to prepare a good talk alerts students to the demands of teaching.

The Senior Honors Thesis is an option in all majors, and is typical of many schools' honors programs. Because it usually involves original research, the thesis is of tremendous benefit to students going on to graduate school. The thesis is one of a senior's four regular courses in the fall semester and the sole obligation during the 4-week January term. More ambitious seniors may extend their theses into the spring term as well, if their advisor approves. Successful completion of an honors thesis along with some other accomplishments give the student the degree with Honors in Mathematics.

The honors thesis culminates in a formal paper written by the student and a thesis defense given orally. The paper gives the student's results along with appropriate background and bibliography. Students present their results in the talks, during and after which faculty members may ask questions or make suggestions. Revisions to the thesis may result though they are usually minor. The student receives a regular course grade determined by the faculty advisor. The department as a whole determines whether the thesis warrants Highest Honors, Honors, or, in rare cases, neither. Some students' work is exceptional enough to publish. Clearly the senior thesis is a good foretaste of research in graduate school. The Honors Program is evaluated in basically the same way as the Senior Colloquium: through student interviews, discussions with MSAB, and faculty discussion at department meetings.

The SMALL Undergraduate Research Project is a 9-week summer program in which students investigate open research problems in mathematics and statistics. The largest program of its kind in the country, SMALL is supported in part by a National Science Foundation grant for Research Experiences for Undergraduates and by the Bronfman Science Center of Williams College. Over 140 students have participated in the project since its inception in 1988. Students work in small groups directed by individual faculty members. Many participants have published papers and presented talks at conferences based on work done in SMALL. Some have gone on to complete Ph.D's. Due to funding constraints, about 60% of SMALL students come from Williams; the rest come from colleges and universities across the country. (To answer the obvious question, "SMALL" is an acronym for the faculty names on the original grant.)

SMALL was initiated in order to give talented students an opportunity to experience research in mathematics. We hoped those with particularly strong potential who were inspired by the experience would then choose to attend graduate school and be that much better prepared for it. (I should point out that during the early years of SMALL there were many other efforts undertaken in the department to increase interest in mathematics in general. The best evidence of our success is the increase in majors from about 10 per year to an average over 30, and the increase in enrollments in mathematics courses from around 600 to over 1000 per year.)

Evaluation of SMALL has two in-house components: one by student participants and one by faculty. Each summer the students conduct their own evaluation in the last two weeks of the project. Two or three students oversee the process of designing the questionnaire, collecting responses, and assembling a final report in which comments are presented as anonymously as possible. Each faculty supervisor gets a copy of the evaluation. (The students themselves each receive a written evaluation from their faculty supervisor.) Student comments cover many aspects of SMALL, including the structure of research groups, amount of direction from faculty, number and type of mathematics activities, length of the program, and various extracurricular matters such as housing and social life. Some of the changes motivated by these comments will be presented in the next section.

Faculty evaluate SMALL more informally through discussions during the summer and school year. We have debated many issues, including how to attract and select participants, expectations and work structure for research groups, and the value of attending conferences and other activities.

Because participants work on open problems, SMALL is an excellent means for students to test their graduate school mettle. Though the circumstances are somewhat idealized (summer in the Berkshires, no classes or grades), how one responds to doing mathematics all day every day is a very basic factor in the decision to go to graduate school. More important is one's response to the challenge, frustration, and, hopefully, exhilaration that comes from tackling a real research problem.

Both the SMALL Project and the Honors Program also provide feedback on graduate school preparation in a more indirect but still important way. The number of our most talented and dedicated students who choose to apply to SMALL and go on to do an honors thesis reflects how well we prepare, advise, and encourage students regarding graduate study. Of course, there will always be variations from year to year in the number of honors students and SMALL applicants, and Williams mathematics majors often have strong interests in other fields, but if we notice particularly talented students choosing not to pursue these options, we are forced to examine our role in such a decision. Did we miss the opportunity to create a department mentor for the student? Did the student have some negative experience we could have avoided? Is the student well-informed about the benefits of both SMALL and the Honors Program even if they are unsure about their commitment to mathematics? There may be many reasonable explanations; not all our strong students will choose graduate school or our programs in undergraduate research. In fact, given the current job market, we must be cautious and honest with our students about graduate study. Nonetheless, it is extremely helpful for us and for our students to have programs in which participation is instructive in so many ways.

Findings

Near the end of the spring term, seniors are interviewed by a senior faculty member about their experience as a mathematics major. A question like "What do you think of the Senior Colloquium?" is always included. Most seniors consider the colloquium a highly valuable experience; some feel it is one of the best parts of the major, even though it lies outside the formal curriculum.

Evaluation of the Senior Colloquium from the faculty point of view occurs mainly through discussions at department meetings. Over the past several years, we have discussed balancing advising work among faculty, responding to recalcitrant students (ill-prepared, canceling talks, etc.), evaluating talks (in a different sense than that above), and the process by which we select the best colloquium (for which we have a prize each year).

We feel our rather informal evaluation procedures are more than adequate in large part because the overall quality of colloquium talks has increased. Failures have always been rare, but there used to be several (3-4) borderline cases each year. Now it's more likely to have only one or two such cases, and they are almost always not poor enough to fail but rather just weak or disappointing.

As mentioned previously, the department has grown a lot in the last ten or more years. In particular, the number of students involved in research and the level to which we encourage such activity has increased dramatically. Where most honors theses used to be expository, now most involve original research. Not every student produces interesting results, but the experience of doing mathematics rather than just learning about mathematics is still significant. We believe this change is due in large part to the existence of the SMALL Project and the department's conviction that undergraduates can indeed do research in mathematics.

Several external agencies provide a more indirect evaluation of SMALL. The NSF and other granting agencies expect annual reports. More significantly, by renewing the REU grant three times over the last 9 years, the NSF has signaled its approval of our project. Otherwise, the NSF has provided no specific findings resulting in changes to SMALL. Further indirect evaluation comes from those journals to which research results are submitted for publication. (Journals publishing SMALL papers include The Pacific Journal of Mathematics, The American Mathematical Monthly, and the Journal of Undergraduate Research.)

Use of Findings

In discussion with students about the Senior Colloquium, concerns raised over the years include attendance, how talks are evaluated, and the procedures for preparing a talk. On the first matter, keeping track of attendance has been awkward and unreliable in the past. We now have a system that works very well. At the front of the room we hang a poster-sized attendance sheet for the entire semester with the heading "Colloquium Attendance Prize." (We do give a small prize at the end of the year to the student who attends the most talks.) The prize is a reasonable cover for the attendance sheet which, being posted publicly, encourages students to be responsible about coming to talks. We have also relaxed the requirement to specify a certain number of talks each semester, rather than one each week as we used to require.

Faculty used to evaluate a talk immediately afterwards, leaving students to chat over tea. Though usually we would disappear into another room for only a few minutes, by the time we returned most students would be gone, leaving only the speaker and a few loyal friends. Now we delay our meeting a bit, talking with students about the colloquium or other matters. Though the speaker has to wait a little longer for our verdict, we all benefit from the relaxed time to interact over cookies and soda. On the preparation of talks, wide variations in faculty approaches to advising a talk used to confuse many students. We now have, to some extent, standardized the timetable under which talks are prepared, so students know, for example, that they have to pick an advisor at least four weeks before the talk and give at least one practice talk.

Some of the issues that continue to be revisited include balancing colloquium advising work among faculty and responding to recalcitrant students (ill-prepared, canceling talks, etc). In some years, by their own choice students seem to distribute themselves pretty well amongst different faculty. In other years, however, a few faculty members have found themselves deluged with requests to advise colloquium talks. It's hard to say no, especially for junior faculty. Rather than set specific limits on the number of talks one can advise, we prefer to remind ourselves annually that an imbalance may occur, and then keep an eye on each other and especially on nontenured colleagues, encouraging them to say no if they feel overburdened advising talks.

Our internal evaluations of the SMALL Project have resulted in some significant changes over the years. The project now lasts nine weeks instead of ten. In response to requests from both students and faculty for a more flexible schedule, there is now only one morning "convocation" per week for announcements and talks rather than daily gatherings. (Weekly afternoon colloquia and Friday afternoon teas are still standard.) A more significant change affected the project format. The original format had students work in two groups of four to five on problems supervised by two different faculty members. Under this structure students and faculty often felt that student efforts were spread too thin. Furthermore, fostering efficient and comfortable group dynamics among five or six students could be quite a challenge. Participants now work with only one faculty member in one group of three students. Though there is some risk that students might not find their one group engaging, for the most part students now seem more focused, productive, and, ultimately, satisfied.

The Senior Colloquium has affected some aspects of our teaching in a way not solely connected to graduate school preparation. Seeing the benefits students gain by giving a colloquium talk, some faculty have added a presentation component to their more advanced courses. This not only enhances students' experiences in that course, but can also in return help prepare them for the more demanding Senior Colloquium which may lie ahead.

Success Factors

Our Honors Program is typical of that found at many other schools. Many of the benefits of such a program have already been mentioned: gives talented students an experience in independent work and original research, offers students and faculty a chance for extended one-on-one work, fosters interest in graduate study, helps prepare students for some of the rigors of graduate study. The costs tend to be what one would suppose: the challenge of finding appropriate projects, the extra workload for faculty, the potential for meandered dabbling rather than the structured learning the student would find in a regular course. There is a bit of an ironic downside as well. We have learned through student interviews that sometimes majors who are not doing honors theses feel they are less interesting to the faculty or less important to the department. Such a circumstance reflects the broader challenge any department faces in trying to meet the needs of all its students.

A full-fledged REU program like the SMALL Project is rarer than an Honors Program. It requires a strong commitment by the department as well as the privilege of an NSF grant or other outside funding and, in our case, substantial institutional resources. The costs and benefits of such a program are extensive. On the plus side are many of the same benefits as an Honors Program as mentioned above. In terms of cost, a summer research program requires faculty who are both willing to commit most of their summer and able to engage students in research. While not all faculty at Williams do SMALL every summer, and those who do also spend time on other projects, the fact remains that working in SMALL can take away substantially from other work faculty may wish or need to do. The project also requires a large commitment from the College. Students get subsidized housing and over half of all participants have their stipends paid by the College. (Williams has a strong tradition of supporting student summer research, especially in the sciences. Many faculty receive outside funding and the College returns a portion of the resulting overhead to the sciences to help support student research.) Further information about SMALL may be found in [1].

Though the department is fully committed to it, the Senior Colloquium is an expensive undertaking. Advising a talk is time consuming. Typically majors require some assistance selecting a topic, two or more meetings to digest the reading(s) they have found or which the advisor has given them, further discussion on the design of the talk, and then at least one practice talk. Over the course of three or four weeks one can spend eight or ten hours advising a talk. It's not unusual for some faculty to advise five talks in a year, so the time commitment is nontrivial. Then there is the colloquium series itself in which visitors speak as well as seniors. All faculty try to attend at least one of the two talks each week, often both. Since we also have a weekly faculty research seminar, the time spend attending talks is also nontrivial. All of this effort would probably not be well-spent if its only benefit were to students preparing for graduate school. The overall benefit to all our majors, however, makes the effort well worthwhile. Seniors gain independence, poise, and confidence even as they learn a healthy respect for the effort required to prepare a good presentation. We think the Senior Colloquium is one of the best aspects of our major.

Reference

[1] Adams, C., Bergstrand, D., and Morgan, F. "The Williams SMALL Undergraduate Research Project," *UME Trends*, 1991.

Part II

Assessment in the Individual Classroom

Introduction

Bonnie Gold
Monmouth University

When Sandy Keith first invited me to join her in putting together this book, my motivation was the section on assessing the major. In our last accreditation review, Wabash College, where I began this work, had been told that it was only being provisionally accredited, pending adoption of an acceptable assessment plan. The administration then told each department to come up with a plan for assessing its activities. With little guidance, we each put together a collection of activities, hoping they would be sufficient to satisfy the accrediting association. Having wasted a lot of time trying to figure out what assessment meant and what options we might consider, I felt that by writing a book such as this one, we could save many other colleagues a similar expenditure of time. Thus, I began work on this book more from a sense of duty to the profession than from any excitement over assessment.

My attitude toward assessment reversed dramatically when, on Sandy's suggestion, I read Angelo and Cross's, *Classroom Assessment Techniques*. This book contained ideas which could help me in the classroom, and immediately. I began putting some of these techniques to use in my classes and have found them very worthwhile.

Most of us have had the experience of lecturing on a topic, asking if there are any questions, answering the few that the better students have the courage to ask, assigning and grading homework over the topic, and yet when a question on that topic is asked on an examination, students are unable to answer it. Classrooms Assessment Techniques (CATs) are ways to find out whether students understand a concept, and what difficulties they are having with it, *before* they fail the test on it. They also include ways to find out how well students are reading, organizing the material, and so on. Traditional assessment is *summative* assessment: give students a test to find out what they've learned, so they can

be assigned a grade. CATs are *formative* assessment: find out where the problems are so students can learn more and better.

Not all of the assessment techniques in this section of our book are CATs; many of them can be used for both formative and summative assessment, and some are mainly summative. However, what they all have in common is that in the process of assessment, students are learning by participation in the assessment technique itself. We've always given lip service to the assertion that on the best tests, students learn something taking the test; the techniques in this section make this homily a reality. This section doesn't try to duplicate Angelo & Cross, and we still recommend that you look at that book, since many of the techniques which they don't specifically recommend for mathematics can be adapted to use in our classrooms. However, we present here a good variety of assessment techniques actually being used in mathematics classrooms around the country to improve student learning.

This section begins with variations on the traditional **tests** and **grades**. Sharon Ross discusses ways in which traditional tests can be modified to examine aspects of student understanding, beyond the routine computations which form the core of many traditional examinations. One variation on tests is Michael Fried's Interactive Questionnaires, a form of quiz taken electronically and sorted automatically to give large classes semi-individualized detailed responses to their work. Then William Bonnice discusses allowing students to focus on their strengths by choosing what percent (within a range selected by the instructor) of their grade comes from each activity.

Next comes a collection of **CATs**, ways of finding out how well students understand a concept or section which has been taught. David Bressoud discusses adapting Angelo & Cross' One-Minute Paper to mathematics classes, large or small. It takes a very small amount of classroom time to find out how productive a class period has been, and what topics need further clarification. Dwight Atkins also adapts an Angelo & Cross technique, Concept Maps, to mathematics classes to learn how well students understand connections between related concepts. John Koker has his students discuss the evolution of their ideas as they work on homework sets, which gives him insight into how much they actually understand and what they have learned by working the problems. John Emert offers three quick CATs: a brief version of the class mission statement (which Ellen Clay discusses, later in the section, in more detail), a way to check how well the course is going *before* the end of the semester, and a quick way to assess how groups are working. Sandra Keith describes a number of "professional assessment techniques," ways to encourage students to take a more professional attitude toward their studies, going beyond the individual course.

Several articles discuss helping students **review and clarify concepts**. Janet Barnett uses true/false questions, but has students justify their answers. These justifications bring out confusions about the concepts, but are also the beginning, for calculus students, of writing mathematical proofs. Joann Bossenbroek helps developmental and precalculus students become comfortable with mathematical language and learn how to use it correctly. Agnes Rash encourages students to look for their own applications of the methods they're studying while reviewing the material for a test.

At the opposite extreme from most CATs, which are fairly short and give helpful information on individual details, are methods used in research on teaching and learning. We have not tried to cover that field in this section, but do have one article, by Kathleen Heid, on using videotaped in-depth interviews to **assess student understanding**. While very time-consuming, it is a method of finding out just what students have understood, and where confusions lie.

There has been considerable attention in the last several years to having students engage in larger **projects** and **write about mathematics** as a route to learning it better. Several other Notes volumes (especially #16, *Using Writing to Teach Mathematics*, and #30, *Problems for Student Investigation*) discuss these processes in more detail; here we present additional guidance in this direction. Annalisa Crannell discusses assigning and assessing writing projects. Charles Emenaker gives two kinds of scales which can make the process of assessing project reports less tedious. Brian Winkel explains how having students work on large projects during class time gives multiple opportunities to redirect students' formation of concepts before they are set in stone in the students' minds. Alan Knoerr and Michael McDonald use two kinds of portfolios to help students aggregate and reflect on concepts studied. Dorothee Blum uses student papers in an honors, non-science majors' calculus course to integrate ideas studied. Alvin White discusses using student journal writing to expand students' vistas of mathematics.

Patricia Kenschaft uses short papers and students' questions to have general education students think carefully about the readings.

As many of us move away from complete dependence on the lecture method, the issue of assessing these alternative teaching techniques arises. Nancy Hagelgans responds to the concern, when using **cooperative learning**, of how to ensure that individuals take responsibility for their own learning. Catherine Roberts discusses effective ways of giving tests in groups. Carolyn Rouviere discusses using cooperative groups for both learning and assessment. Annalisa Crannell explains how to give and grade collaborative oral take-home examinations. (See also the article by Emert earlier in this section for a quick CAT on discerning how well groups are working.) Sandra Trowell and Grayson Wheatley discuss assessment in a **problem-centered** course.

In recent years we have become more aware that not all students are 18-22-year-old white males. Some of these other students don't "test" well, but can demonstrate what they've learned when alternative assessment methods are used. Two articles discuss these **special-needs students**. Regina Brunner discusses what techniques she's found effective in her all-female classes, and Jacqueline Giles-Giron writes of strategies to assess the adult learner.

Turning from assessment of individual parts of a course to assessment of how a **course as a whole** is going, Ellen Clay helps set the tone of her courses at the beginning by developing a "class mission statement," which can be revisited as the course progresses to assess progress toward meeting course goals. William Bonnice, David Lomen, and Patricia Shure each discuss methods of finding out, during a semester, how the course is going. Bonnice uses questionnaires and whole-class discussion. Lomen uses student feedback teams. Shure has an outside observer visit the class, who then holds a discussion with the class in the absence of the instructor, and provides feedback to the instructor. Janet Barnett and Steven Dunbar discuss alternatives to the standard course questionnaire for summing up a course once it's finished. Their methods provide more useful information than do student evaluation questionnaires for improving the next incarnation of the course. Barnett has students write letters to friends describing the course. Dunbar collects data throughout the semester into a course portfolio, which can then be used by that instructor or passed on to others.

Keep in mind that as you consider using one of these techniques, even if your course and institution are very similar to that of the author, to make the technique work well you will need to modify the technique to fit your personality and the tone and level of your class. But for most of your assessment problems, there should be at least one technique from this assortment which can help.

What Happened to Tests?

Sharon Cutler Ross
DeKalb College

For students to believe that we expect them to understand the ideas, not just be able to do computations, our tests must reflect this expectation. This article discusses how this can be done.

Background and Purpose

Tests — timed, in-class, individual efforts — are still the primary assessment tool used by mathematics faculty at DeKalb College, but comparing current tests to tests of ten, or even five, years ago show that perhaps the characteristics given above are the only things that are still the same. And even these three descriptors are often wide of the mark. A variety of other types of assessment are used on a regular basis at DeKalb College, but this essay is primarily a report on how tests have changed.

DeKalb College is a large, multi-campus, public, commuter, two-year college in the metropolitan Atlanta area. We have approximately 16,000 students, many of whom are non-traditional students. The curriculum is liberal arts based and designed for transfer to four-year colleges. There are only a handful of career programs offered by the college. The mathematics courses range from several low-level courses for the career programs through a full calculus sequence and include discrete mathematics, statistics, linear algebra, and differential equations courses. In each course we follow a detailed common course outline, supplemented by a course-specific teaching guide, and use the same textbook in each class. At regular intervals we give common final examinations to assess the attainment of course goals, but otherwise individual instructors are free to design their own tests and other assignments. A comprehensive final examination is required by the college in all courses. In practice, mathematics faculty often work together in writing tests, assignments, and final exams.

Mathematics teachers at all levels joke about responses to the dreaded question, "Will this be on the test?" But students are reminding us with this question that in their minds, tests are the bottom line in their valuation of material. Most students value ideas and course material only to the extent that they perceive the instructor values them, and that is most clearly demonstrated to them by inclusion of a topic on a test. Our students come to us with very strong ideas about what tests are and what they are not. Changing tests means changing our messages about what we value in our students' mathematical experiences. Acknowledging this means we have to think about what those messages should be. At DeKalb College, we see some of our entry-level courses primarily as skill-building courses, others as concept-development courses. Even in courses that focus on skill building, students are asked to apply skills, often in novel settings. We want all students to be able to use their mathematics. Also, fewer skills seem critical to us now; in time it will be hard to classify a DeKalb College mathematics course as skill building rather than concept building.

Technology was one of the original spurs for rethinking how tests are constructed; writing common finals was the other spur. One of the first issues we tackled in using technology was how to ask about things that faculty valued, such as students' knowing exact values of trig functions. Those discussions and their continuations have served us well as we use more and more powerful technology. Some of the questions that recur are: Is this something for the technology to handle? at this stage in the student's learning? ever? Is this topic or technique as important as before? worth keeping with reduced emphasis? Common finals mean we have to discuss what each course is about, what is needed for subsequent courses, and what can really be measured.

The driving force now is our evolving philosophy about our courses and their goals. Changes in course goals have resulted in changes in instruction. We use lots of collaborative work, lab-type experiences, extended time out-of-class assignments, and writing in a variety of settings. All of these require appropriate types of assessment, and this affects what our traditional tests must assess.

Method

Assessing Skills

When we do want to evaluate mathematical skills, we ask directly for those things we think are best done mentally. The experience of choosing appropriate technology including mental computations and paper-and-pencil work is a necessary part of testing in a technology-rich environment. We try to be careful about time requirements as we want to assess skill not speed. As skill assessments, the next two examples differ very little from textbook exercises or test items from ten years ago, but now items of this type play a very minor role in testing.

Example: Give the exact value of $\sin \frac{\pi}{3}$.

Example: (a) Give the exact value of $\int_0^1 e^x \, dx$.

(b) Approximate this value to three decimal places.

Another type of question for skill assessment that gives the student some responsibility is one that directly asks the student to show that he or she has acquired the skill.

Example: Use a 3×3 system to demonstrate that you know and can apply Cramers rule.

Example: Describe a simple situation where the Banzhaf power index would be appropriate. Give the Banzhaf index for each player in this situation.

Assessing Conceptual Understanding

Good ways to measure conceptual understanding can be more of a challenge to create, but there are some question styles we have found to be useful. Here are examples of (1) creation items, (2) reverse questions (both positive and negative versions), (3) transition between representations, (4) understanding in the presence of technology, and (5) interpretation.

(1) Creation tasks are one way to open a window into student understanding. Often a student can identify or classify mathematical objects of a certain type, but can not create an object with specified characteristics.

Example: Create a polynomial of fifth degree, with coefficients 3, -5, 17, 6, that uses three variables.

Example: Create a function that is continuous on $[-5,5]$, whose derivative is positive on $[-5,3)$ and negative on $(3,5]$, and whose graph is concave up on $[0,2]$ and concave down elsewhere.

(2) One thing we have learned is that students are always learning. Whether they are drawing the conclusions the instructor is describing is an open question however. Reversing the type of question on which the student has been practicing is a good way to investigate this question. These two examples come from situations where exercises are mostly of the form: Sketch the graph of .... (The graphs have been omitted in the examples.)

Example: Give a function whose graph could be the one given.

Example: Why can the graph given not be the graph of a fourth degree polynomial function?

(3) Being able to move easily between representations of ideas can be a powerful aid in developing conceptual understanding as well as in problem solving, but students do not make these transitions automatically. Test items that require connecting representations or moving from one to another can strengthen these skills as well as reveal gaps in understanding.

Example: What happens to an ellipse when its foci are "pushed" together and the fixed distance is held constant? Confirm this using a symbolic representation of the ellipse.

Example: The columns of the following table represent values for f, f', and f'' at regularly spaced values of x. Each line of the table gives function values for the same value of x. Identify which column gives the values for f, for f', and for f''. (Table omitted here.)

(4) When technology is used regularly to handle routine computations, there is a possibility that intuition and a "feel" for the concept is being slighted. If this is a concern, one strategy is to ask questions in a form that the available technology can not handle. Another is to set a task where technology can be used to confirm, but not replace intuition.

Example: Let $f(x) = ax^2$ and $g(x) = bx^3$ where $a > b$. Which function has the larger average rate of change on the interval $\left(0, \frac{a}{b}\right)$? Support your conclusion mathematically.

Example: Change the given data set slightly so that the mean is larger than the median.

weight (g)	2	3	4	5	6
frequency	10	9	17	9	10

What is another result of the change you made?

(5) Another source of rich test items is situations that ask students to interpret a representation, such as a graph, or results produced by technology or another person.

Example: The following work has been handed in on a test. Is the work correct? Explain your reasoning.

$$\int_{-1}^{3} \frac{dx}{x-2} = \left[\ln|x-2|\right]_{-1}^{3} = \ln(1) - \ln(3) = -\ln(3)$$

Example: The following graph was obtained by dropping a ball and recording the height in feet at time measured in seconds. Why do the dots appear to be further apart as the ball falls? (The graph is omitted here.)

Assessing Problem Solving

Because developing problem-solving skills is also a goal in all of our courses, we include test items that help us measure progress to this goal. Questions like this also tell our students that they should reflect on effective and efficient ways to solve problems.

Example: Our company produces closed rectangular boxes and wishes to minimize the cost of material used. This week we are producing boxes that have square bottoms and that must have a volume of 300 cubic inches. The material for the top costs $0.34 per square inch, for the sides $0.22 per square inch, and for the bottom $0.52 per square inch. (a) Write a function that you could use to determine the minimum cost per box.
(b) Describe the method you would use to approximate this cost.

Example: (a) Use synthetic division to find $f(2.7)$ for the function $f(x) = 2x^5 - 3x^3 + 7x - 11$. (b) Give two other problems for which you would have done the same calculations as you did in part (a).

Findings

In addition to the broad categories described thus far, questions that require students to describe or explain, estimate answers, defend the reasonableness of results, or check the consistency of representations are used more often now. All these changes in our tests are part of a rethinking of the role of assessment in teaching and learning mathematics. We understand better now the implicit messages students receive from assessment tasks. For example, tests may be the best way to make it clear to students which concepts and techniques are the important ones. Assessing, teaching, and determining curricular goals form an interconnected process, each shaping and re-shaping the others. Assessment, in particular, is not something that happens after teaching, but is an integral part of instruction.

Use of Findings

As our tests have changed, we have learned to talk with students about these changes. Before a test students should know that our questions may not look like those in the text and understand why they should be able to apply concepts and do more than mimic examples. Students should expect to see questions that reflect the full range of classroom and homework activities—computations, explorations, applications — and to encounter novel situations. We suggest that students prepare by practicing skills (with and without technology, as appropriate), by analyzing exercises and examples for keys to appropriate problem-solving techniques, and by reflecting on the "big" ideas covered since the last test. After a test, students should have opportunities to discuss what was learned from the test, thought processes and problem-solving strategies, and any misunderstandings that were revealed. These discussions, in fact, may be where most of a student's learning takes place.

Success Factors

The world provides many wonderful and clever problems, and newer texts often contain creative exercises and test questions. But our purpose is to learn what the student knows or does not know about the material, not to show how smart we arc. So we examine course goals first, then think about questions and tasks that will support those goals. We suggest also reconsidering as well the grading rubric for tests. Using an analytic or holistic grading rubric can change the messages sent by even a traditional test. Consider the proper role of tests in a mix of assignments that help students accomplish course goals. And ask one final question: do you really need to see timed, in-class, individual efforts?

Interactive E-Mail Assessment*

Michael D. Fried
University of California at Irvine

This article discusses how internet technology can be harnessed to give students semi-automated individualized help. The intended reader appreciates how little problem solving guidance students get in class, and how they are on their own to handle giving meaning to learning an overstuffed curriculum.

Background and Purpose

Vector calculus is difficult for *all* students. It combines algebra and geometry in ways that go beyond high school training. Students need frequent and consistent responses. Without assurances they are learning a model supporting the course's ideas, students revert to memorizing details and superficial imitation of the instructor's blackboard topics. Many bog down, lose interest, and drop the course.

I needed more resources to give a class of 65 students the help they were clamoring for. So, I set out to develop a system of e-mail programs that would give more effective interaction with students. My goals were to:

1. Retain minority students
2. Enhance interaction between professors and students through project and portfolio creation
3. Find out why — despite good evaluations — fundamental topics didn't work
4. Continue to live a reasonable life while accomplishing goals 1 and 2.

I chose technology because it is the only resource we have more of than five years ago. We have less money and less time. Further, a Unix e-mail account is cheap and powerful, once you learn how to use it. Here are basic elements of the program, which was funded by the Sloan Foundation:

- Establish quality student/instructor e-mail communication
- Gather student data through e-mail Interactive Questionnaires (**IQ**s)
- Develop a system of automated portfolios
- Create a process for developing individual and team projects
- Reinforce student initiative toward completion of projects
- Collect weekly comment files for students and teaching assistants

These elements required designing and programming an office system of seven modules. Complete documentation will appear in [3]. [2] has more discussion of what to expect from on-screen use of the programs. Each module incrementally installs into existing e-mail Unix accounts. These are now available to others at my institution. For example, having teaching fellows use parts of the system improved their performance and communication with me. Using the system requires training on basics, like how to move to a directory in a typical e-mail account. In this article I concentrate on **IQ**s and automated portfolios.

This e-mail system allowed efficient fielding of 30–40 quality interchanges a day. Further, it kept formatted

*Partial support: Sloan Foundation 1994–96, NSF #DMS 9622928.

electronic portfolios tracking each student's interactions with me, in several forms.

Method

Interactive Questionnaires (**IQs**). An **IQ** is an enhanced interactive quiz, (controlled by a computer program) offering various aids and guidance to students who receive it by e-mail. Students take **IQs** at computer terminals — any terminal with access to their mail accounts. When the student finishes, the **IQ** automatically returns the student-entered data to the instructor by e-mail for (automatic) placement in the student's portfolio. The **IQ** reader software sorts and manipulates evaluative data entered by the student and returned by the **IQ**.

Teaching the modularity of mathematics is exceptionally hard. **IQs'** highly modular structure assists this process. They help students focus on analyzing one step at a time. Data from an **IQ** is in pieces. You view pieces extracted from a student's (or many students') response(s) to any **IQ** with the **IQ** reader. Therefore, **IQs** help an instructor focus on small conceptual problem elements in a class.

Here is a sample **IQ** from vector calculus, divided into parts, to help students analyze lines in 3-space. (Numbers after each part show the point value that part contributes to the 45 points.)

Comparing lines. (45 points) Consider two lines $L_1 = \{(2,1,3) + t(v_x, v_y, v_z) \mid t \in R\}$ and $L_2 = \{(x_0, y_0, z_0) + s(2,1,0) \mid s \in R\}$. You choose the vector (v_x, v_y, v_z) and the point (x_0, y_0, z_0) to construct lines with certain properties.

a) **7**: For what vectors (v_x, v_y, v_z) is L_1 parallel to L_2?

b) **13**: Suppose $(x_0, y_0, z_0) = (2,1,3)$. Let V be the set of nonzero vectors (v_x, v_y, v_z) which make $L_1 \perp L_2$. Find vectors $\mathbf{v}_1, \mathbf{v}_2 \in V$ such that $\mathbf{v}_1$ is perpendicular to $\mathbf{v}_2$.

c) **10**: Suppose $(v_x, v_y, v_z) = (1,3,2)$, and L_1 and L_2 meet at a point. Find a vector $\mathbf{w}$ and a point (x_1, y_1, z_1) so the plane $P = \{(x,y,z) \mid \mathbf{w} \cdot ((x,y,z) - (x_1, y_1, z_1)) = 0\}$ contains all points of both L_1 and L_2.

d) **15**: Suppose $(v_x, v_y, v_z) = (1,3,2)$. For what points (x_0, y_0, z_0) is there a plane P that contains both L_1 and L_2?

An **IQ** presents each problem part, with help screens, separately. We (now) use simple TeX notation (see discussion of this in Findings). Numbers after help screen menu items inform students of how many points they will lose for using that help option. The **IQ** tracks student use of help screens, returning that data to the instructor via e-mail. Here is the help screen from part (b) of the problem above.

Help for Part b: Choose from among these if you feel you need help.

[0.0]: No help needed; I'm ready to answer the question.
[1.2]: Choosing an example vector in V.
[2.2]: Checking if two vectors in 3-space are perpendicular.
[3.3]: Describing V as a set given by equations.

[4.3]: Solving equations for $\mathbf{v}_1$ and $\mathbf{v}_2$.

The student enters a menu choice. Choosing 0 lets the student type written answers into the **IQ**. These answers are automatically sent to the instructor and put in the instructor's portfolio for that student. Any use of other help items before the student sends in the answer is also included in the information the instructor gets. Students can retry menus any time. After returning the **IQ** to the instructor, a student can go through it again, this time using the help screens to check the work. As a developer of **IQs**, I work on enhancing the quality of evaluation and interaction.

Polling Portfolios. To *poll* a portfolio is to collect **IQ** and other interaction data from the student portfolios. Polling portfolios gives an instructor an instantaneous screen or file report on how a collection of 65 students answered a specific question. Unless I choose otherwise, only I have access to the portfolio.

The phrase *automated portfolios* refers to creating (additions to) portfolios semiautomatically from interactions. Each piece of each problem in an **IQ** has a tag. This simplifies moving material from a student's **IQ** responses (e-mail messages), to the student's portfolio, and gathering the responses of all students to a given question so the instructor can evaluate them. It is as if the instructor had ripped the bluebook apart to allow grading all students' responses to a given item together. Placing items correctly into the portfolio is what allows easy polling (creation of reports about the portfolio contents). **IQs** are much easier to grade than similarly enhanced paper and pencil exams. The viewer program for that **IQ** can create a report across the student portfolio collections using responses to that tagged piece. **IQ** technology allows the following activities, under the rubric we call *polling* portfolios.

• Automatic placing/formatting of **IQ** responses into student portfolios
• Batched evaluating and commenting on specific pieces from an **IQ**
• Automatic mailings of commented **IQ** portions to portfolio owners [2]

Below are the answers the **IQ** reader collected from students with mail-names Akoines and Gmoore (two of my 60+ students) to part b of the problem above, followed by my response. Student responses to problem b showed two types of difficulty among all 60+ answers. Other students tried to start the problem without using the help screens. The response to them showed what they would have gained by doing so.

The problem asked:

Suppose $(x_0, y_0, z_0) = (2,1,3)$. Let V be the set of nonzero vectors (v_x, v_y, v_z) which make $L_1 \perp L_2$. Find vectors $\mathbf{v}_1, \mathbf{v}_2 \in V$ such that $\mathbf{v}_1$ is perpendicular to $\mathbf{v}_2$.

Akoines responded: "$(x_0, y_0, z_0) = (2,1,3)$, so $L_2 = \{(2,1,3) + s(2,1,0)\ s \in R\}$." Akoines then paraphrased what Help told him, and proceeded with the problem: "As L_1 and L_2 both go through (2,1,3), they are perpendicular. This means their directions are perpendicular. So, (v_x, v_y, v_z) is perpendicular to (2,1,0). Dot product gives $2v_x + v_y = 0$. I didn't get what *general solution* means."

Akoines admitted not understanding an idea in the last sentence. Though he didn't use it, Help had a phrase on this point: Let v_z be anything; solve the relation between v_y and v_x. Gmoore tackled this point without needing Help.

Gmoore responded, "As $(x_0, y_0, z_0) = (2,1,3)$, $L_2 = \{(2,1,3) + s(2,1,0)\}$. Since L_1 and L_2 go through (2,1,3), they are perpendicular if their vectors are perpendicular. The product $(v_x, v_y, v_z) \cdot (2,1,0) = 2v_x + v_y$ is 0. So, v_z is anything, and v_x and v_y satisfy this equation."

The **IQ** reader also told me that Akoines had used help screens 1 and 3, at a total cost of 5 points out of the 13 for that part of the problem, and Gmoore had used different help screens, with a different resulting score.

My response to Akoines was "List vectors (v_x, v_y, v_z) with $2v_x + v_y = 0$. Take (v_x, v_y, v_z) from $V = \{(u, -2u, v) \mid u, v \in R\}$ as in b3 Help: $u = 0$, $v = 1$, and $v = 0$, $u = 1$ (as in b.4 Help) to give $\mathbf{v}_1 = (0,0,1)$ and $\mathbf{v}_2 = (1, -2, 0)$."

Gmoore and every other student with difficulties similar to Akoines got the same response I sent to Akoines. I only had to write this response once, and then tag which students that response should be sent to. Solving part b required using the solution of one linear equation in two unknowns. Even with help screens, this was a hard analysis for students. So, the same grader response appeared in many **IQ**s. The viewer lets the grader — seeing many answer parts juxtaposed on the screen prior — *simultaneously* add this one response to many **IQ**s. This typifies how a batch response to problem pieces cuts grading time. My upcoming book [3] features many applications of the system for batch and personalized merge mail, including returning **IQ**s to students.

Findings

Enhanced e-mail interactions brought more contact with students in one course than I had in my previous 20 years of teaching. Without this system, these interactions and associate evaluations would have overwhelmed me.

A ten-week course covers a few succinct ideas. Instructors know where the road goes. Portfolio data shows few students know there is a road. Day to day, students lose their way, and we march on without them. Many faculty are aware students get too little feedback, and that giving more without some aid would overwhelm instructors. I now routinely retain close to 100% of my students. Often students report to me from the next class, on how some particular idea improved their ability to use my course in the next.

Further, these portfolios started the process of documenting the value added by the instructor. Side messages from students — also put into their portfolios — show specific changes of attitude. Epiphany occurs when students realize it's not luck if an exam suits their studying. Rather than being lucky to have memorized a problem rubric, they see they need more concentration to maintain an analyzing mode. My web site [2] quotes a student likening her negative approach to analyzing problems with her difficulties using a modem.

Using TeX helped my whole class, for the first time ever, use precision in their mathematical writing. Computer shorthand for mathematics formulas (via TeX) aids effective use of mathematics e-mail. The instructor can teach this incrementally. Further, running e-mail exchanges through a TeX compiler translates symbols to magnificent output, a reward for students.

Use of Findings

Suppose in a particular class I find students show they have forgotten material from two weeks before. I use portfolios to show us all what is happening. I ask the polling program to list questions from previous **IQ**s. When I find one that illustrates (say from an **IQ** two weeks ago) the same questions students couldn't answer in today's class, I create a report on the old responses to that question. That takes no time: I just choose "create a report" from the menu in the program. The report of students' responses several weeks ago astounds them. Most would have asserted we never had that material before. It shows them they must consider anew the effectiveness of their study habits, and other aspects of how they respond to classroom data.

Student portfolios contain interactions, including with classmates, which lay out their intellectual problems. Until I used **IQ**s and portfolios, I didn't realize how often students lose what they have learned. Despite years of excellent interactions with students, by the early 90's I felt students were deceiving me. They appeared to be masters at convincing me they understood material that in truth, they did not. Using **IQ**s showed what was happening.

[2] gives detailed screen shots from **IQ** sessions. A graphic with it illustrates polling portfolios. This shows the dynamics of students learning and losing core material. Students *had* a tenuous grasp on material soon after a preliminary introduction. Then, weeks later, they *lost* it. The viewpoint from high school kept reclaiming territory from university ideas. This happened with characteristic material repeatedly (example: falling back on point/slope lines in 3-space). Polling student portfolios simplified catching this and interceding.

My Sloan Technology activities have enjoyed some campus support (for trying to handle 18,000 students). Steve Franklin and Leonard Meglioli [1] at UCI's Office of Academic Computing have replicated simple **IQ**s in HTML forms under the name "ZOT dispatch." OAC offers training in it to induce faculty to get students' comments on their Web offerings. It is a valid, though limited, entrance to the whole portfolio system.

Success Factors

Any use of paper, like students responding to **IQ**s on paper rather than on-line, bogged down recording the effort and responding for the teaching fellow or me. My upcoming book [3] discusses at length why *office hours* can't come close to the effectiveness of **IQ**s. Especially: The Sloan system gives an electronic record — put to many uses — and it has all students participate.

References

[1] Franklin, S. and Meglioli, L., III, *Documentation for Zot Dispatch*, comment system for EEE-class resources at UCI, http://www.oac.uci.edu/X/W6/forms/zot-dispatch/, 1996.

[2] Fried, M.D. "Electronic Portfolios: Enhancing Interactions Between Students and Teachers," essay and graphic at http://www.oac.uci.edu/indiv/franklin/doc/mfried, 1994.

[3] Fried, M.D. *Interactive E-mail; Enhanced Student Assessment*, in preparation, 6 of 10 chapters complete as of August 1997.

Flexible Grade Weightings

William E. Bonnice
University of New Hampshire

Using a computer spreadsheet to compute grades, it's easy to let students (even in a large class) focus on their strengths by choosing what percent (within a range selected by the instructor) of their grade comes from each required activity.

Background and Purpose

I facilitate cooperative classrooms where students are supportive of one another and where much of the learning takes place in class discussions and with students working in groups. To deepen student involvement and to inculcate a sense of their responsibility, I work with the class in making decisions about course operation and learning processes. One day, when the class was arguing about how much weight to place on the various factors that contribute to their grade, it came to me: why not let each student decide for him or herself from a given range of choices? This way students can weight heavily factors which they like or in which they are strong, and they can minimize the weight of areas in which they are weaker or which they dislike. For example, some very mature and conscientious students claim that they study more efficiently if they make their own choices as to what homework to do. One could respond constructively to such students by allowing zero as the lower bound of the homework range. These students then could choose not to turn in any homework and weight other factors correspondingly higher.

Assessment methods for ascertaining grades should be more broadly based, yet the number of different tasks we can expect a student to undertake is limited. Ideally each student should be able to choose his or her own method of assessment to demonstrate competence and learning. The flexible weighting method, to be described here, is a simple versatile extension of standard grading methods. It accommodates the diverse strengths, interests, and desires of the students.

Method

Early in the course, a form which lists the ranges for each factor that contributes toward the grade is handed out to the students. They are given some time to experience the course and gauge the teacher before they must make their decisions. Usually shortly after the first exam and the first project are returned, the students are required to make their choices of specific weights, sign the form, and turn it in. At the end of the semester, I record the computation of the students' grades right on this form, in case they want to check my computations and the weights that they chose.

A typical grading scheme I now use is:

Self-Evaluation	5%
Teacher-Evaluation	5%
Board Presentations	5 to 10%
Journal	5 to 20%
Projects	10 to 30%
Homework	5 to 20%
Hour Exams	30 to 45%
Final Exam	15 to 25%

The following simplified example illustrates the possibilities. The first column of figures gives the ranges offered and the subsequent columns show possible selections of percentages by different students.

	Ranges Offered	Exam Hater	Exam Lover	Homework Hater	Portfolio Hater	Balanced Student
Homework	0 – 25	25	10	0	25	25
Quizzes	25 – 40	25	40	40	40	25
Final Exam	20 – 40	20	40	30	35	25
Portfolio	0 – 30	30	10	30	0	25

We see that the "Exam Hater" may choose to put maximum weight on homework and the portfolio, whereas the "Exam Lover" may choose to put maximum weight on the exams. The "Homework Hater" may choose that the homework not count at all and the "Portfolio Hater" may choose not to keep a portfolio. We also see that the quizzes and final exam combined must account for a minimum of forty-five percent of a student's grade; thus homework and the portfolio combined can account for at most fifty-five percent of a student's grade. If one did not want to make homework optional, one would simply change the lower bound on the homework range to be greater than zero.

Findings

Since I have been using flexible weighting, students have always reacted to it favorably. Often students ask, "Why don't other teachers use this method?" In view of current technology, that is a legitimate question.

Because I want to foster cooperation, not competition, in my classes, I set grade categories at the beginning of the course: 90% or above receives an A, 80% to 89% receives a B, 70% to 79% gets a C, 60% to 69% gets a D, and below 60% receives an F. When students are able to focus on their strengths and interests their learning seems to improve.

Although originally I made some factors optional by giving them a lower bound of zero, lately I have been using five percentage points as the smallest lower bound. This gives the students the message that all methods of assessment that we use are important but that if they really don't want to use one of them, the loss of the five points might possibly be made up by utilizing the freed up time to work more on a different, preferred area.

Use of Findings

Using this flexible weighting scheme has influenced me to include more factors in my grading than I had in the past. For example 5% of a student's grade comes from a "self-evaluation." What part of this 5% students receive depends on the *quality* of their self-evaluation. Students could say that they deserves an A in the course and receive a zero or close to 0 (out of a possible 5%) because they did a poor job

of justifying the A. Similarly students could receive the full 5% of the self-evaluation factor for doing an excellent job of justifying a poor grade in the course. These self-evaluations have enabled me to know and understand my students much better. They make very interesting reading. My present thought is that they ought to be done earlier in the course, around mid-semester, so that students would have time to make behavioral changes based on my feedback.

Note that five percentage points may come from "teacher-evaluation." On the same day that the students turn in their self-evaluations, I give them my teacher-evaluation of their work and learning. Writing the teacher-evaluation is quite demanding and requires getting to know each student. For each student I keep a separate sheet of paper in my notebook on which I record significant observations as they occur during the semester. I record examples of effort and accomplishment such as class participation, willingness to help classmates, intelligent comments and contributions, well-done homework or projects, etc. On the flip side, I record lapses and lack of effort and participation, especially instances of times when students were unprepared or did not do their assigned work. If students have no negative comments on their record, I give them the full five points and tell them some of the good things I recorded about their work during the semester.

Note that in the typical scheme, board presentations have been given a minimum of five percentage points. Some students are terrified of presenting in front of the class. They may choose not to present and the loss of five points won't kill them. However my experience has been that, in order not to lose those five points, students overcome their fear and make a presentation. It is surprising how many of the fearful students go on to "flower" and become regular presenters at the board.

Note also that a minimum of five percentage points has been given to keeping a journal. This motivates most students to perform the useful process of writing about the mathematics that we are learning. Yet the recalcitrant student may still choose, without undue penalty, not to keep a journal.

This flexible grade method has led me to give more respect to students as individuals with their own strengths, weaknesses, preferences, and interests. It has stimulated me to move away from being an authority in the classroom toward looking for other ways to turning responsibility and authority for learning over to the students.

Success Factors

After the students have read the first day's handout describing the method, they are usually enthusiastic about it and the following in-class discussion takes less than twenty minutes. It must be made clear that, after they have chosen their percentages, their choices may not be changed later in the semester.

I have written a program on my calculator which I use to compute each student's final numerical grade. If one keeps the class grades on a computer spreadsheet, a formula can be entered which will immediately calculate each student's final numerical grade, once her/his percents (weights) have been entered into the spreadsheet.

Because the method of flexible weightings makes grading more agreeable for both teachers and students, it is worth the extra time and effort that it entails.

The One-Minute Paper

David M. Bressoud
Macalester College

This quick technique helps the instructor find out what students have gotten out of a given day's class, and works well with both large and small classes.

Background and Purpose

No matter how beautifully prepared our classroom presentation may be, what the student hears is not always what we think we have said. The one-minute paper (described in Angelo and Cross, *Classroom Assessment Techniques*) is a quick and easy assessment tool that helps alert us when this disjuncture occurs, while it also gives the timid student an opportunity to ask questions and seek clarification. I have used the one-minute paper in large lecture classes at Penn State where I would choose one of the ten recitation sections to respond and where it helped keep me informed of what the students were getting from the class. I have also used it in the relatively small classes at Macalester College where it is still helpful, alerting me to the fact that my students are never quite as fully with me as I would like to think they are.

Method

In its basic format, the instructor takes the last minute (or, realistically, three minutes) of class and asks students to write down short answers to two questions:

- What was the most important point made in class today?
- What unanswered question do you still have?

Responses can be put on 3×5 cards that are I hand out, or on the student's own paper. Students can be allowed to respond anonymously, to encourage them to admit points of confusion they might hesitate to put their name to, or they can be asked to write their names so that the instructor can write a brief, personal response to each question or encourage thoughtful answers by giving extra credit.

The questions can be modified in various ways, but they should remain open-ended. In one variation described by Angelo and Cross, the instructor asked each student to name five significant points that had been made in that session. This can be especially useful in identifying the range of perceptions of what has been happening in class. By spending some time early in the semester discussing these perceptions and how they relate to what the instructor hopes that the students will see as the central ideas of the class, students can learn how to identify the central themes in each lecture.

In the large lecture classes at Penn State, students were required to write their names on their papers. After class, it would take me less than 30 minutes to go through the thirty or so papers that would be turned in, check the names of those who had turned them in (a bonus amounting to 1% of the total grade was given to those who turned them in regularly), and then write a one-sentence response to each question. These were returned to the students by their recitation instructors.

Findings

Many of the students in large lecture classes viewed the one-minute paper as simply a means of checking on whether or not they attended class, and, in fact, it did help keep class attendance up. It kept me abreast of what students were getting out of the lectures and helped establish some personal contact in the very impersonal environment of the amphitheater lecture hall. But even in the small classes at Macalester College where I now teach, where students are less hesitant to speak up, and where it seems that I can sense how well the class is following me, the one-minute paper often alerts me to problems of which otherwise I would not be aware until much later in the semester.

Use of Findings

Since the purpose of the one-minute paper is to identify and clarify points of confusion, I start the next class with a few minutes spent discussing student answers to the first question and explaining the misunderstandings that seemed to be shared by more than one student.

For example, after a class in which I introduced exponential functions of the form b^x and explained how to find their derivatives, I discovered that there was still a lot of confusion about how the derivative of 3^x was obtained. For many students, using $(3^{x+.01} - 3^{x-.01})/.02$ as an approximation to the derivative of 3^x was more confusing than useful because they thought that an approximation to the derivative should look like $(3^{x+h} - 3^x)/h$. Some students were thrown by my use of the phrase "in terms of x" when I spoke of "the derivative in terms of x." They knew about derivatives, but had never heard of derivatives in terms of x. This provided an opening the next day to come back to some of their continuing misunderstandings about functions. More than one student thought that the most important point of this class was how to find the derivative of e. This forewarned me that some of my students considered e to be the name of a function. Only four students picked out what I thought I had been emphasizing: that the significance of e is that it provides a base for an exponential function that is its own derivative. Almost as many thought that the identification of ln with $\log_e$ was the most important thing said all period. Knowing the principal points of confusion about derivatives of exponentials, I was able to start the next class by clearing up some of them and using selected questions to motivate the new material I wanted to introduce.

To use the one-minute paper as a learning tool, it is essential that you be consistent and regular and spend time early in the course clarifying what you want. It can also be employed simply as a periodic check on how accurate are your perceptions of what students are learning and what unanswered questions remain at the end of each class. The beauty of this tool lies in its simplicity and flexibility.

Success Factors

It is not easy to get students to identify important points or to ask good questions. They will tend to retreat into generalities. Following the class on derivatives of exponential functions, most students wrote that the most important idea was how to take derivatives of exponential functions. A few simply identified "taking derivatives" as the most important idea. Similarly, many students will either write that they have no questions or tell you only that they are confused. You can correct this by refining the question ("What was the most surprising or enlightening moment in class today?") or by spending time early in the course discussing examples of the kinds of specific observations that you would like them to be able to make. You may also want to talk about how to formulate questions during a lecture or presentation as a means of sharpening attention, illustrate this with examples — taken from students in the class — of questions that demonstrate this kind of attention, and then give some small bonus credit to those students who can consistently end each class with a question.

Reference

[1] Angelo, T.A., and Cross, K.P. *Classroom Assessment Techniques*, 2nd ed., Jossey-Bass, San Francisco, 1993, pp. 148–153.

Concept Maps

Dwight Atkins
Surry Community College

By drawing concept maps, students strengthen their understanding of how a new concept is related to others they already know.

Background and Purpose

Concept maps are drawings or diagrams showing connections between major concepts in a course or section of a course. I have used concept maps in precalculus and calculus classes at Surry Community College, a medium sized member of the North Carolina Community College System. About one-third of our students transfer to senior institutions. This includes most of the students who enroll in precalculus and nearly all of those who take calculus.

Concepts maps can be a useful tool for formative assessment. They can also provide an added benefit of helping students organize their knowledge in a way that facilitates further learning. I first learned about concept maps from Novak [3]. Further insight about concept maps and their uses in the classroom can be found in [1] and [2].

Method

Concept maps are essentially drawings or diagrams showing mental connections that students make between a major concept the instructor focuses on and other concepts that they have learned [1]. First, the teacher puts a schematic before the students (on a transparency or individual worksheet) along with a collection of terms related to the root concept. Students are to fill in the empty ovals with appropriate terms from the given collection maintaining the relationships given along branches. From the kinds of questions and errors that emerge, I determine the nature and extent of review needed, especially as related to the individual concepts names that should be familiar to each student. I also give some prompts that enable students to proceed while being careful not to give too much information. This is followed by allowing the students to use their text and small group discussions as an aid. The idea at this stage is to let the students complete the construction of the map. Eventually they all produce a correct concept map. Once this occurs, I discuss the concept map, enrich it with other examples and extensions in the form of other related concepts.

This strategy seems to work particularly well when dealing with several related concepts, especially when some concepts are actually subclasses of others. I have used this procedure in precalculus in order to introduce the concept of a transcendental number. Students chose from the list: integers, real, 3/5, rational, irrational, e, algebraic, π, transcendental, $\sqrt{2}$, and natural to complete the concept map on the top of page 90.

As students proceeded individually, I walked around the classroom and monitored their work, noting different areas of difficulty. Some students posed questions that revealed various levels of understanding. Among the questions were "Don't these belong together?" "Does every card have to be used?" and "Haven't some groups of numbers been left out?"

I have also used this procedure in calculus to introduce the gamma function. I wanted to introduce the gamma function by first approaching the idea of a non-elementary function. This prompted me to question what students understood about the classifications of other kinds of functions. I presented a schematic and gave each student a set of index cards. On each card was one of the following terms: polynomial, radical, non-elementary, trigonometric,

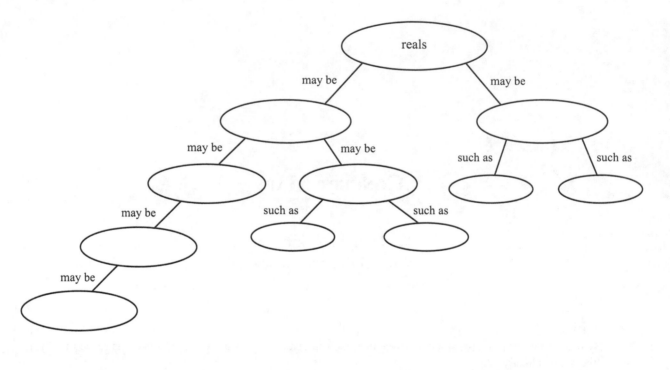

inverse trigonometric, hyperbolic, and gamma. The students were told to arrange the cards in a way that would fit the schematic.

Use of Findings

The primary value of using concept maps this way is that it helps me learn what needs review. They provide a means for detecting students misconceptions and lack of knowledge of the prerequisite concepts necessary for learning new mathematics.

Success Factors

The primary caveats in this approach are that the teacher should be careful not to let the students work in groups too quickly and not to give away too much information in the way of prompts or hints. One alternative approach is to have the students, early in their study of a topic, do a concept map of the ideas involved without giving them a preset schematic. Then you gather the concept maps drawn and discuss the relations found: what's good about each, or what's missing. You can also have students draw a concept map by brainstorming to find related concepts. Then students draw lines between concepts, noting for each line what the relationship is.

References

[1] Angelo, T. A. and Cross, K. P. *Classroom Assessment Techniques, A Handbook for College Teachers* (2nd ed., p. 197). Jossey-Bass, San Francisco, 1993.

[2] Jonassen, D.H., Beissneer K., and Yacci, M.A. *Structural Knowledge: Techniques for Conveying, Assessing, and Acquiring Structural Knowledge*. Lawrence Erlbaum Associates, Hillsdale, NJ, 1993.

[3] Novak, J.D. "Clarify with Concept Maps: A tool for students and teachers alike," *The Science Teacher*, 58 (7), 1991, pp. 45–49.

If You Want To Know What Students Understand, Ask Them!

John Koker

University of Wisconsin Oshkosh

To gain insight into how much students actually understand, and what they have learned by working the problems, have them discuss the evolution of their ideas as they work on homework sets.

Background and Purpose

The University of Wisconsin Oshkosh is both a major undergraduate and regional graduate campus in the statewide University of Wisconsin System with a student population of about 10,000. Many of the students I work with are pre-service elementary and middle school teachers. I teach an Abstract Algebra class for prospective middle school mathematics teachers. This course is specifically designed as part of a mathematics minor for those majoring in elementary education.

In my classes I assign non-routine, challenging problems that are to be handed in and evaluated. These problems usually go a step beyond the material discussed in class and are designed to set up situations where students can discover some mathematics on their own. Most students are successful in working exercises similar to those done in class and examples in the text, while few hand in "good" solutions to the challenging problems. I found myself giving low scores to what looked like poor attempts and meager results.

Anyone familiar with the history of mathematics can think of several examples of mathematicians who have worked on a given problem for many years. Perhaps the original problem was never solved, but the amount of mathematics learned and developed in working on the problem was astounding. I applied this situation to my students. What is more important: that they solve a given problem, or they learn some new mathematics while struggling with a problem? I was looking for papers which were organized, logical, thoughtful and demonstrated students had engaged themselves with the course material.

Method

In many of the problems I assign in my Abstract Algebra course, I expect students to compare properties of abstract mathematical structures with more familiar structures (integers and rationals). However, these students are not yet ready to respond to this kind of assignment with a clear, textbook-like essay. To help them understand what I do expect of them, I hand out "Guidelines for Written Homework," which explains

> Your papers can be as expository as you want to make them. Tell how you arrived at your particular solution. You can discuss how the problem relates to material in class and possible applications of the problem. Do you understand the class material better now by working on this problem? Explain. If you could not arrive at a solution tell what you tried and why it didn't work, or describe what you think may work. Vent frustrations! Identify obstacles you encountered and what you think you need to overcome these. **The most important thing is to demonstrate to me that some thought went into your work and that you learned something!**

I add that an expository paragraph describing their experience with the problem **must** be included with the solution.

For example, the following homework assignment was given while we were discussing groups, before they were aware that a group satisfies the cancellation property.

HOMEWORK #3

When working with numbers, we "cancel" without even thinking. For example, if $3 + x = 3 + y$, then $x = y$, and if $3x = 3y$, then $x = y$. The purpose of this homework problem is to get you thinking about this "cancellation process."

a. Let a, b and c be elements of $\mathbf{Z}_6$ (integers modulo 6). Then if $a \oplus b = a \oplus c$, is it necessarily true that $b = c$? Justify your answer.

b. Let a, b and c be elements of $\mathbf{Z}_6$. Then if $a \otimes b = a \otimes c$, is it necessarily true that $b = c$? Justify your answer.

c. Let $(G, *)$ be a mathematical structure and suppose that a, b and c are elements of G. What properties must $(G, *)$ satisfy so that $a * b = a * c$ implies that $b = c$?

Findings

The expository portions of the first two assignments were quite unrevealing. Comments like, "the problem was hard," "I really didn't understand what you wanted," "I worked 2 hours on this." "this one wasn't too bad," etc., were common. I used the "power of the grade" to obtain more thoughtful responses. If, in my opinion, the expository section was too shallow, I would deduct a few points. I would also include a question or two which they could consider for the next assignment. Those who did not include any expository section, lost several points. After a few assignments, I began to get some interesting comments.

Assignment #3 described above was the first one on which I received good comments. Comments I received included:

I first thought in part (a) you just subtract a from each side and in part (b) and (c) you just divide by a on each side. When I mentioned this to you, you asked me if subtraction and division are defined in these structures. This helped me in part (b), because we did a problem in class about division in $\mathbf{Z}_n$. But I am stuck with part (c). How can I define division in $(G, *)$? I am sure that when a, b and c are in $\mathbf{Z}_n$, $a \otimes b = a \otimes c$ implies that $b = c$ if a is not a zero divisor because I can then define division. You said I am on the right track, but I still don't see it.

I didn't really have any trouble understanding parts a and b. I simply made the tables and checked to see if each element of $\mathbf{Z}_6$ appeared at most one time in each row of the table. If this happens the element can be canceled. For example in $\mathbf{Z}_6$, $3 \otimes 0 = 3 \otimes 4$, but $0 \neq 4$. In the "3" row, the number 0 appears more than once. However, part c was a different story. I now see what you mean when you say that it is different to show something which is an example and something is true

in general. I could just make a table for $\mathbf{Z}_6$ but how in the world do you make one for $(G, *)$? When I came and talked to you, you told me to think of the algebraic properties involved. I didn't know what you meant by this. You then asked me what are the "mechanics involved in going from $3x = 3y$ to $x = y$." I then started to see the picture. Is this what abstract algebra is like? Trying to decide what properties will allow certain results?

Neither of the students who wrote the above paragraphs above solved part c. But because of what they wrote, I know that the first student has a better understanding of what it means for division to be defined and the second appreciates the difference between working examples and justifying statements in general. Both of these tell me the student has learned something.

I realized that this was the first assignment where the student was given an open-ended problem in which they needed to make a conjecture. I didn't require proof. They needed to convince themselves that their guess was accurate. I also got something that I didn't expect. Here is what one student wrote about part c:

I feel that the examples we did in class on Friday helped serve as a guide for this problem. It wasn't really too difficult. I didn't spend very much time on it.

By itself this seems like a pretty insignificant comment. However the solution the student handed in was completely incorrect! Some students submit incorrect solutions knowing they are incorrect. That, in itself, shows that the student knows something. It is quite different, however, when a student believes an incorrect solution is correct. I simply wrote a note to this student and we discussed this problem in my office.

Later in the semester I gave an assignment which was designed to have the student conjecture when a linear equation $ax + b = 0$ (with a and b elements of a ring R) has a unique solution. In the first part, students were asked to find examples of a linear equation which has more than one solution, a linear equation which has a unique solution, and a linear equation which has no solution in each of the rings $\mathbf{Z}$, $\mathbf{Q}$, $\mathbf{Z}_4$, $\mathbf{Z}_5$, and $\mathbf{Z}_6$ (if possible). Then they were asked to give a general condition which will imply that a linear equation $ax + b = 0$ over a ring R has a unique solution.

Most students could completely handle the integer and rational cases. Additionally, only a few struggled with the linear equations over $\mathbf{Z}_4$, $\mathbf{Z}_5$, and $\mathbf{Z}_6$. Discovering the algebraic properties involved was the heart of the problem and the difficult part. One interesting comment I received indicated that the student never thought about the number of solutions an equation has and that this is somehow connected to the elements and properties of the structure. She just solved them without much thought. She also wondered if the same problem could be solved for quadratic equations.

A second student thought that a linear equation always had at most one solution. This proved to be true in $\mathbf{Z}$ and $\mathbf{Q}$. She described trying to use "regular" algebra in $\mathbf{Z}_4$ by subtracting b from both sides and then dividing by a. This caused problems when she found that division is not always possible in $\mathbf{Z}_4$. So she chose numbers and substituted them for a and b in the equation $ax + b = 0$. She solved the problem in $\mathbf{Z}_4$ by using the operation tables. She then went on to explain that this is where she figured out that if an element doesn't have an inverse, there can be more than one solution. She concluded that this is why in $\mathbf{Z}$ and $\mathbf{Q}$ there are unique solutions.

I feel like the first student's interest was sparked by this problem. The second has some misconceptions. She says that there are multiple solutions when an element is not invertible and comments that this is why equations over the integers have unique solutions (when solutions exist). Yet only two integers are invertible.

Several students told me they learned a lot by completing this assignment. That is insufficient. I want to know what they have learned.

For the most part, the student reaction to writing about their experience with a problem was quite positive. At the end of the semester I asked for students to comment on this process. Students said that the expository section of the homework really helped them to think about what they did and why it did or did not work. In general they felt they had an alternate route to demonstrate learning was taking place. They also appreciated getting credit even if their solution was not entirely correct.

Use of Findings

I have been learning about how students approach and think about problems. Before, I could tell whether or not a student could solve a problem, but that is all. I couldn't even tell if they knew if they were right or wrong. Students see this technique as a nonthreatening way to express their understanding. I use their paragraphs to construct new problems, to prepare for class, and to instigate classroom discussions. For example I wrote a problem asking when the quadratic equation will work in $\mathbf{Z}_4$, $\mathbf{Z}_5$, and $\mathbf{Z}_6$ based on the student comment above. After grading a problem with poor results, I go over it in class. These paragraphs helped

me to focus on what the students did and didn't understand. I could use their ideas to change a condition in the problem and asked them how the results change. We found ourselves asking "What if …?" a lot.

In addition, I now reward process, not just results. I try to assess the solution together with the exposition, giving one score for the combination. I try to convince my students that the most important part of their paper is to demonstrate to me that substantial thought went into their work. The paragraphs which addressed a student's own learning were particularly helpful with my assessment. In addition, students were using these papers to ask questions about material that was not clear. The method created a nonthreatening situation for the student to express confusion.

Success Factors

It is important to realize that this expository writing will not work for all kinds of problems, e.g., routine symbol manipulating problems. I had my greatest success on open-ended problems. Initially, I didn't get any really good responses, and some students never really cooperated. In the future I may include specific questions I want answered with a given problem. Some example of questions on my list are:

- Do you understand more or less about this topic after this problem?
- Explain how you arrived at your solution. Discuss trials that didn't work.
- Discuss possible alternate solutions.
- Discuss possible applications of this problem.
- Discuss the connections between this problem and our work on _____ in class.
- Describe what you were thinking and feeling while working on this problem.
- Can you think of a better problem to assign on this topic? If so, what is it?

These questions may be helpful at the beginning of the semester. Students will be encouraged to add additional comments beyond responding to the specified question. Then, depending on the results, I can allow more freedom as the class moves on and the students' confidence increases.

Improving Classes with Quick Assessment Techniques

John W. Emert

Ball State University

This article discusses three quick techniques which can alert the instructor to potential problems. The first helps students understand course goals, the second evaluates the effectiveness of group work, and the third is a general way of finding out how things are going in time to make changes.

As I continue to experiment with various informal classroom assessment techniques, I have come to favor assessment tools which use no more than ten minutes of class time and require no more than an hour after class to tabulate and form a response. The three techniques included here provide quick ways to find out what is going well and what is not, and allow me to address any problems in a timely manner.

Ball State University is a comprehensive community with approximately 17,000 undergraduates. The University offers a strong undergraduate liberal and professional education and some graduate programs. Most University courses are taught in small classes of 35 or fewer students by full-time faculty.

My classes tend to sort into three categories: content courses for departmental majors, colloquia courses for our Honors College, and a mathematics topics course that fulfills a University general studies requirement. While each class develops its own unique personality, each of these categories brings its own challenges.

Technique 1: Setting Course Goals

Background and Purpose

A section of our general studies topics course generally includes 35 students, mostly first-year, with widely varying mathematics abilities and declared majors in the creative arts, communications, humanities, social sciences, and architecture. Students in this course often enter the course annoyed that a mathematics class is required by the University. I want my students to examine the role of this course in their college curriculum.

Method

At the first meeting I follow a brief course introduction by distributing copies of the General Studies Program Goals, as found in the Undergraduate Catalogue.

General Studies Program Goals The General Studies program is designed to help you develop knowledge, skills, and values that all graduates of the university are expected to share. The program of General Studies has the following goals:

1. An ability to engage in lifelong education by learning to acquire knowledge and to use it for intelligent ends.
2. An ability to communicate at a level acceptable for college graduates.
3. An ability to clarify one's personal values and to be sensitive to those held by others.
4. An ability to recognize and seek solutions for the common problems of living by drawing on a knowledge of historical and contemporary events and the elements of the cultural heritage related to those events.
5. An ability to work with others to solve the common problems of living.
6. An ability to assess one's unique interests, talents, and goals and to choose specialized learning experiences that

will foster their fulfillment. This program is made up of core requirements and distribution requirements, which are groups of courses from which you choose. All students graduating with baccalaureate degrees must complete the 41-hour requirement in General Studies.

After reviewing these six goals, I ask student to select the goals which they believe are important to all graduates, and of these, to choose one which they think the course will help them to achieve. I then ask students to write a paragraph identifying the goal and why they think it is important to all graduates.

Findings

Through this activity, I am able to focus the class on the general studies mission of the course. While this method effectively forces a positive reaction from each student, it allowed me to solicit student reactions and learn what they might expect from this class. This activity also creates an expectation for additional classroom assessment activities in the course.

Use of Findings

I tabulate the responses and share them with the class the next day. All six goals are generally included among the responses. As the course progresses through the term, when examples or illustrations could support these goals, I can to refer back to this list and the supporting student paragraphs. This activity helps me effectively combat the nagging question of *relevance* which can often plague a general studies course.

Success Factors

Courses for majors can also benefit from an initial overview of the course's intent, purpose, and breadth. You can also solicit the students' perceived course objectives and compare them to yours and those of the university.

Technique 2: Evaluating How Groups Work

Background and Purpose

Our departmental majors, including undergraduates and master's candidates in actuarial science, mathematics, mathematics education, and statistics, range broadly in ability and motivation. These students take our Discrete Systems course after one term of Calculus. Group laboratory projects have developed in recent years to become part of the Calculus sequence, but the experience can vary greatly from instructor to instructor. Since Discrete Systems includes several topics that can naturally support group exploration, I elected to develop similar projects for this course.

Method

After the second project was completed, I wanted to learn how my students' perception of this activity compared to my own. Therefore, I developed an evaluation form and solicited responses from each student.

Group Projects Evaluation Form

1. Overall, how effectively did your groups work together on the project? (Poorly, Adequately, Well, Extremely Well)
2. In an effective working group, each person should be an active participant. How well did your groups meet this goal? (Poorly, Adequately, Well, Extremely Well)
3. Give one specific example of *something you learned from a group* that you probably would not have learned alone.
4. Give one specific example of *something another group member learned from you* that he or she probably would not have learned alone.
5. What is the biggest challenge to group projects? How could this challenge be overcome?

Findings

After tabulating the responses, I found that over 75% of the class rated their ability to work together in groups "Well" or "Extremely Well" in the first two questions. In fact, about 45% of the class responded "Extremely Well" to both questions. In response to questions 3 and 4, positive aspects of group work included: different points of view (45%), better understanding of class topics (40%), improved library skills (35%), improved computer skills (15%) and group commitment (10%). The comments to the final question included several versions of "Finding time when we both can work on the project was the biggest challenge."

Use of Findings

I returned copies of these tabulated responses to the class the following day, along with the following paragraph:

Based on your responses, most of you view the group projects as a valuable part of the course. In addition to focusing on specifically assigned "group questions," the group structure promotes discussion of class topics, improves study skills, and helps to develop class rapport. Because I agree that group work is a valuable part of this class, and since the scheduled class time offers a guaranteed opportunity for groups to meet, I intend to devote parts of future class times for group discussion. Though the projects will not be completed during such short meetings, these times may allow groups to share ideas, report progress, and plan strategy. I intend to use a similar evaluation instrument following the final report. —JWE

Success Factors

I have found that the third and fourth questions can be adapted as an effective evaluation tool, as well as a prompt for classroom discussion. For example, when another Discrete Systems class recently visited our Annual Student Symposium, a poster session for student research projects, I asked the following three questions:

1. What is something you learned from a presenter at the symposium that you might not have otherwise learned?
2. What is something that someone else at the symposium (another student or a presenter) learned because of you that he or she might not have otherwise learned?
3. Should I encourage other mathematics classes to attend similar symposia in the future? Please elaborate.

Technique 3: Mid-Term Adjustments

Background and Purpose

My Honors College colloquia tend to attract a broad spectrum of curious upper class students, including majors in architecture, art, computer science, mathematics and music. Each Colloquium course develops in a different way. Even when a topic list is repeated, class dynamics can change drastically from term to term.

I recently taught a colloquium course on "Common themes" using Douglas Hofstadter's book, *Gödel, Escher, Bach* [2], to examine traits common to Bach's music, Gödel's mathematical logic, and Escher's graphic art. Having taught this course several times, I have learned the necessity of providing sufficient foundational knowledge in each of these areas so that effective discussions can occur in class. This time, I found that after the first few weeks the class discussions and activities had drifted precipitously from my initially announced goals and I knew that a few students were somewhat frustrated with this shift. While I had used other informal assessment tools in this course (such as the minute paper), the appropriate assessment activity for this occasion was much more focused.

Method

I asked the students to anonymously answer the following three questions:

1. What's one specific way that this class has addressed a goal? That is, "what worked?"
2. What's something that shouldn't have happened? That is, "what didn't work?"
3. What's something missing or unnoticed that should happen? That is, "what's missing, so far?"

Findings

These are difficult questions, both for the student and for the instructor. However, my class had a healthy atmosphere, I asked sincerely for their feedback, and I received specific and helpful responses from each student. Some students observed that the videotapes I had used at times were "boring." A few students thought that discussions had drifted too far into mathematics without sufficient foundation. Students also observed that some alternative activities, such as creating examples of self-reference in literature and the arts, were very effective. Most importantly, the third item provided a valuable resource for additional ideas: another guest speaker, an experiment using video feedback, and a culminating group project.

Use of Findings

This activity gave me the chance to solicit and immediately react to my students' concerns and desires, rather than waiting until "the next term." I summarized the results and reported them to the class the following week, along with my intentions: to arrange a tour to the Art Museum by the Curator of Education, to greatly reduce the use of videos, and to continue a significant amount of class discussions. I also solicited volunteers to set up the video feedback experiment for the class. The class collected their favorite examples and illustrations and compiled a "Proceedings" which was distributed to each participant at the conclusion of the course.

Success Factors

This should not be the first assessment tool you use in a course. Your students must know by your example that you value their sincere responses and that you will react appropriately to their comments.

Final Words

As Angelo and Cross [1] often say, "Don't ask if you don't want to know." An effective classroom assessment should attempt to confirm or refute your suspicions. Before you solicit student responses, identify what you want to learn, what you think you will find, and what you intend to do with the results. Without feedback of some sort, these activities lose their *assessment* quality, and become quizzes and study tools. For me, techniques that are quick to develop and administer tend to be used most often. One must remember, however, to allot the time to evaluate and respond to the students.

References

[1] Angelo, T.A., and Cross, K.P. *Classroom Assessment Techniques*, second edition. Jossey-Bass, San Francisco, 1993.

[2] Hofstadter, D. *Gödel, Escher, Bach: an eternal golden braid*. Basic Books, New York, 1979.

Creating a Professional Environment in the Classroom

Sandra Z. Keith

St. Cloud State University

This article describes four activities which can help students develop a more professional attitude toward their coursework: looking directly at their expectations, revising goals after a test, finding applications of the course in their chosen field, and preparing a résumé.

Background and Purpose

St. Cloud State University serves approximately 13,000 students. Since most students are self-supporting, the job ethic is very strong. The state pays 2/3 of tuition costs, expecting in return at least an eight-hour day of classes and studying from students, but most students maintain part-time and even full-time jobs (often at night). Worse than the effect of this on studying habits is the fact that our students often associate greater value to their jobs, and this can translate into excuses for missing class or turning in assignments late. Over the years I have found that some of these attitudes can actually be turned to a learning advantage — especially when one explains education situations in terms of jobs. For example, the student who complains that a take-home exam is too difficult will understand my point when I ask how a boss would react to the sentiment, "the job is too difficult."

The assignments described here (which I might call PATs, for "Professional Assessment Techniques") are designed to take advantage of the situation, and to encourage students to think of themselves as pre-professionals in the educational environment. My work with PATs preceded my discovery of Angelo and Cross's Classroom Assessment Techniques (CATs, [1]) and the assessment movement, tracing back to a Writing-Across-the-Curriculum faculty development workshop in the early 80's, during which I became interested in using exploratory writing assignments to enhance learning [3]. Angelo & Cross, however, helped me to think more systematically about using student writing for classroom research and for shaping learning. Some of the techniques I use are explained in [2].

Method

I use some combination of the following assignments in a variety of courses. I tell the class in advance of an assignment how many points it will count.

1. Examining Teacher/Student Roles.

Assignment: List (overnight) ten characteristics of a good teacher and ten characteristics of a good student.

Findings. The combined list for the teacher tends to be highly diverse, mostly featuring friendly qualities such as "smiles a lot, understands when we don't get something, is available, doesn't assign too much homework." In fact, this list inevitably sounds like the preferred qualities of an instructor on faculty evaluation forms, raising the question of exactly what comes first. The list of features of the good student, however, is by comparison very short and incomplete. A good student "smiles a lot, asks questions, is helpful to others, tries to come to class as often as possible, keeps up as much as possible." I have never received the response, "reads the book." After compiling and reviewing these two lists, I draw up my own expectations, and in the next class we spend a half-hour comparing our lists on an overhead projector. I highlight comments that strike me as interesting.

I wish I could say that this exercise is enjoyable. Rather, coming at the beginning of the term, with our expectations on a collision course, it's more likely to cause students to feel that they are being "set up." But I don't expect this exercise to enhance my popularity. It is instead intended to let students reflect on their own learning objectives, and in particular, to make some expectations extremely clear; e.g., "to come to class as often as possible" is not acceptable. ("How would that sound to your boss?") More congenially, throughout the course, I can remind them, for example, that "good students ask questions."

2. Revising Goals.

Assignment: (a) Resubmit your exam with revisions and corrections to anything "in red." (b) Complete the Post-Test Survey. (c) Review your initial goals and create five new ones, based on your experiences in this class to date.

Some questions I use for the Post-Test Survey:
(1) How prepared did you feel for this test?
(2) Did you do as well as you expected?
(3) Was the test fair?
(4) On what questions did you make your mistakes and why? What did you feel were your weak areas? Be explicit.
(5) How many hours per day do you actually study?

Findings: Many industries and academic institutions, including my own, ask for goal reports. While many teachers set goals for their classes and expect the same from students, there is value in having students ASSESS and REVISE their goals. Students write five learning goals at the outset of the course; then after each exam they submit a folder consisting of corrections to their tests, the Post-Test Summary, and a revision of their goals, based on the test performance. I consider the revising process an important part of student self-assessment. It is gratifying to see goals become more specific over time, with instructor feedback. For example, a typical goal from students at the outset is, "to get an A." I may write that this is my goal for the student as well, but HOW shall we achieve it? For example, how will we know when the work will produce A's on tests? Subsequent goals become more sophisticated — one student amended this goal to "During this class I wish to obtain better study habits by organizing myself and the notes I take. I see from this test that I will probably have to study longer on certain things that I do not understand in class and go to the tutoring sessions if it is necessary for me to do." Perhaps this student is learning to write longer sentences, but he is also becoming more aware of his share of responsibility in learning.

I also collate responses to selected questions on the Post-Test Summary and show them on overhead projector. If 24 students felt the test was fair, while four did not, reasons can be discussed. Occasionally I compare study patterns of some anonymous A-students with those of low performers. Test performance is generally a private thing, and when tests are opened up to discussion, students become very curious about what their peers are saying. Any tendencies of students to regard themselves as victims of the test, the teaching, or the

book dissolve in the more serious discussion about what we can all do productively to improve learning.

The folders draw me into the students' lives, and I find myself making extensive comments. Learning flows from engagement. Red-flagging a rationalization exactly when the student is engaged with it provides the best opportunity for catching the student's attention. Many students will claim to have made "careless" mistakes, which allows me to discuss why these are nevertheless a problem. Or I might flag an excellent student, demoralized by an exam, whose goal is now "to settle for a C." The interactive format allows me to respond to issues and questions the students themselves have raised, and thus provides coaching in a more personalized framework.

3. Getting into the real world.

Assignment: Go to the library and photocopy a page, from an ADVANCED text or journal in the field of your major interest, that includes some calculus. Document your source. Write a 1-page explanation about what the mathematics is saying, in language that will be understandable to another student in this class. (Anyone doubting calculus will be used in his or her field should look at an advanced statistics or economics text.)

Findings. A universally expressed goal among students is "to understand where this material will apply!" Nevertheless, the examples from applied fields in a calculus book never seem to have an impact. Possibly these examples go by too fast, or seem rigged, or they are not the type to be on tests. In this writing assignment, students explore the section of the library devoted to books in their chosen field; usually they have not been to this section of the library. They must choose a page that includes some calculus (perhaps use of the integral) and explain, as best they can, the background of the problem being discussed, as well as the steps of the mathematics. Many texts will be too advanced for the students at their current level of understanding, and they may simply have to admit as much, but chances are, with some searching, they will find a page they can minimally interpret. Since this assignment takes students out of the normally sequenced learning of both mathematics and their major fields, it is probably unlike anything they have ever done before.

To help students understand what I expect on this project, I show the work of former students (good result, poor result). Most students react positively; they are proud of the level of difficulty of the fields they plan to enter. And while students may not always provide the best of explanations, they at least can witness first-hand that the mathematics in their chosen field is probably much tougher and even more theoretical than what is currently being expected of them. This assignment is also a wonderful opportunity to observe how students perceive the meanings of the mathematics they have learned. For instance, students may accept some equation in physics as a given rule or law, without any sense that the rule could be derived with the mathematics we have just studied. For example, it is a rule, to them, that total charge is the integral of current over time; it does not seem to be something about which they should question, "why?" or "how?" On overhead projector, I show all the xeroxings of the text pages, briefly explaining in what context we are seeing the integral, for example. This gives a very streamlined overview of the ubiquity of calculus in applied fields.

Since the grading for this assignment is somewhat problematic, generally I grade on effort. A more in-depth grading is not easy if one must make comments on the mathematics. The following system for giving fast but personal responses helps: since I type fast, I create a form letter on the word-processor, citing the directions; below I add my personal responses. This way it somehow becomes easier to explain where the student has succeeded or fallen short of expectations.

4. Looking Good on a Resume.

Assignment: Write a 2-page self-introduction to a potential employer presenting your activities as a student (including mathematics classes) as evidence for your potential as an employee.

Findings. At some point during the term, I spend a half-hour discussing with the class how they can build their resumes. Even at the sophomore level, some students are still coasting on their high school athletic achievements. We discuss joining clubs, doing student research, becoming a tutor, going to conferences, and setting up a file in the placement office. Many students at our institution (which is looking into its general advising policies) have never heard these sorts of suggestions before.

This writing assignment may directly impact on their jobs: our institution, for example, when hiring, first asks interested individuals to write a letter introducing themselves (with a list of publications), and on this basis, determines who should receive an application form. Since many students feel embarrassed about presenting themselves in a positive or forceful way, I usually show what former students have written as well as what I have written as letters of recommendation for students (names erased, of course). In this way, students can see that the work we do on the board, in groups, and even the very writing we do on this assignment allow me to add more information on my recommendation letters for them.

When I give this assignment, I explain that I like to tack onto my letters of recommendation some exceptional work of the student (this might be the interpretive library assignment above or a take-home exam), conjoining once again the work we do in class with the students' future jobs.

Findings and Use of Findings

The credit given for these assignments constitutes a small percentage of the grade. Students seem to appreciate any take-home work that provides balance to test grades, although they may initially be puzzled by these assignments, which tend to be unlike any they have seen in mathematics classes before. A major benefit of these projects is a heightened sense of collegiality among the students in class as they find interesting examples of applications of calculus in different fields to share, and as the goals and autobiographical material provide an opportunity for more personal interaction than math classes typically offer. Most importantly, these projects heighten student awareness of how to shape their prospects for jobs and careers, and this is a value they recognize with no difficulty at all.

For my own portfolios, I copy selected strong/medium/weak responses to the assignments to represent the range of class responses. I keep these portfolios to read for reflection and to mine for evidence of effective teaching. Whether these teaching portfolios are seriously studied by university evaluators or not, I do not know, but the massive binders I have collected over the years allow me to study how students learn, and refresh me with ideas for future classes.

Success Factors

Some students respond trivially to these assignments early in the term, and I may request reworkings. Generally students come to appreciate the assignments when they see the serious responses of other students. I show transparencies of much of the students' output on an overhead projector; this provides students with response to their input — important in any assessment activity and invaluable in discussing "student professionalism."

References

[1] Angelo, T.A., and Cross, K.P. *Classroom Assessment Techniques*, 2nd ed., Jossey-Bass, San Francisco, 1993.

[2] Keith, S.Z. "Self-Assessment Materials for Use in Portfolios," *PRIMUS*, 6 (2), 1996, pp. 178–192.

[3] Keith, S.Z. "Explorative Writing and Learning Mathematics," *Mathematics Magazine*, 81 (9), 1988, pp. 714–719.

True or False? Explain!

Janet Heine Barnett
University of Southern Colorado

One way to find out what students understand is to ask them true/false questions, but have them justify their answers. These justifications bring out confusions about the concepts, but are also the beginning, for calculus students, of writing mathematical proofs.

Background and Purpose

My initial experimentation with modified true/false questions was prompted by frustration with my students' performance in the first-year calculus courses which I was responsible for teaching as a graduate assistant at the University of Colorado, Boulder. Although my students' ability to perform the various algorithms of the course was largely acceptable, the proofs which they were producing (especially on tests) left much to be desired. As a replacement for these "proofs," I began to use true/false questions addressing course concepts on exams; to avoid mere guessing, I also required students to write a short explanation of their response. These written explanations proved to be extremely valuable as a method of identifying student misconceptions and learning difficulties, thereby allowing me to modify my instruction to address those problems.

After joining the faculty at the University of Southern Colorado, I began to use modified true/false questions both for on-going assessment and on exams in a variety of lower division courses, including calculus, college algebra, liberal arts mathematics courses, and content courses for pre-service elementary teachers. I have also used them as one means to assess conceptual understanding on upper division course exams. A regional state institution with moderately open enrollment, USC limits class size in mathematics to 45 students, and in many classes the size is closer to 30. This allows me to use modified true/false and other writing assignments as often as once a week so that students become familiar with both the

format and my expectations of it. Regular use of these questions provides me with instructional feedback, and serves as a mechanism to encourage student reflection on concepts.

Method

The statements themselves are designed to test students' understanding of specific theoretical points, including definitions, properties, theorems, and relations among concepts. (See below for examples.) By casting these statements in the form of implications, it is possible to draw attention to the logical form of mathematical theorems, such as the distinction between the hypothesis and the conclusion, and the relation between a statement, its contrapositve and its converse. The form of the question is thus conducive to a wide variety of concepts, making it readily adaptable to different topics. The critical step in designing a good question is to identify those aspects of the concept are (i) most important and (ii) most likely to be misunderstood. The purpose should not be to trip students up on subtle technical points, but to call their attention (and reflection) to critical features of the concepts. Some examples:

> Determine if each of the following is true or false, and give a complete written argument for your response. Your explanation should be addressed to a fellow student whom you are trying to convince of your reply. Use diagrams, examples, or counterexamples as appropriate to supplement your written response.

If $6 \mid n$ and $4 \mid n$, then $24 \mid n$.

If a is a real number, then $\sqrt{a^2} = a$.

If f is differentiable at a, then $\lim_{x \to a} f(x)$ exists.

If $f(c)$ is a local maximum of f, then $f'(c)=0$.

If the series $\sum a_n$ converges, then $a_n \to 0$.

A typical true/false writing assignment consists of two - four questions addressing the same basic concept or relation; sample instructions for the students are shown above. The stipulation that students write their explanation to a fellow student is made to encourage students to provide a convincing, logical argument consistent with their current level of formal understanding. (In upper division courses, more formalism and rigor is expected.) The goal is to have students communicate their understanding as precisely and clearly as possible, but to articulate this understanding in their own terms, rather than using formal terms or conventions which they don't understand. When employing writing assignments as a part of the course grade, the role of this audience is explained and emphasized to the students at the beginning of and throughout the semester. Students are allowed three to five days to complete the assignment, and are encouraged to discuss the questions with each other and with me during this time. Students who wish to discuss the assignments with me must show that they've already made a significant attempt, and they find that I am not likely to give them "the answer."

In evaluating individual responses, credit is given primarily for the explanations, rather than the true/false response. In fact, a weak explanation for a correct response may earn less credit than a "good" explanation for a wrong response. Both composition and content are considered, with the emphasis placed on content. Typically, I read each response twice: the first time to make comments and place papers into rough piles based on either a three or five point scale, the second time to verify consistency within the piles and assign the score. When using true/false assignments as part of the course grade, a generic version of the scoring guide is shared with students early in the semester. This generic guide is tailored to specific assignments by setting specific content benchmarks for each score (for two types of such scales, see the article by Emenaker in this volume, p. 116). The scores from these assignments can then be incorporated into the final course grade in a variety of ways depending on overall course assessment structure and objectives. Methods that I have used include computing an overall writing assignment grade, incorporating the writing assignment scores into the homework grade, and incorporating the writing assignment scores into a course portfolio grade.

Findings

Students tend to find these assignments challenging, but useful in building their understanding. Their value in building this understanding has led me to use writing increasingly as an instructional device, especially with difficult concepts. The primary disadvantage of this assignment is the fact that any type of writing requires more time for both the instructor (mostly for grading) and the student (both for the physical act of writing and the intellectual challenge of conceptual questions). I have found that modified true/false questions require less time for me to evaluate than most writing assignments (as little as one hour for three questions in a class of 25). Because these questions focus on a specific aspect of the concept in question, most students are also able to respond fairly briefly to them, so that including them on exams is not unreasonable provided some extra time is allowed.

Use of Findings

The instructional feedback I've obtained from student responses to these questions has led me to modify my teaching of particular concepts in subsequent courses (see [1]). For instance, the responses I initially received to questions addressing the role of derivatives in locating extrema suggested that students were confused, rather than enlightened, by class presentations of "exceptional" cases (e.g., an inflection point that is also an extremum). Accordingly, I now use true/false assignments (e.g., "If $(c, f(c))$ is an inflection point, then $f(c)$ is not a local maximum of f.") as a means to prompt the students to discover such examples themselves, rather than including them in class presentations. This worked well in calculus, where students have enough mathematical maturity for such explorations, but was less successful in college algebra. In response, I began to modify the question format itself in order to force the students to confront conceptual difficulties. For example, rather than ask *whether* $\sqrt{a^2} = a$ for all reals a, I may tell the students this statement is false, and ask them to explain why.

A more global instructional change I made as a result of a true/false assignment occurred when one of the best students in a first semester calculus course repeatedly gave excellent counterexamples in response to false statements, while insisting that the statements themselves were not false since he knew of other examples for which the conditions of the statement did hold. I now regularly explore the distinction between universally and existentially quantified statements in introductory courses, whereas I had previously not considered it an issue.

Success Factors

Early in a semester, one should expect the quality of responses to be rather poor, both in terms of content and composition. Quality improves dramatically after the first few assignments as students become familiar with its requirements. To facilitate this process, it is important to provide each student with specific written comments on their work, and to encourage revisions of unacceptable work. I have also found that sharing samples of high quality student responses helps students learn to distinguish poor, good, and outstanding work. For some students (non-native English speakers, learning disabled, etc.), the writing requirement (and the fact that composition does count) has posed problems not solved by these measures. Since the number of these students is small at USC, I have always been able to arrange an alternative for them, such as an oral interview or a transcriber.

References

[1] Barnett, J. "Assessing Student Understanding Through Writing." *PRIMUS* (6:1), 1996. pp. 77–86.

Define, Compare, Contrast, Explain …

Joann Bossenbroek
Columbus State Community College

These short writing assignments help students clarify concepts, and show the instructor where more work is needed.

Background and Purpose

I teach mostly at the precalculus level at Columbus State Community College and have used the following technique in my Intermediate Algebra, College Algebra, and Precalculus classes. Our students come to us very often with extremely weak mathematical backgrounds, with a history of failure at mathematics, with a significant amount of math anxiety, and frequently after being out of school for several years. I feel that my students need a variety of assessment techniques in order to learn the material and also to communicate to me what they have learned. Often my students have only a surface understanding of a particular concept. The assessment technique described in this paper helps students learn how to use mathematical language precisely. It helps me check whether they have absorbed the distinctions that language is trying to establish and the details of the principal concepts.

Method

One week before each test I hand out ten or twelve questions which require students to explain, describe, compare, contrast, or define certain mathematical concepts or terms. The goal is to have my students discuss these concepts using correct terminology. I expect the students to answer the questions using correct sentence structure, punctuation, grammar, and spelling. This strategy also serves as a review of the material for the students as it covers concepts that are going to be on my traditional test.

Examples:

From an Intermediate Algebra class:

- In your own words *define* "function." Give an example of a relationship that is a function and one that is not and *explain* the difference.
- *Explain,* without using interval notation, what $(-\infty, 7]$ means. What is the smallest number in the interval? The largest?

From a College Algebra class:

- If $x = 2$ is a vertical asymptote of the graph of the function $y = f(x)$, *describe* what happens to the x- and y-coordinates of a point moving along the graph, as x approaches 2.
- *Compare/contrast* the difference between a vertical asymptote and a horizontal asymptote.

From a Precalculus class:

- *Explain,* referring to appropriate geometric transformations, and *illustrate,* using graphs, why $\sin 2x \neq 2 \sin x$.

Emphasizing why $\sin 2x \neq 2 \sin x$ by referring to geometric transformations and graphs helps students when they must learn the double angle formulas. It is so tempting for students to assume $\sin 2x = 2 \sin x$.

In order to receive full credit for this example, students must explain that the function $y = \sin 2x$ has a horizontal compression by a factor of 2 which makes the period of the function π and leaves the range at $[-1, 1]$ whereas the function $y = 2 \sin x$ has a vertical stretch by a factor of 2 which makes the range $[-2, 2]$ but leaves the period at 2π. They must also sketch the graphs of each function. Requiring students to both verbalize and visualize a concept is an excellent way to strengthen understanding.

- *Define* a logarithm. *Explain* how a logarithm relates to an exponential function both algebraically and graphically.

Findings

I have been using this strategy for more than six years. Many students have told me that these assignments help them make connections and gain a deeper understanding of the concepts that are being taught. One student recently told me that I forced him to think; his other mathematics instructors only made him work problems. Students who go on to take higher level mathematics courses frequently come back to thank me for the rigor that this type of strategy demands. They claim that the higher level courses are easier for them because of this rigor. At Columbus State half of the final exam in College Algebra and Precalculus is a department component which all instructors must use. I compared the results on the department component from my students for the last eight quarters with the results from the entire department and found that, on average, my students scored 4% higher than the students from the department as a whole. I believe that this assessment strategy is a contributing factor for this difference. A few examples will illustrate the type of misconceptions that I have found:

From Intermediate Algebra: When I ask the students to explain the meaning of $(-\infty, 7]$, I am reminded that students frequently confuse the use of (and [. I am also reminded how fuzzy students ideas of the meaning of ∞ and $-\infty$ are. One student told me that "the smallest number in the interval was the smallest number that anyone could think of."

From College Algebra: When asked to respond to "If $x = 2$ is a vertical asymptote of the graph of the function $y = f(x)$, *describe* what happens to the x- and y-coordinates of a point moving along the graph, as x approaches 2," one student responded, "as x gets closer, nothing specific happens to the y-value but the graph approaches $-\infty$." This kind of an answer gives me insights that I would never get by asking a traditional test question.

Use of Findings

When evaluating the results, I make extensive comments to my students trying to help clarify the concepts that have confused them and to deepen their understanding. I rarely just mark the result incorrect. These assignments have helped me improve my teaching. They have given me a better understanding of what it is about a specific concept that is confusing. I then carefully tailor my examples in class to address the ambiguities. So for example, when I discuss the vertical asymptotes of the graph of a rational function, I very carefully find the x- and y-values for a number of points on the graph for which the x-value is approaching the discontinuity. I have found certain concepts that I thought were obvious or easy to grasp needed more emphasis such as when to use "(" and when to use "[." I give more examples in class to illustrate the difference between using "(" and "[." When I return the papers, I only discuss the responses with the class if a number of students have made the same error or if a student requests more clarification.

Success Factors

- Students need to be cautioned to take these questions seriously. They are not accustomed to writing in a mathematics class. These questions only require short written statements for the most part, but students sometimes forget to use correct sentence structure, grammar, spelling, and punctuation.

- Students need to be reminded that these questions may not be easy to answer. Students often find it easier to work a problem than to describe a concept. They should be cautioned not to try to finish the assignment in one evening.

- Students need to be encouraged to be very complete and thorough. They have had very little experience in explaining mathematical concepts, and so, at first, they tend to be very superficial with their explanations. I find that students do improve their ability to answer these types of questions as the quarter progresses.

- At first I found these assignments to be quite time consuming to correct, but I have become more efficient. Because I use them frequently, I have had practice making decisions as to how to grade them.

- When I grade these assignments, I do point out grammar and spelling errors, but I do not count off heavily for these mistakes. I am primarily interested in the students' understanding of mathematics.

Student-Created Problems Demonstrate Knowledge and Understanding

Agnes M. Rash
Saint Joseph's University

Having students write problems shows the instructor where the gaps in their understanding are at the same time that it has students review for an upcoming test. Further, it can help students find the relevance of the course to their own interests.

Background and Purpose

Saint Joseph's University is a small, comprehensive institution with majors in the humanities, business, social sciences and physical and life sciences. Every student is required to complete two semesters of mathematics. The courses vary from one division to another.

In any course, we are challenged to construct tests that measure what students have learned. Introspectively, we often acquire a deeper understanding of a concept by teaching it, by explaining it to another, or by trying to construct problems for a test. The inadequacy of in-class tests for assessing the learning and understanding of my students led me to explore new methods of appraising their knowledge and reasoning. I wanted to ascertain student perception of which concepts are important and how each individual thinks about problems. Using student-created problems (and solutions) as an assessment procedure reveals both the thinking processes and interests of the students.

Student-created problems allow the instructor to determine the level of each student's comprehension of topics and her/his ability to apply the concepts to new problems. In the process, students hone communication skills, since problems and solutions are submitted in writing and often explained orally in class as well. With several days to create a problem, speed is not an issue; each student works at her/his own rate. This process fosters a commitment to personal achievement, and encourages the student to become more self-directed. Although some students submit only routine

problems, each knows the grading scale in advance and so shares control in the teaching-learning environment. The difficulty level of the problems is entirely in the hands of the students, and hence they set their own personal goals for achievement.

The method of using student-created problems can be implemented in any course. I have used the technique successfully in freshmen level courses (with a maximum enrollment of 30) and in upper division courses with smaller enrollments. In this article, we demonstrate a use of the technique in a course for social science majors.

Method

In the courses for social science majors, the textbook *For All Practical Purposes* [1] is used. Broadly stated, the concepts include Euler and Hamiltonian circuits, scheduling, linear programming, voting strategies, population growth, bin packing, optimization, and statistics. The topics are current and relevant to today's world. During the course, each student is required to design a word problem and its solution for each of the major concepts. The instructor collects the problems, organizes them, and checks for accuracy. The corrected problems are used as a study guide and review aid for tests and the final exam.

- For each of the five to seven units discussed in a semester, each student constructs a word problem and its solution. Within a unit, students are asked to construct problems for different concepts. For example, in a unit on graph theory, some students construct a problem using Euler circuits; others prepare a problem on Hamiltonian circuits or minimum-cost spanning trees, etc.

- Students are divided into small groups of four to six members. All members of a group work on the same topic but each student designs her/his own problem on the concept. After completing a topic, students are given 15 minutes at the end of a period for discussion of ideas. Questions can be asked of the instructor at this time. Students are given several days (for example Monday through Friday, or a long weekend) to construct and submit a correct word problem and its solution. (Working individually may be as effective in some classes.)

- Problems and solutions are collected, read, graded and marked with comments by the instructor. Grammatical as well as mathematical errors are indicated when the problems are returned to the students.

- When a problem needs correction or additional work, the student is given the opportunity to correct and resubmit a revised version, usually within three days. (Problems not relevant to the current topic are returned also. A suitable problem may be resubmitted within the time limit.)

As one would expect, the student-created problems run the gamut from mundane to exciting. Interesting problems are presented in class. A set of correct problems (without solutions) is given to the class as a review. Questions can be asked in a later class. Each student then becomes an authority on her/his problem and its solution, and can be called upon to explain or solve the problem in class. In practice, members of the class actively engage in questioning each other about the solution to a particular problem. Lively discussions take place. The grade on a problem becomes one part of the evaluation of a student's knowledge in the course. The instructor acts as an advisor in the development of problems and solutions of high quality.

In my classes, the collection of problems is given the weight of one test, which is 25% of the final grade. (There are two in-class tests, a collection of problems, and a final examination. I include student-constructed problems on the tests.)   Problems are graded on the basis of correctness, relevance to the topic, and originality. If a student elects to correct a problem, the numerical value it receives is the same as if it had been submitted correctly initially.

Findings

Analysis of a problem allows me to determine if a student understands a concept. Possible errors include misusing the definition, confusion among the concepts, applying a concept to a problem for which the method is inappropriate, or in extreme cases, inability to apply the concept to a new situation. The interests of the students appear in the problems constructed and I use this information in designing examples for other topics in the course. Since my tests consist of routine exercises and problems drawn from the students' work, students' complaints about the test questions, when the test is returned, are virtually eliminated.

Since resubmission is encouraged for poorly posed problems or incorrect solutions, students tend to be more careful in designing the initial problem. Complimenting students on innovative problems, even if they are not completely accurate on the initial submission, improves students' self-esteem and self-confidence. They take more responsibility for the learning process, and develop pride in their work. Because students are actively engaged in the creative process, listening, writing, seeing and thinking are all involved. Furthermore, students learn to read the textbook carefully and to reread examples in order to develop problems.

Students are actively engaged in a cooperative activity and view me as a resource in developing problems rather than as the controller of their academic destiny. Each person has an opportunity to show the depth of her/his understanding without a time constraint. Since students may discuss ideas with each other and with the instructor before submitting a problem, more student-faculty contact, greater cooperation among students, and proactive learning result. As students internalize a concept, they can apply the concept to situations of personal interest. The instructor shows respect for the diverse interests, talents, and modes of learning among the students.

The outcomes of this process are well worth the effort. It is gratifying to observe the students as they find creative ways to use the concepts. There is a great variety in the problems, some of which are quite ingenious and truly enjoyable to read. An unanticipated benefit of the technique is that the instructor obtains a broader range of applications to discuss in the class. Below are a few examples of students' work.

A woman is campaigning for mayor. She needs to go door to door to gather votes. If she parks her car at 61st Street and Lancaster Avenue, design a path for her to use that minimizes deadheading. (Included with the problem is a map of a 3 block by 4 block section of the city.)

A chocolate chip cookie company ships cookies to a convenience store. The bags of cookies come in four different sizes: 1/2 lb., 1 lb., 2 lb., and 4 lb. The company packs these bags into cartons. Each carton holds at most ten pounds. The quantities of cookies that must be shipped are: ten 1/2-lb. bags, fifteen 1-lb. bags, ten 2-lb. bags, and two 4-lb. bags. Using the best-fit decreasing heuristic, place the 37 bags into as few cartons as possible.

In making a decision about which college to attend, the following tasks must be completed: (a) make a list of schools to consider (2 days), (b) send for information and applications (14 days), (c) research the schools (60 days), (d) apply to schools (30 days), (e) visit the schools (5 days), (f) receive acceptances/rejections (90 days), (g) make final decision about the school to attend (30 days). Construct an order requirement digraph and find the critical path.

Course evaluation forms from students indicate that the assignments are worthwhile, increase understanding, and (sometimes) are fun to do. A student's time commitment in designing and solving a problem exceeds that which a student usually spends solving exercises in the textbook. Course evaluations indicate that students spend more time and effort in this course than they usually devote to studying mathematics taught in a traditional style.

Use of Findings

This alternative to in-class testing is enlightening to me as well as beneficial to the student. I am more open to student ideas, and use a formative approach to the development of concepts as a result of this practice. Sharing control in the teaching-learning environment means that the topics discussed can be modified to reflect the interests of the students. When submitted problems have common errors or exhibit similar misconceptions about the application of a topic, I know I have not made the topic clear and delineated its parameters adequately. I need to find another approach or to give more complete examples. Usually, I respond by more carefully explaining a new example and examine how this problem fits the assumptions of the technique. In the following year, I know to be more thorough when explaining this topic. I also can tailor my examples to the interests of the students.

Success Factors

Using this strategy to determine a student's knowledge involves a considerable investment of time and effort on the part of the instructor, particularly in reading and commenting on the work submitted by the students. I estimate that it probably takes three times as long to read and return student-created problems as test problems. Grading the problems and checking the accuracy of the solutions can be a slow process. Solving each problem and checking for accuracy takes longer than checking routine problems. On an ordinary test, all students are solving the same problems. While grading a test, one becomes familiar with the common errors and easily sees them. With student-created problems, each one is different, so an instructor solves 20–30 problems rather than 8–12 problems. On the other hand, time is saved by not having to construct good test questions. Furthermore, class time is saved by eliminating at least one test, and perhaps another class used to return and review the test.

Depending on the availability of readers and/or typists, you may be performing the task of organizing and collating the problems alone. Compiling and editing can be completed more simply when the problems are composed on a computer and submitted electronically than when they are submitted typed on paper.

Some helpful hints:

- Adopt a policy that no problems are accepted late, lest you find yourself cornered at the end of the semester with more evaluation to do than you can handle.

- Prepare in advance the topics you want the students to explore. Determine the number of problems you will collect. Explain to students the purpose of the problems and how the questions will be used.

- Demonstrate what you want students to do. Give examples of problems in each of the grading categories for the students to use as guidelines. Involve the class in a discussion of the grade each example should receive.

- Give students ample time to create a problem.

- Provide feedback in a timely fashion. Grade and return the problems so that they can be corrected and used. Even if you have insufficient time to organize the problems completely, return the papers with comments.

- Make use of the problems, as a study guide for an upcoming test, as test questions, as examples in class, or as problems on the final examination.

- Compliment the students on their work.

- If possible, have students submit problems electronically, on disk or by e-mail. (This may not be possible with all problems.) Those not in this form can be photocopied for distribution.

For other assistance in getting started and further references see Rash [2].

References

[1] COMAP. *For All Practical Purposes*, W. H. Freeman and Company, New York, 1991.

[2] Rash, A.M. "An Alternate Method of Assessment, Using Student-Created Problems." *Primus*, March, 1997, pp. 89–96.

In-Depth Interviews to Understand Student Understanding

M. Kathleen Heid
The Pennsylvania State University

Students' answers on tests don't always show their true level of understanding. Sometimes they understand more than their answers indicate, and sometimes, despite their regurgitating the correct words, they don't understand what they write. This article discusses a method to probe what they actually understand.

Background and Purpose

Drawing on and expanding techniques developed by Jean Piaget in his studies of the intellectual development of children, researchers in mathematics education now commonly use content-based interviews to understand students' mathematical understandings. What started as techniques for researchers to assess student understanding has more recently been used by mathematics instructors who are interested in understanding more deeply the nature of their students' mathematical understandings. Although it is possible to use interviews for the purpose of grading, that is not the intention of this article. If the instructor approaches the interview by informing the student that the goal is to understand how students think instead of to assign a grade, students are generally quite ready to share their thinking. Many seem fairly pleased that the instructor is taking this kind of interest in them. The primary purpose of the interviews described in this article is the enhancement of the mathematics instructor's understanding of students' understandings with its subsequent improvement of mathematics teaching.

My first experiences with using interviews to understand student understanding came when I interviewed high school students to learn about what kinds of activities made them think deeply in their high school classes. A few years later, I wanted to explore differences between good and poor students' understandings of calculus, so I began by interviewing an "A" student and a "D" student (as reported by their instructor). One of the questions I asked each of them was "What is a derivative?" The "A" student answered in the following way: "A derivative? I'm not sure I can tell you what one is, but give me one and I'll do it. Like, for x^2, it's $2x$." Further probing revealed no additional insight on the part of the student. The "D" student gave an explanation of the derivative as the limit of the slope of a sequence of tangent lines, complete with illustrative drawing. This experience first suggested to me the potential for interviews as vehicles to reveal students' mathematical understandings in a way that grades do not.

Since my first experiences with interviews, I have used interviews in a variety of circumstances. The goals of interviews like the ones I have conducted are to find out more about each student's personal understanding of selected mathematical concepts and ideas and to assess the depth and breadth of students' mathematical understandings. Many assessment techniques are designed to inventory students' abilities to execute routine skills or their abilities to demonstrate a pre-identified set of understandings. This type of interview, however, is designed to uncover a network of understandings that may or may not be part of a pre-identified list. The interview is not just an oral quiz or test but rather a way to dig more deeply into the complexities of students' mathematical understandings.

Method

Using interviews to examine student understanding usually involves several steps.

Identifying the goal(s) of the interview. Before beginning to construct an interview schedule, the interviewer needs to clarify his or her goals for the interview. Interviews may be directed toward goals like describing the interviewee's concept image of derivative, determining the extent to which the student sees connections among the concepts of limit, derivative and integral, or determining the depth of a student's knowledge of a particular set of concepts.

Designing an interview schedule of questions to be asked. An interview schedule is a set of directions for the interview, including questions that the interviewer plans to ask, directions for how to follow-up, and tasks to be posed during the course of an interview. The schedule should include a core set of questions or tasks that will be posed to every interviewee and a set of potential follow-up questions or tasks — items whose use would depend on the interviewee's initial set of responses. The schedule should also include a plan for what the interviewer will do under different circumstances. For example, the interviewer will want to plan ahead of time what to do in case the interviewee gets stuck on particular facets of the task, or in case the interviewee asks for assistance, or in case the interviewee brings up an interesting related point. Finally, attached to each copy of the interview schedule should be copies of the core set of questions or problem tasks in large enough type to be visible by interviewer, interviewee, and the video-camera.

Piloting and revising the interview schedule. Just as for any assessment instrument, interview schedules need to be piloted and revised based on the success of those pilots in achieving the goals of the interview. I recommend taping a pilot interview or two to enable more accurate analysis. After conducting a pilot interview, ask yourself questions such as:

- Were the questions understood as intended?
- Were the questions adequate catalysts for finding out about the student's mathematical understandings?
- Were the planned follow-up questions useful?
- Are there additional follow-up questions that should be included?
- Was the sequence of questions appropriate for the purpose of the interview?
- For this interview, was it more appropriate to focus the camera on students' work or on the computer or calculator images (which can be projected behind the students) or on the students?

Preparing for and conducting the interviews. I have found it useful to prepare, before each round of interviews, an "interview box" which contains a copy of the interview schedule, tapes, batteries, an accordion pocket folder for each interviewee, a pen, blank paper, graph paper, a straightedge, and an appropriate calculator. Knowing that needed supplies are ready allows me to focus my attention on the interview.

Conducting the interviews. I will illustrate my method of conducting an interview by describing a particular set of interviews. (See sample interview schedule at the end of the article.) In a yearlong curriculum development, teaching, and research project, I examined the effects of focusing an applied calculus course on concepts and applications and using a computer algebra system for routine skills. Central to my understanding of the effects of this curricular approach was a series of content-based interviews I conducted both with students from the experimental classes and with students from traditionally taught sections. These interviews were loosely structured around a common set of questions. As catalysts with which to start students talking about their understanding of calculus concepts, these interviews used questions like: "What is a derivative?", "Why is it true that 'If $f(x) = x^2$, then $f'(x) = 2x$.'?", "How could you find the slope of a tangent line like the one shown in the sketch?", "Could you explain in your own words what a second derivative is?", and "How can you estimate the area under a given curve?" In these and other interviews I have conducted, I continued to probe as long as the probing seemed to produce additional information about the interviewee's mathematical understandings. I asked the interviewees to share their rationales for each answer, regardless of the "correctness" of the response. At every perceived opportunity as interviewees responded to these initial questions, I encouraged them to talk openly and freely about their understandings of course concepts. When their answers contained phrasing that appeared to be uniquely personal, I probed so that I could understand its meaning. When they made statements that resembled the language of their textbook or of my classroom explanations, I asked them to explain their ideas in another way. When they made generalizations, I asked them to give instances and to explain them. Perhaps my most useful guideline is when I think I understand what the student is saying, I probe further. Many times in these cases, the student's response turns my initial interpretation on end. Interviews like these were designed to allow each student's personal understandings to emerge.

Analyzing the results of the interviews. When possible and reasonable, I have found it useful to watch interview tapes with a colleague who is willing and able to engage in an in-depth discussion of what the tapes seem to indicate about the interviewee's mathematical understandings. Because we have a videotape, we can see what the student is writing and entering into the calculator or computer while making a particular statement. The question we continually ask ourselves is "How does this student make sense of the mathematics he or she is using?" rather than "Is the student's answer right or wrong?"

Findings

Content-based interviews can reveal a variety of aspects of students' mathematical understandings that are not visible through many other methods of assessment. The interrelatedness of one student's understandings of various mathematical topics was revealed in an interview I designed to tap understanding of limit, derivative, and integral. Early in the interview, I asked the interviewee to use the word limit to describe a few graphical representations of functions which I had provided. In each case, the interviewee's response suggested a view of the limit as the process of approaching and never reaching. When asked to explain what was meant by a derivative, the interviewee said that the derivative approached, but never reached, the slope of a tangent line. And finally, the interviewee spoke of the definite integral as approaching the area under a curve, never exactly equal to that area. Further probing in each of these cases uncovered a strongly held and consistent belief based on the notion of limit as the process of approaching rather than as the number that is being approached. The interview allowed me to discover this aberrant belief and to test its strength and consistency in a way that other assessment techniques could not.

Use of Findings

As mathematics instructors learn more about their students' mathematical understandings through interviews, several benefits arise. First, the experience of designing and conducting content-based interviews can help instructors to listen with a new attention and ability to focus on the student's personal interpretations and ways of thinking. Second, an instructor who is aware of possible pitfalls in reasoning can construct examples that are likely to pose cognitive conflicts for students as they struggle with refining the ways they are thinking about particular aspects of mathematics. These cognitive conflicts are helpful in inducing a more useful and robust way to think about the concept in question. For example, after recognizing that students were viewing the limit as a process of approaching rather than as the number being approached, I have centered class discussion on the difference between thinking of limit as a process and thinking of limit as a number. I have been more attentive to the language I use in describing functions, and I have identified examples that would engage students in thinking about limit as a process and as a number. For example, a discussion of whether $.\overline{9} = 1$ is likely to bring this issue to the fore.

However, even the best discussion is not enough to move a student who is not ready beyond a procedural understanding of limit.

Success factors

Key to success in using this method is a deep-seated belief on the part of the interviewer that each student's understanding is unique and that this understanding is best revealed through open-ended questions and related probes. Also key to the success of interviewing as a way to assess student understanding is the belief that each student constructs his or her own mathematical knowledge and that this knowledge cannot be delivered intact to students no matter how well the concepts under consideration are explained.

Perhaps just as important as understanding what interviews are is understanding what they are not. Interviews are *not tutoring sessions* because the goal is not for the interviewer to teach but to understand what the interviewee is thinking and understanding. A successful interview may reveal not only what the interviewee is thinking about a piece of mathematics but also why what he or she is thinking is reasonable to him or her. Interviews are *not oral examinations* because the goal is not to see how well the interviewee can perform on some fixed pre-identified set of tasks but rather to be able to characterize the interviewee's thinking. And this sort of interview is *not usually a teaching experiment* because its goal is not to observe students learning but to access what they think in the absence of additional purposeful instruction.

For further examples of what can be learned about student understanding from content-based interviews, the reader can consult recent issues of the journals, *Educational Studies in Mathematics* and the *Journal of Mathematical Behavior*, and David Tall's edited volume on *Advanced Mathematical Thinking* [1].

The interview schedule on the next page was developed by Pete Johnson (Colby-Sawyer College) and Jean Werner (Mansfield University), under the direction of the author.

Reference

[1] Tall, D. (Ed.). *Advanced Mathematical Thinking*. Kluwer Academic Publishers, Dordrecht, 1991.

INTERVIEW ON REPRESENTATIONAL UNDERSTANDING OF SLOPES, TANGENTS, AND ASYMPTOTES[*]

Goal for this interview: To assess interviewee's numerical, graphical, and symbolic understanding of slopes, tangents, and asymptotes and the connections they see between them.

The interview schedule consisted of four parts: one on slope, one on tangent, one on asymptote, and one on the connections among them. Below are sample questions used in the "tangent" and "connections" parts.

Part II TANGENT

1. DESCRIBE OR DEFINE

In your own words, could you tell me what the word "tangent" means?

[If applicable, ask] You said earlier that the slope of the function is the slope of the tangent to the function. Now I want you to explain to me what you mean by "tangent." [If they say something like slope of a function at a point, ask them to sketch a picture.]

I have a graph here [Show figure.], with several lines which seem to touch the curve here at just one point. Which of these are tangents?

What does the tangent tell you about a function?

2. ROLE OF TANGENT IN CALCULUS

Suppose it was decided to drop the concept of tangent from Calculus. How would the study of Calculus be affected?

3. SYMBOLIC REPRESENTATION

Suppose I gave you a function rule. How would you find the tangent related to that function?

Can you find the tangent in another way?

Discuss the tangent of the function $f(x) = \sqrt{x}$. [Then ask,] What is the tangent at the point $x = 0$? [Follow up with questions about the limit to the right and left of 0.]

4. GRAPHICAL REPRESENTATION

Can you explain the word "tangent" to me by using a graph? Explain.[Show a sketch of $f(x) = \dfrac{1}{x}$.] Discuss the tangents of this function.

Part IV - CONNECTIONS

1. Some people say that the ideas of asymptote and slope are related. Do you agree or disagree? Explain.

2. Some people say that the ideas of asymptote and tangent are related. Do you agree or disagree? Explain.

---

[*] This interview schedule was developed by Pete Johnson (Colby-Sawyer College) and Jean Werner (Mansfield University), under the direction of the author.

Assessing Expository Mathematics: Grading Journals, Essays, and Other Vagaries

Annalisa Crannell
Franklin & Marshall College

This article gives many helpful hints to the instructor who wants to assign writing projects, both on what to think about when making the assignment, and what to do with the projects once they're turned in.

Background and Purpose

This article will attempt to illuminate the issues of how, when and why to assign writing in mathematics classes, but it will focus primarily on the first of these three questions: how to make and grade an expository assignment.

I have been writing and using writing projects for just over four years at Franklin & Marshall College, a highly selective liberal arts college in bucolic Lancaster, Pennsylvania. The majority of our students hail from surrounding counties in Pennsylvania, Delaware, and New Jersey, but we also draw some 15% of our students from overseas. We have no college mathematics requirement, but there is a writing requirement. Those of our students who choose to take mathematics usually begin with differential or integral calculus; class size is generally limited to 25 students. Our students tend to be more career-oriented than is usual at liberal arts colleges (an implication of which is that post-class wrangling about the distinction between a B and B+ is common).

I have worked with instructors who come from a variety of geographic regions and from a variety of types of institutions: liberal arts colleges, community colleges, branch campuses of state universities. I have found that their experiences with having students write mathematical essays are remarkably similar to my own. Therefore, although I will use my own projects as specific examples, what I would like to discuss in this paper are general guidelines that have helped me and my colleagues. I will present some of the key ingredients to first assigning and then grading a writing experience which are both beneficial to the students and non-overwhelming to the instructor.

Method

The first question to address when creating an assignment is: why assign writing in the first place? What is the goal of the assignment, and how does it fit into the larger course? Other articles in this volume will address this question in greater detail, so I will risk oversimplifying the treatment of this question here. Let me then posit several possible reasons for assigning writing: (1) to improve students' mathematical exposition; (2) to introduce new mathematics; (3) to strengthen understanding of previously encountered mathematics; or (4) to provide feedback from the student to the instructor. Closely related to this question of "why do it?" is the question of audience: to whom is the student writing? Students are most used to writing to an omniscient being: the instructor. Other possibilities include a potential boss, the public, a fellow student, or the student him/herself.

Deciding for yourself the answers to these two questions, and then *explaining these to the students*, is a very important part of creating an assignment. It focuses the student on your intentions, it reduces the student's level of confusion and angst, and it results in better papers.

Knowing the answers to the above questions, you can design your assignments to match your goals. For example,

I want my Calculus students to explain mathematics to those who know less mathematics than they do, and I want their writing experience to reinforce concepts being taught in the course. The projects I assign take the form of three letters from fictional characters who know little mathematics but who need mathematical help. My students are expected to translate the problems from English prose into mathematics, to solve the problems, and to explain the solutions in prose which the fictional characters could understand. I require that these solutions be written up on a word processor. (An example of one such problem can be found in [2]).

Once you have decided the over-arching purpose behind your assignment, you should try to pin down the specifics. The "details" that matter little to you now matter a great deal to the students—and will haunt you later. Do you want the papers to be word processed or handwritten? What is a reasonable length for the final product (one paragraph/one page/several pages)? What is the quality of writing that you expect: how much do the mechanics of English writing matter? How much mathematical detail do you expect: should the students show every step of their work, or instead provide an overview of the process they followed? Will students be allowed or required to work in groups? How much time will they be given to complete the paper? What will the policy on late assignments be? What percentage of class grade will this assignment be?

Especially for shorter papers, students find it very helpful if you can provide an example of the type of writing you expect. I have made it a habit to ask permission to copy one or two students' papers each semester for this purpose, and I keep my collection of "Excellent Student Papers" on reserve in the library for the use of future students.

Findings

After you assign and collect the papers, you face the issue of providing feedback and grades. There is a considerable literature which documents that instructors provide poor feedback to our students on writing. What students see when they get their papers back is a plethora of very specific comments about spelling and grammar, and only a few (confusing) comments on organization and structure. As a result, students do *not* in general rewrite papers by thinking about the process of writing. Instead, when rewriting is permitted they go from comment to comment on their draft, and change those things (and only those things) that their instructor pointed out. For this reason, I avoid assignments that require rewriting whenever I can: the instructor does a lot of work as an editor, and the student learns a minimal amount from the process.

One of my favorite assessment tools is the *rubric*, "a direction or rule of conduct." My own rubrics take the form of a checklist with specific yes or no questions, which I list here:

Does this paper:

1. clearly (re)state the problem to be solved?
2. state the answer in a complete sentence which stands on its own?
3. clearly state the assumptions which underlie the formulas?
4. provide a paragraph which explains how the problem will be approached?
5. clearly label diagrams, tables, graphs, or other visual representations of the math (if these are indeed used)?
6. define all variables used?
7. explain how each formula is derived, or where it can be found?
8. give acknowledgment where it is due?

In this paper,

9. are the spelling, grammar, and punctuation correct?
10. is the mathematics correct?
11. did the writer solve the question that was originally asked?

Certainly I am not advocating that this is the only form a cover sheet could take. Other rubrics might be more general in nature: for example, having one section for comments on mathematical accuracy and another for comments on exposition, and then a grade for each section. At the other extreme, there are rubrics which go into greater detail about what each subdivision of the rubric entails, with a sliding scale of grades on each. There is a growing literature on the use of rubrics—see Houston, *et al* [3], or Emenaker's paper in this volume, p. 116.

Use of Findings

Students get a copy of my checklist even before they get their first assignment; they are required to attach this sheet to the front page of their completed papers. I make almost all of my comments directly on the checklist, with a "yes" or "no" next to each question. The total number of "yes's" is their grade on the paper.

The main reason I like the rubric as much as I do is that it serves as both a teaching and an assessment tool. Students *really* appreciate having the guidance that a checklist provides, especially on their first paper. Moreover, the fact that the checklist remains consistent from one paper to the next (I assign three) helps the student to focus on the overall process of writing, rather than on each specific mistake the instructor circled on the last paper.

I said that my foremost reason for using the checklists is pedagogical idealism, but running a close second is practicality. Using this checklist has saved me a considerable amount of time grading. For one thing, it makes providing feedback easier: I can write "X?" next to question (6) above instead of the longer "What does X stand for?" Moreover, the checklist makes it less likely that the student will neglect to describe "X." Using the checklist gives the appearance of objectivity in grading; I have had very little quibbling over scores that I give—this is unusual at my college. I describe this checklist in greater detail in [2]; and I refer the interested reader there.

Success Factors

If an instructor does decide to have students rewrite their work, one way to provide valuable commentary but still leave responsibility for revision in the students' hands is to ask students to turn in a cassette tape along with their draft. A simple tape player with a microphone allows for much more detailed comments of a significant kind: "In this paragraph, I don't understand whether you're saying this or that ... you jumped between three topics here without any transitions ... oh, and by the way, you *really* need to spell-check your work before you turn it in again." I've found that students much prefer this kind of feedback. It means less writing for me; and it results in significantly changed (and better) final drafts.

If you assign a long-term project, it is *extremely* helpful to mandate mini-deadlines which correspond to relevant subtasks. These help to reduce student anxiety and to avoid procrastination. For example, when I assign a semester-long research paper, I have a deadline every week or two, in which students are expected to turn in:

- a thesis topic;
- a partial bibliography;
- a list of mathematicians with the same last names as they have (this teaches them to use *Science Citation Index* or *Math Reviews*, from which they can expand their bibliography);
- a vocabulary list relevant to their paper, including words that they don't understand;
- a rough draft of an introductory paragraph;
- an outline with a revised bibliography;
- a rough draft of the paper; and finally
- the final paper.

One last piece of advice: As you prepare your assignments, make sure that you have resources available on campus to help the anxious or weak student. One of these resources will necessarily be you, but others might include math tutors, a writing center, and books or articles on writing mathematics (such as [1] or [4]).

References

[1] Crannell, A., *A Guide to Writing in Mathematics Classes,* 1993. Available upon request from the author, or from http://www.fandm.edu/Departments/Mathematics/Writing.html

[2] Crannell, A., "How to grade 300 math essays and survive to tell the tale," *PRIMUS* 4 (3), 1994.

[3] Houston, S.K., Haines, C. R., Kitchen, A. , *et. al.*, *Developing Rating Scales for Undergraduate Mathematics Projects*, University of Ulster, 1994.

[4] Maurer, S., "Advice for Undergraduates on Special Aspects of Writing Mathematics," *PRIMUS*, 1 (1), 1990.

Assessing Modeling Projects In Calculus and Precalculus: Two Approaches

Charles E. Emenaker

University of Cincinnati
Raymond Walters College

Once you've assigned a writing project and collected the papers, how are you going to grade it without spending the rest of the semester on that one set? This article outlines two grading scales which can make this more efficient.

Background and Purpose

Projects that require students to model a real-life situation, solve the resulting mathematical problem and interpret the results, are important both in enabling students to use the mathematics they learn and in making the mathematics relevant to their lives. However, assessing modeling projects can be challenging and time-consuming. The holistic and analytic scales discussed here make the task easier. They provide the students with a set of guidelines to use when writing the final reports. They also provide me with a structured, consistent way to assess student reports in a relatively short period of time.

Currently I use these scales to assess projects given in precalculus and calculus classes taught at a two-year branch of a university. I have used these scales over the last seven years to assess projects given in undergraduate as well as graduate-level mathematics courses. The scales have worked well at both levels.

Method

I provide each student with a copy of the assessment scales the first class meeting. The scales are briefly discussed with the students and care is taken to point out that a numeric solution without the other components will result in a failing grade. I review the scale when the first group project is assigned. This helps the students to understand what is expected of them.

The analytic scale (p. 117, adapted from [1]) lends itself well to situations where a somewhat detailed assessment of student solutions is desired. When the analytic scale is used, a separate score is recorded for each section: understanding, plan, solution, and presentation. This allows the students to see the specific strengths and weaknesses of the final report and provides guidance for improvement on future reports.

The holistic scale (p. 119, adapted from [2]) seems better suited to situations where a less detailed assessment is required. It often requires less time to apply to each student report and can be used as is or adjusted to meet your individual needs and preferences. For some problems it might be useful to develop a scale that has a total of five or six points possible. This will require rewriting the criteria for each level of score.

Findings

One of the easiest ways to demonstrate how the scales work is to actually apply them. Figure 1 contains a proposed solution to the problem given below. The solution has been evaluated using the analytic scale.

The Problem: You want to surprise your little brother with a water balloon when he comes home, but want to make it look "accidental." To make it look accidental you will call, "Watch out below!" as you release the balloon, but you really don't want him to have time to move. You time the warning call and find that it takes 1 second. You have noticed that it takes your brother about 0.75 seconds to respond to warning calls and you know sound travels about 1100 feet per second. If your brother is 5 feet tall, what is the greatest height you could drop the balloon from and still be certain of dousing him?

Using the holistic scale on p. 119 to assess this proposed solution results in a score of "3a." The "a" is included to provide the students with guidance regarding the shortcoming(s) of the report.

Use of these scales, especially the analytic scale, has led to reports that are much more consistent in organization and usually of a higher quality. Students have repeatedly made two comments regarding the use of the scales. First they appreciate the structure. They know exactly what is expected of them. Second they are much more comfortable with the grade they receive. On more than one occasion students have expressed concern regarding the consistency of assessment in another class(es) and confidence in the assessment based on these scales.

Use of Findings

One issue I focus on is how well the students are interpreting what is being asked of them. After reading many reports where the students are first asked to restate the problem, it has become clear that I need to spend time in class teaching the students how to read and interpret problems. They may be literate, but they often lack the necessary experience to understand what is being described physically.

I also focus on the students' mathematical reasoning. In the problem presented above, a number of students run into difficulty developing an equation for the brother's reaction time. They will add the times for the warning call and the reaction while completely ignoring the time for sound to travel. Although it does not significantly alter the solution in this problem, it is important for the students to realize that sound requires time to travel. Prior to giving this assignment, we now spend time in class looking at situations where the time needed for sound to travel affects the solution.

Success Factors

When using the holistic scale, I include a letter (the "a" of "3a" above) to indicate which criterion was lacking. Students appreciate this information.

When new scales are developed it is well worth remembering that as the number of points increases so does the difficulty in developing and applying the scale. In the case of the holistic scale, it is often best to start with four points. As you gain experience, you may decide to develop more detailed scales. I have always found five or at most six points sufficient for holistic scales.

To avoid problems with one or two students in a group doing all of the work, I have them log the hours spent on the project. I warn the class that if a student in a group spends 20 minutes on the project while each of the other members averages four hours, that person will receive one-twelfth of the total grade. There has only been one case in seven years where this type of action was necessary.

A final note about using these scales. It often seems a huge task to evaluate the reports, but they go rather quickly once started. First, there is only one report per group so the number of papers is reduced to one-third of the normal load. Second, having clearly described qualities for each score reduces the time spent determining the score for most situations.

An analytic scale for assessing project reports

(Portions of this are adapted from [1]. See Figure 1 for a sample.)

Understanding

3 Pts The student(s) demonstrates a complete understanding of the problem in the problem statement section as well as in the development of the plan and interpretation of the solution.

2 Pts The student(s) demonstrates a good understanding of the problem in the problem statement section. Some minor point(s) of the problem may be overlooked in the problem statement, the development of the plan, or the interpretation of the solution.

1 Pt The student(s) demonstrates minimal understanding of the problem. The problem statement may be unclear to the reader. The plan and/or interpretation of the solution overlooks significant parts of the problem.

0 Pt The student(s) demonstrates no understanding of the problem. The problem statement section does not address the problem or may even be missing. The plan and discussion of the solution have nothing to do with the problem.

Plan

3 Pts The plan is clearly articulated AND will lead to a correct solution.

2 Pts The plan is articulated reasonably well and correct OR may contain a minor flaw based on a correct interpretation of the problem.

1 Pt The plan is not clearly presented OR only partially correct based on a correct/partially correct understanding of the problem.

0 Pt There is no plan OR the plan is completely incorrect.

Solution

3 Pts The solution is correct AND clearly labeled OR though the solution is incorrect it is the expected outcome of a slightly flawed plan that is correctly implemented.

The Water Balloon Problem
Project 2

Feb 25, 1997

The Problem: I need to find how high I can climb a tree and still drop a water balloon on my brother without him having time to move. The warning call takes 1 second and my brother takes .75 seconds to respond to warning calls.

You need to include information about the speed of sound

The balloon will be pulled down by gravity so I want to use the formula

$$s(t) = -16t^2 + vo(t) + so \quad \checkmark$$

My brother is 5 feet tall so $s(t) = 5$ and $vo = 0$ since the balloon is being dropped and not thrown. The calculations for the balloon to drop look like:

$$5 = -16t^2 + so \quad \checkmark \longrightarrow \quad t = \frac{\sqrt{s_0 - 5}}{4}$$

To find the time to react you add the time for the warning call and the time to react, so the time to react is 1.75 seconds. Using $t = 1.75$ seconds you get

$$5 = -16(1.75)^2 + so$$
$$5 + 49 = so$$
$$so = 54 \text{ ft}$$

The maximum height you can drop the balloon from is 54 feet.

U - 2
P - 1
S - 3
P - 1
___
7

This reaction time also involves time for sound to travel:

$$\frac{s_0 - 5}{1100}$$

$$T_{REACT} = 1.75 + \frac{s_0 - 5}{1100}$$

Setting $T_{FALL} = T_{REACT}$

$$\frac{\sqrt{s_0 - 5}}{4} = 1.75 + \frac{s_0 - 5}{1100}$$

Solving for s_0 either algebraically or graphically $\longrightarrow s_0 \approx 56.7$ ft

Figure 1. A proposed solution evaluated using the analytic scale.

2 Pts Solution is incorrect due to a minor error in implementation of either a correct or incorrect plan OR solution is not clearly labeled.

1 Pt Solution is incorrect due to a significant error in implementation of either a correct or incorrect plan.

0 Pt No solution is given.

Presentation

1 Pt Overall appearance of the paper is neat and easy to read. All pertinent information can be readily found.

0 Pt Paper is hard to read OR pertinent information is hard to find.

A Holistic Scoring Scale

(This is an adaptation of a scale from [2].)

4 Points: Exemplary Response

All of the following characteristics must be present.

- The answer is correct.

- The explanation is clear and complete.

- The explanation includes complete implementation of a mathematically correct plan.

3 Points: Good Response

Exactly one of the following characteristics is present.

a The answer is incorrect due to a minor flaw in plan or an algebraic error.

b The explanation lacks clarity.

c The explanation is incomplete.

2 Points: Inadequate Response

Exactly two of the characteristics in the 3-point section are present OR

One or more of the following characteristics are present.

a The answer is incorrect due to a major flaw in the plan.

b The explanation lacks clarity or is incomplete but does indicate some correct and relevant reasoning.

c A plan is partially implemented and no solution is provided.

1 Point: Poor Response

All of the following characteristics must be present.

- The answer is incorrect.

- The explanation, if any, uses irrelevant arguments.

- No plan for solution is attempted beyond just copying data given in the problem statement.

0 Points: No Response

- The student's paper is blank or contains only work that appears to have no relevance to the problem.

References

[1] Charles, R., Lester, F., and O'Daffer, P. *How to Evaluate Progress in Problem Solving*, National Council of Teachers of Mathematics, Reston, Va, 1987.

[2] Lester, F. and Kroll, D. "Evaluation: A New Vision," *Mathematics Teacher* 84, 1991, pp. 276–283.

Formative Assessment During Complex, Problem-Solving, Group Work in Class

Brian J. Winkel
United States Military Academy*

During student efforts to attack and solve complex, technology-based problems there is rich opportunity for assessment. The teacher can assess student initiative, creativity, and discovery; flexibility and tolerance; communication, team, and group self-assessment skills; mathematical knowledge; implementation of established and newly discovered mathematical concepts; and translation from physical descriptions to mathematical models.

Background and Purpose

Complex, technology-based problems worked in groups during class over a period of several days form a suitable environment for assessment of student growth and understanding in mathematics. The teacher can make assessment of student initiative, creativity, and discovery; flexibility and tolerance; communication, team, and group self-assessment skills; mathematical knowledge; implementation of established and newly discovered mathematical concepts; and translation from physical descriptions to mathematical models. Viewing students' progress, digression, discovery, frustration, formulation, learning, and self-assessing is valuable and instant feedback for the teacher.

During student efforts to attack and solve complex, technology-based problems there is rich opportunity for assessment. Active and collaborative learning settings provide the best opportunity for assessment as students interact with fellow students and faculty. The teacher sees what students know and use and the students receive immediate feedback offered by the coaching faculty.

Student initiative is very easy to measure, for the students voice their opinions on how they might proceed. Creativity, discovery, and critiquing group member's offerings come right out at the teacher as students blurt out ideas, build upon good ones, and react to ideas offered by others. Listening to groups process individual contributions will show the teacher how flexible and tolerant members are of new ideas. The write-up of a journal recording process and the final report permits assessment qualities such as follow-through, rigor, and communication.

I illustrate success using this assessment strategy with one problem and direct the reader to other problems. I used the problem described below at Rose-Hulman Institute of Technology several times [1] as part of a team of faculty teaching in an Integrated First-Year Curriculum in Science, Engineering, and Mathematics (IFYCSEM). In IFYCSEM the technical courses (calculus, mechanics, electricity and magnetism, chemistry, statics, computer programming, graphics, and design) are all put together in three 12 credit quarter courses [2]. This particular problem links programming, visualization, and mathematics.

This activity fits into a course in which we try to discover as much as possible and reinforce by use in context ideas covered elsewhere in this course or the students' backgrounds. Further, it gives the teacher opportunities to see the students in action, to "head off at the pass" some bad habits, to bring to the attention of the class good ideas of a given group through impromptu group presentations of their

---

*This work was done at the author's previous institution: Rose-Hulman Institute of Technology, Terre Haute, IN 47803 USA.

ideas, and to have students "doing" mathematics rather than "listening" to professorial exposition.

The activity is graded and given a respectable portion of the course grade (say, something equal to or more than an hourly exam) and sometimes there is a related project quiz given in class after the project has been completed.

In IFYCSEM we used *Mathematica* early in the course as a tool for mathematics and as the introductory programming environment. Thus we were able to build on this expertise in a number of projects and problems throughout the course. I have also used such projects in "stand-alone" calculus courses at West Point and they work just as well.

Method

A suitable class size for "coaching" or "studio modeling" such a problem is about 24 students — eight groups of three. It is possible to visit each group at least once during an hour, to engage the class in feedback on what is going on overall, and to have several groups report in-progress results. I have used this problem with two class periods devoted to it at the start, one class period a week later to revisit and share progress reports, and one more week for groups to meet and work up their results.

One has to keep on the move from group to group and one has to be a fast read into where the students' mindsets are, to try to understand what they are doing, to help them build upon their ideas or to provide an alternative they can put to use, but not to destroy them with a comment like, "Yes, that is all well and good, but have you looked at it this way?" (Translated into adolescent jargon — "You are wrong again, bag your way, do it my way!")

An Example

The problem is to describe what you can see on one mountain while sitting on an adjacent mountain. See Figure 1. In a separate paper [1] I discuss how students, working in groups, attack the problem and the issues surrounding the solution strategies. This problem develops visualization skills, verbalization of concepts, possibly programming, and the mathematical topics of gradients, projection, optimization, integration, and surface area.

The problem stated

For the function

$$f(x,y) = \frac{x^3 - 3x + 4}{x^4 + 5y^4 + 20}$$

suppose your eye is precisely on the surface $z = f(x, y)$ (see Figure 1) at the point $(2.8, 0.5, f(2.8, 0.5))$. You look to the left, i.e., in the direction (roughly) $(-1, 0, 0)$. You see a mountain before you.

(a) Determine the point on the mountain which you can see which is nearest to you.

(b) Describe as best you can the points on the mountain which you can see from the point $(2.8, 0.5, f(2.8, 0.5))$.

(c) Determine the amount of surface area on the mountain which you can see from the point $(2.8, .5, f(2.8, 0.5))$.

I introduce the above problem in a room without computers so the students will visualize without turning to computers to "crunch." I give the students the statement of the problem with the figure (see Figure 1) and say, "Go to it!" Working in groups, they turn to their neighbor and start buzzing. Circulating around the class, I listen in on group discussion.

Findings

Typically students offer conjectures which can be questioned or supported by peers, e.g., the highest point visible on the opposite mountain can be found as the intersection of a vertical plane and the opposite mountain. Students with hiking experience will say you cannot see the top of the mountain over the intervening ridges. A recurring notion is that when placing a "stick" at the point of the viewing eye and letting it fall on the opposite mountain this stick is tangent to the opposite mountain. Other students can build upon this vision, some even inventing the gradient as a vector which will necessarily be perpendicular to the stick at the tangent point on the opposite mountain. Others can use this latter tangency notion to refute the vertical plane notion of the first conjecture.

I also assign process reporting (summative evaluation) at the end of the project in which students describe and reflect upon their process. This is part of their grade. They keep notes on out of class meetings, report the technical advances and wrong alleys, and reflect on the workings of the group. Assigned roles (Convenor, Recorder, and Reactor) are changed at each meeting.

Use of Findings

The formative assessment takes place in the interaction among students and between students and teacher. Basically, the students "expose" their unshaped ideas and strategies, get feedback from classmates on their ideas, hone their articulation, and reject false notions. In so doing they clarify and move to a higher level of development. Observing and interacting with students who are going through this problem-solving process is an excellent way for the teacher to assess what students really understand.

Reading the journal accounts of progress gives the teacher insight into the learning process as well as group dynamics. The former helps in understanding the nature of student

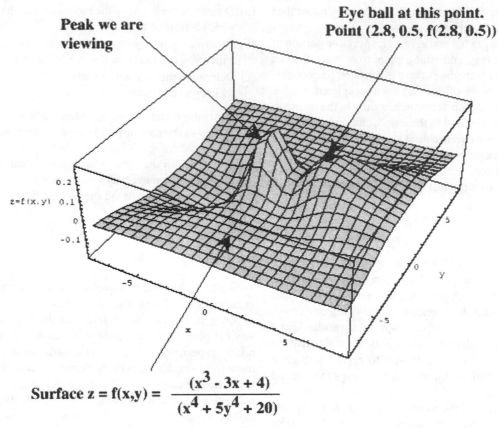

Peak we are viewing

Eye ball at this point.
Point (2.8, 0.5, f(2.8, 0.5))

$$\text{Surface } z = f(x,y) = \frac{(x^3 - 3x + 4)}{(x^4 + 5y^4 + 20)}$$

Figure 1

difficulty. The latter can help the teacher see why some groups are smoother than others and thus be aware of potential problems in the next group activity assigned.

Success Factors

The only real difficulty in this entire approach is how to address the many conjectures and convictions the students bring forth in the discussion. One has to be thinking and assessing all the time in order to give meaningful, constructive feedback. The teacher has to watch that one idea (even a good one) does not dominate from the start and that all ideas are given fair consideration. The entire process never fails if the problem is interesting enough.

A *Mathematica* notebook, ASCII, and HTML version of this problem, with solution and comments is available (under title "OverView") at web site http://www.rose-hulman.edu/Class/CalculusProbs. This is a part of a larger National Science Foundation project effort, "Development Site for Complex, Technology-Based Problems in Calculus," NSF Grant DUE-9352849.

References

[1] Winkel, B.J. "In Plane View: An Exercise in Visualization," *International Journal of Mathematical Education in Science and Technology,* 28(4), 1997, pp. 599–607.

[2] Winkel, B.J. and Rogers, G. "Integrated First-Year Curriculum in Science, Engineering, and Mathematics at Rose-Hulman Institute of Technology: Nature, Evolution, and Evaluation," *Proceedings of the 1993 ASEE Conference, June 1993,* pp. 186–191.

Student Assessment Through Portfolios

Alan P. Knoerr and Michael A. McDonald
Occidental College

Reflective portfolios help students assess their own growth. Project portfolios identify their interests and tackle more ambitious assignments.

Background and Purpose

We discuss our use of portfolio assessment in undergraduate mathematics at Occidental College, a small, residential liberal arts college with strong sciences and a diverse student body. Portfolios are concrete and somewhat personal expressions of growth and development. The following elements, abstracted from artist portfolios, are common to all portfolios and, taken together, distinguish them from other assessment tools:

- work of high quality,
- accumulated over time,
- chosen (in part) by the student.

These defining features also indicate assessment goals for which portfolios are appropriate.

Our use of portfolios is an outgrowth of more than six years of curricular and pedagogical reform in our department. We focus here on two types of portfolios — reflective portfolios and project portfolios — which we use in some second-year and upper-division courses. We also describe how we integrate portfolios with other types of assessment.

Reflective Portfolios. In a reflective portfolio, students choose from a wide range of completed work using carefully specified criteria. They are asked to explicitly consider their progress over the length of the course, so early work which is flawed may nonetheless be included to illustrate how far the student has progressed. A reflective portfolio helps students assess their own growth.

The collection of portfolios can also help a teacher reflect on the strengths and weaknesses of the course. It can point out strong links made by the students, and indicate struggles and successes students had with different topics in the course. It can further open the window to student attitudes and feelings.

Project Portfolios. Project portfolios are one component of a "professional evaluation" model of assessment [1, 2]. The other components are "licensing exams" for basic skills, small but open-ended "exploration" projects, and targeted "reflective writing" assignments. These different components, modeled on activities engaged in by professionals who use and create mathematics, have been chosen to help students develop a more mature approach to their study of mathematics.

Students choose projects from lists provided for each unit of the course. Completed projects are included in their project portfolio. This method of assessment helps students identify their interests, produce work of high quality, and tackle more ambitious assignments which may take several weeks to complete. Collecting work from the entire course encourages students to look beyond the next test or quiz.

Method I. Reflective Portfolios

To increase students' self-awareness of how their understanding has developed requires that information about this understanding be collected by the students as the course progresses. Keeping a journal for the course is a useful supplement to the usual homework, classwork, quizzes, and tests. For example, in a real analysis course, students are asked to make a journal entry two or three times each week as they, among other possibilities: reflect on their readings or problem sets, discuss difficulties or successes in the course, clarify or connect concepts within this course or with those in other courses, or reflect on their feelings toward this course or mathematics in general. The teacher collects and responds to the journals on a regular basis.

Near the end of the semester, students are asked to prepare a reflective portfolio. The assignment says:

> The purpose of this portfolio assignment is to allow you to highlight your own selections of your work and give an analysis of them in your own words. I will focus on two specific things in evaluating your portfolio: (1) an understanding of some key concepts in real analysis and (2) a self-awareness of your journey (where you started from, where you went, and where you are now).
>
> Select three pieces of work from this semester to include in your portfolio. These pieces can include journal writing, homework assignments, tests, class worksheets, class notes, computer experiments, or any other pieces of work you have produced in this class. Your analysis should explain your reasons for picking these pieces. As examples, you might consider a selection which shows the development of your understanding of one key concept, or a selection which shows your growing appreciation of and proficiency with formal proofs, or a selection which shows your connection of two or more key concepts.
>
> The most important part of the portfolio is your reflection on why you chose the pieces you did, how they show your understanding of some key concept(s), and how they show a self-awareness of your journey through this class. You should definitely write more than one paragraph but no more than five pages. The portfolio is 5% of your final grade.

Method II. Project Portfolios

The project portfolio corresponds to papers and other completed projects a professional would include in his or her curriculum vitae. Two sorts of themes are used for projects in a multivariable calculus course with a linear algebra prerequisite. For example, "Vector Spaces of Polynomials" concerns fundamental ideas of the course, while "Optimization in Physics" is a special topic for students with particular interests. All students are required to complete certain projects, while in other cases they choose from several topics. Here is an example of a portfolio project assignment.

Vector Spaces of Polynomials. What properties define a vector space over $\mathbf{R}$? Let P_3 denote the space of polynomials over $\mathbf{R}$ of degree less than or equal to three. Show that P_3 is a vector space over $\mathbf{R}$ with natural definitions for "vector addition" and "scalar multiplication." Find a basis for P_3 and show how to find the coordinate representation of a polynomial in P_3 relative to this basis. Show that "differentiation" is a linear transformation from P_3 to P_2, where P_2 is the space of polynomials over $\mathbf{R}$ of degree less than or equal to 2. Find a matrix representation for this transformation relative to bases of your choice for the input space P_3 and the output space P_2. To check your work, first multiply the coordinate representation of a general polynomial $p(x)$ in P_3 by your matrix representation. Then compare this result with the coordinate representation of its derivative, $p'(x)$. Discuss what you learned from this project. In particular, has this project changed how you think about vector spaces, and if so, in what way?

Project reports must include: a cover page with a title, author's name, and an abstract; clear statements of problems solved, along with their solutions; a discussion of what was learned and its relevance to the course; acknowledgement of any assistance; and a list of references. Reports are usually between five and ten pages long. They are evaluated on both mathematical content and quality of presentation. First and final drafts are used, especially early in the course when students are learning what these projects entail.

For the typical project, students will have one week to produce a first draft and another week to complete the final draft. Students may be working on two different projects in different stages at the same time. Allowing for test breaks and longer projects, a completed course portfolio will comprise seven to ten projects. Students draw on this portfolio as part of a self-assessment exercise at the end of the course.

Findings

The collection of reflective portfolios can serve as feedback of student mathematical understanding and growth through the course. It can sometimes also give us a glimpse of the joy of learning, and thus the joy of teaching. One student wrote:

> Without hesitation I knew the selections I was going to analyze. I think one of the reasons why this worksheet and these two journal entries came to mind so quickly is because they reflect a major revelation I had in the course. Not often in my math classes have I felt so accomplished .... These selections represent something I figured out ON MY OWN! That's why they so prominently came to mind.

The first drafts of projects for the project portfolio are a rich source of information on how students are thinking about important topics being covered in the course. The final drafts reveal more clearly their depth of understanding and degree of mastery of mathematical language. We find that these projects help students achieve a better conceptual understanding of important aspects of the course and become more mature in presenting mathematical arguments. The discussion sections of their reports offer students some chance for reflection. For example, one student wrote the following in his report on vector spaces of polynomials:

> I am fascinated to see how much we can deduce from an abstract space without being able to visualize it. While the concept of a polynomial vector space still amazes me it is very interesting to see how we can make this abstraction very tangible.

Use of Findings

Reflective portfolios document the development of mathematical knowledge and feelings about mathematics for each student in a particular course and class. This information has been used to realign class materials, spending greater time on or developing different materials for particularly difficult concepts. In addition, the evaluation of student journals throughout the semester allows for clarifying concepts which were not clearly understood. As an example, one student discussed an in-class worksheet on uniform continuity of a function $f(x)$ and did not understand the role of x in the definition. Seeing this, class time was set aside to clarify this concept before more formal evaluation.

First drafts of project portfolios may highlight difficulties shared by many students and thus influence teaching while a course is in progress. Reflecting on completed portfolios can also lead to changing how we teach a course in the future. The projects themselves may be improved or replaced by new ones. Special topics which were previously presented to the entire class may be treated in optional projects, leaving more class time for fundamental ideas. These projects can also lead to more radical reorganization of a course. For example, one year a project on Fubini's Theorem revealed that students had trouble understanding elementary regions for iterated integrals. The next year we first developed line integrals, then introduced elementary regions through Green's Theorem before treating iterated integrals. Teaching these topics in this unusual order worked quite well.

Success Factors

The reflective portfolio, journal writing, and the project portfolio are all writing-intensive forms of assessment. Clarity, time and patience are required of both teachers and students in working with assignments like these which are unusual in mathematics courses.

The reflective portfolio is little more than a short paper at the end of the semester. The time is put in during the semester as the teacher reads and responds to the journals. Cycling through all the journals every few weeks is probably the most efficient and least burdensome way of handling them.

Some students enjoy reflective writing while others may feel awkward about it. In the real analysis class, for example, the female students used the journal more consistently and produced deeper reflections than did the males. Teachers must also allow for reflections which may not be consistent with the outcomes they desire.

Several techniques can make correcting first drafts of project portfolios more efficient and effective. Comments which apply to many reports for a given project can be compiled on a feedback sheet to which more specific comments may be added for individual students. Peer review will often improve the quality of written work. A conference with a student before his or her report is submitted may make a second draft unnecessary. Students often need to complete several projects before they fully appreciate the degree of thoroughness and clarity required in their reports. Sharing examples of good student work from previous years can help communicate these expectations.

References

[1] Knoerr, A.P. "A professional evaluation model for assessment." Joint Mathematics Meetings. San Diego, California: January 8–11, 1997.

[2] Knoerr, A.P. "Authentic assessment in mathematics: a professional evaluation model." Ninth Annual Lilly Conference on College Teaching — West, Lake Arrowhead, California: March 7–9, 1997.

Using Writing to Assess Understanding of Calculus Concepts

Dorothee Jane Blum

Millersville University of Pennsylvania

Student write expository papers in an honors, non-science majors' calculus course to integrate the major ideas they're studying.

Background and Purpose

Millersville University is one of fourteen institutions in the State System of Higher Education in Pennsylvania. It has an enrollment of approximately 6,300 undergraduate students, about 280 of whom participate in the University Honors Program. Each student who plans to graduate in the Honors Program must complete a stringent set of requirements, including a course in calculus. For honors students seeking degrees in business or one of the sciences, this particular requirement is automatically met because calculus is required for these majors. However, most honors students major in the humanities, fine arts, social sciences, or education. To accommodate these students, the mathematics department designed a six-credit, two semester sequence, Honors Applied Calculus. There are usually between 25 and 30 students in each course of the sequence.

At Millersville University, courses are not approved as honors courses unless they include a well-defined writing component. At its inception, this requirement was not enforced for Honors Applied Calculus because no one at any level in the university associated writing assignments with undergraduate mathematics courses. The lack of an honors writing component troubled me, so I attended workshops such as "Writing Across the Curriculum" in order to obtain ideas about incorporating writing into the two courses. As a result, not only was I able to find ways to add writing assignments to Honors Applied Calculus, but I also decided to replace the usual in-class test on applications of calculus with a portfolio of applications. Writing would be an integral part of the portfolio because the students would be required to preface each type of application with a paragraph explaining how calculus was used to solve the problems in that section.

Method

I use, on a regular basis, four topics for writing assignments in these courses. They are 1) a discussion of the three basic types of discontinuity; 2) a discussion of how the concept of limit is used to obtain instantaneous velocity from average velocity; 3) an analysis of a graph representing a real world situation; and 4) a discussion of how Riemann sums are used to approximate area under a curve. For the Riemann sum paper, I specify the audience that the students should address; usually it is someone who has had minimal exposure to mathematics such as a younger sibling. In the first semester, I assign one of the first two topics during the first half of the semester and the graph analysis paper near the end of the semester. In the second semester, the portfolio of calculus applications is due by spring break, whereas the Riemann sum paper is assigned near the end of the semester. I always allow at least three weeks for students to complete each writing assignment.

Typical instructions for the graph analysis paper are:

Find, in a publication, a graph of a function that represents a real world situation, or make up your own graph of a function that could represent a real world situation. The graph you use for this assignment must exhibit at least six aspects from the list of graph aspects that we compiled in class. You are to make a chart indicating each aspect of the graph, where it occurs, and what it indicates about the function or its derivatives. Then you are to write a narrative in which you explain the connections between each graph aspect and the real world situation it represents. The final paper that you submit must include the graph, the chart, and the narrative.

I believe that this assignment is the definitive assessment tool for the conceptual material that we cover in the first semester. By transferring what they have learned in the course to a situation outside of the course, the students clarify for themselves the significance of such things as extrema, discontinuities, and inflection points. It is important to realize that in this assignment, I have no interest in evaluating computational ability. Rather, I am assessing students' critical thinking, analytical thinking, and interpretative skills.

When I grade writing assignments, my primary concern is accuracy and my secondary concern is organization. No one is required to be creative or original, but most students voluntarily incorporate both attributes in the work they submit. The portfolio project, on the other hand, involves problem solving and computation in addition to writing. Each student includes two or three problems, chosen from a list provided by me, for each type of application (Newton's Method, optimization, differentials, and related rates). Each section of the portfolio corresponds to one of the types of applications and includes the brief written explanation mentioned earlier, followed by the statement of each problem to be solved in the section and its detailed solution, including relevant diagrams. I evaluate the portfolio primarily on correctness of solutions and secondarily on writing and organization.

Findings

Students have reacted favorably to the writing assignments. On the Honors Program course evaluation questionnaire, many students claim that writing about mathematical ideas is a greater intellectual challenge and accomplishment than doing calculations or solving problems. They also have claimed that writing clarifies concepts for them. So far, no student has indicated that he or she would prefer to have tests in place of the writing assignments. However, students have indicated that putting together the portfolio is very time consuming and several have remarked that they would have preferred a test in its place. There is also universal agreement that the writing assignments and portfolio do comprise a more than adequate honors component for the sequence.

It is interesting to note that on the graph analysis assignment, approximately the same number of students choose to make up their graphs as choose to use graphs from

publications. In the latter case, many students use graphs related to their majors. Some examples of situations for which students have created their own graphs are:

1) a graph of Joe Function's trip through the land of Graphia where a removable discontinuity is represented by a trap;
2) a graph of Anakin Skywalker/Darth Vader's relationship with the force where the x-intercepts represent his changes from the "good side" to the "dark side" and back again;
3) a graph of profits for Computer Totes, Inc., where a fire in the factory created a jump discontinuity. The narrative written by the student in this paper was in the form of newspaper articles.

Some examples of graphs students found in publications are:

1) a graph of power levels in thermal megawatts of the Chernobyl nuclear power plant accident where there is a vertical asymptote at 1:23 a.m. on April 16, 1986;
2) a graph of tourism in Israel from 1989 to 1994 with an absolute minimum occurring during the Persian Gulf War;
3) a graph of the unemployment percentage where it is increasing at an increasing rate between 1989 and 1992.

The assignment on Riemann sums has also produced some very innovative work. One student prepared a little book made of construction paper and yarn, entitled "Riemann Sums for Kids." A French major, writing in French, explained how one could amuse oneself on a rainy day by using Riemann sums to approximate the area of an oval rug. Another student wrote a guide for parents to use when helping their children with homework involving Riemann sums. An elementary education major prepared a lesson plan on the subject that could be used in a fifth grade class. Several students submitted poems in which they explained the Riemann sum process.

The portfolio produces fairly standard results because it consists of more "traditional" problems. However, there have been some creative presentations, such as prefacing each section of the portfolio by a pertinent cartoon.

Use of Findings

When I first started using writing assignments in Honors Applied Calculus, I only counted them collectively as 10% of the final course grade. At that time, I did not value the writing as much as I did the in-class computational tests. However, I soon realized that students devoted more thought, time, and effort to the writing assignments than they did to the tests. I also realized that I learned more, through their writing, about how my students thought about and related to mathematics than I did from their tests. As a result, I now give equal weight to writing assignments, the portfolio, tests, and homework when determining the final course grades in the course.

Furthermore, when I began to use writing assignments, I did not feel comfortable evaluating them. This situation resolved itself once I realized that the students, writing about mathematics for the first time, were just as unsure of what they were doing as I was. Now, I encourage the students to turn in rough drafts of their work the weekend before the Friday due date. This provides me with the opportunity to make suggestions and corrections as needed without the pressure of assigning final grades right away. Sometimes I will assign provisional grades so students will have an idea of the quality of their preliminary work.

I also have learned about curriculum from these projects. For example, based on results from the writing assignment on types of discontinuity, I am convinced that the most effective way to introduce limits is within the context of continuity and discontinuity. This is the approach I now use in all my calculus courses.

Success Factors

It would be difficult to imagine better conditions under which to use writing in a mathematics course than those that exist in the honors calculus sequence. The students are among the most academically motivated in the university. They are majoring in areas where good writing is highly valued. The pace of the sequence is determined by the needs of the students and not by a rigid syllabus that must be covered in a specific time frame. As a result, there is ample class time to discuss topics after they have been introduced and developed. For example, in the first semester, we spend twice as much time on the relationship between functions and their derivatives as is typically scheduled in other calculus courses.

Even though I initially introduced the writing assignments to fill a void, I now view them as an indispensable part of learning calculus. Consequently, I am looking for ways to incorporate writing in all my calculus courses.

Journals: Assessment Without Anxiety

Alvin White

Harvey Mudd College

Using journals for formative assessment encourages students to explore topics they might have been intimidated by if they were being graded.

Background and Purpose

A journal is a personal record of occurrences, experiences and reflections kept on a regular basis. In my mathematics classes students keep a journal of their mathematical experiences inside and outside of class. The purpose of journals is not to assign a grade for each entry but to help students find their own voices and to be reflective about the subject. Allowing more informal tentative writing into the classroom encourages students to think for themselves as opposed to only knowing second hand what others have thought before them.

Mathematics is sometimes perceived as stark and unbending. This may be caused by presentations which are strictly definition-theorem-proof, or lack a sense of historical evolution and excitement. "Mathematics has a public image of an elegant, polished, finished product that obscures its human roots. It has a private life of human joy, challenge, reflection, puzzlement, intuition, struggle and excitement. The humanistic dimension is often limited to this private world. Students see this elegant mask, but rarely see this private world, though they may have a notion that it exists." (D. Buerk [1], page 151)

Some professors insist that all answers to homework and exam problems be in full sentences, with severe penalties for violations. The students are given a sheet with instructions and illustrative examples and are left to sink or swim. There is little effort to convince students of the merits and advantages of this demand. The perception is that the professor is not "student friendly."

If the object is to help students learn to express themselves in writing, journals offer a more natural approach, and the perceived relationship with the professor is not so confrontational. When the task is to "solve a problem," students who are already writing in their journals may approach the solution in a more expansive and discursive fashion. Mathematics may thus be elevated above memorization of facts and formulas.

When I asked my 20 to 30 students in Calculus and in Linear Algebra to keep journals that would be read and commented upon by me, they remarked that they were also keeping journals in their humanities, social science and biology courses. Harvey Mudd College is a small (650 students) college of engineering and science with a demanding workload. One-third of the curriculum is in the humanities and social sciences. A Writing Center exists to help students improve their abilities. All first year students take a writing intensive course that is limited to eighteen students per classroom. This course is taught by writing specialists who are part of the Humanities and Social Science Department.

Journals are a form of self-assessment, an opportunity for students to think about their knowledge of the subject and to strengthen their confidence. The journals are not graded; they afford an opportunity for dialogue between each student and the teacher. Grades are based on homework, class participation, quizzes and exams. The journal dialogues allow the students and the teacher to know each other beyond the anonymity of the classroom activities.

Method

I ask my students to keep journals about their encounters with mathematics in or out of class. These may include insights or puzzlements. Some students may be interested in pursuing the history or the implications of some aspect of the subject. Entries should be made at least once a week. The journals are collected every two or three weeks and returned within a week. No length or other guidelines are offered. Students are encouraged to express themselves in writing. The exercise of writing may encourage further discussion among students and between students and teacher. Some entries have nothing to do with mathematics, but come from other concerns of young people.

An early entry for my students is "A Mathematical Autobiography." Students seem to enjoy writing about their experiences with good or not so good teachers, their awards and successes, sometimes starting in the third grade. Other entries may be responses to questions or topics that arise. There may not be enough time to reflect on those ideas in class. Students may also express their puzzlements in their journals. The journal becomes part of the conversation between students and teacher.

A colleague periodically asks his students to turn in a looseleaf binder that contains returned, corrected homework, personal responses to readings in the text and classroom notes. These portfolios may or may not be graded, but will be commented upon in the margins.

A successful alternative or addition to journals that I have tried is to ask students to write a short (3–5 pages) essay on anything related to mathematics:

Is there some subject that you are curious about? Learn something about the subject and write about it.

Findings

Many students seize the opportunity to learn independently and to write about their knowledge. Grades are not mentioned. (The activity might be part of class participation.) All students, with rare exception, participate enthusiastically. Some of the responses have been inspired. An Iranian student could not think of an appropriate topic. I suggested that he learn and write about the Persian poet-mathematician Omar Khayyam. He was so excited by what he learned that he insisted on writing a second paper about his countryman. The activity offered an independent, self-directed task that contributed to knowledge and self confidence. A follow-up for that last activity might be duplicating all of those papers and distributing to the class (and beyond) a stapled or bound set.

Use of Findings

Grading journal entries might inhibit the free exchange of ideas and the viewing of the journals as part of an extended conversation. One might assume that not formally grading them would lead many students to neglect keeping journals. Paradoxically, the absence of grades seems to liberate students to be more adventurous and willing to explore ideas in writing and in the classroom. The usual subjects of the journals are the ideas that are studied and discussed in the classroom or from supplementary reading done by all or part of the class. The students flourish from the intrinsic rewards of expressing themselves about things that matter to them, and having a conversation with the teacher.

Although I do not grade the students' journals, I do correct grammar and spelling, and comment on the content with questions, references and appropriate praise. Students respond to the conversations via journals by speaking in class with more confidence and a greater variety of ideas. Entries made before discussions are rehearsals for class participation. Entries made after class are opportunities for reflection.

Success Factors

How do students acquire knowledge of mathematics? Memorization and solving problems are two routes that may be followed. Constructing personal meaning by reflection and conversation is another route. These routes are not mutually exclusive. Asking students to discuss in their journals some of the questions that were touched on or discussed in class extends the intellectual agenda in a meaningful way. The journal is an opportunity for each student to be personally involved in the agenda of the class.

Students will not be anonymous members of the class if they participate in writing and reflecting as well as speaking.

Reference

[1] Buerk, D. "Getting Beneath the Mask, Moving out of Silence," in White, A., ed., *Essays in Humanistic Mathematics,* MAA Notes Number 32, The Mathematical Association of America, Washington, DC, 1993.

Assessing General Education Mathematics Through Writing and Questions

Patricia Clark Kenschaft
Montclair State University

General education students can learn to read mathematics more thoughtfully and critically by writing questions over the reading, and short response papers—without enormous investment of faculty time grading.

Background and Purpose

Montclair State University was founded as a public two-year normal school in 1908 and has evolved into a teaching university with a Master's Degree program and plans for a doctorate. Almost 13,000 students are enrolled. Most have a mid-range of collegiate abilities, but there are a few super-stars. Most are first generation college students; many are working their way through college.

MSU requires all students to take mathematics, and specially designed mathematics courses support many majors. Students in other majors have a choice of courses to satisfy their mathematics general education requirement. One, "Development of Mathematics" has as its aim "to examine mathematics as a method of inquiry, a creative activity, and language, and a body of knowledge that has significantly influenced our culture and society through its impact on religion, philosophy, the arts, and the social and natural sciences." It attracts majors in religion, philosophy, psychology, the arts, foreign languages, English, journalism, elementary education, pre-law, political science, and history.

I was first assigned this course in 1993. How could one grade students in such a course? Computation doesn't have much of a role in achieving the stated aims, but downplaying computation essentially eliminates traditional methods of mathematics assessment. A friend who teaches collegiate English and gives workshops nationwide for teachers of introductory college writing courses urged me to consider giving short writing assignments and just checking them off,

as is common in introductory writing courses. This flies against our custom in teaching mathematics. When I balked, she admitted she gives "check plus" to those who show special insight in their writing.

Wanting more variety, I decided to include weekly questions on the reading and a term paper as major contributors to the semester grade. A smaller weight is given to a final paper, attendance, and class participation, although (to avoid intimidating the students) I de-emphasize the oral component.

The two aspects that are most unusual in mathematics courses are discussed here: short papers and weekly questions on the reading.

Method

Short Papers. Typically, the weekly 250-word essay counts three points toward the final grade. The syllabus that I hand out on the first day of classes says, "You may explore ideas further (if you want an 'A' in the course) or merely summarize the author's comments (if you are content with a 'B–')." Many students choose to "explore ideas further," and when they do this creatively, they obtain a "check plus," worth four points instead of the usual three.

Some papers, but relatively few, are skimpy, and deserve only two points. The level of most papers is more than adequate. For lateness I subtract points, but I stay free of an algorithm, since some papers are more justified in being late than others.

Weekly Question. A more explicit device for discovering what students want and need to learn is to require students to hand in questions at the beginning of each week on that week's assigned reading. Each question counts one point toward their final grade, and ensures that most students have read the assignment before the week begins. These questions form the basis of much of our in-class discussion (see Use of Findings).

Findings

One of the surprises is the variety in the students' papers as they try to write something that is more than routine. I inevitably learn a great deal, not only about the students, but about new ways of viewing and teaching mathematics, and about other subjects related to mathematics that they know better than I do.

Another surprise has been how easy it is to decide for grading purposes which papers demonstrate originality. Excerpts from one of the first papers I received indicates how one student explored ideas further. (Rudy Rucker is the author of the text.)

....Suddenly, I'm drowning in base 27's and numbers I've never heard of before. I cry out for help as I notice a googol chasing me across the room.

My senses eventually come back and I realize that [the author] has made one major blunder... [He claims] that if we were to count two numbers a second for the better part of six days, one would reach a million.

....One could count two numbers a second very easily if it were up to one hundred. [but]...the numbers in the one hundred thousand to nine hundred ninety-nine thousand, nine hundred ninety-nine range are going to take more than a half a second. They are actually going to take the better part of a second. This does not even take into account that a person will probably get winded somewhere along the line and have to slow down. Fatigue may also factor into this because a person is bound to be tired after not sleeping for six days.

Nevertheless, I did want to make sure that my point was legitimate, so I took different sequences of numbers in a clump of ten and said them, checking the second hand of my watch before and after the set of numbers in the one hundred thousand range. My average was considerably greater than a half second per number. For example, try counting the sequence 789,884 through 789,894.

Mr. Rucker, your question is good. Its answer isn't. I'd tell whoever decided to attempt this to set forth a couple extra days. Six days is not enough.

Such a paper clearly merits an extra point beyond those that merely regurgitate the reading.

Another discovery is the extent to which students reread each week's assignment for their papers. Most read it reasonably carefully before submitting their question at the beginning of the week. Many refer to a second, third, or even fourth reading in their paper written at the end of the week. They are obviously seeking a deeper understanding based on the class discussions. Some students actually chronicle for me how much they comprehended at each reading, a useful account for my future teaching.

Use of Findings

Grading such papers is far more interesting than checking computations; it is worth the time it takes. I feel more in touch with the students than in any other course. Since I try to write at least one comment about the content of each paper, the papers become, in effect, a personal correspondence with each student.

I often read excerpts from the best papers aloud, and it is clear that the students also are impressed. Having one's paper read aloud (but anonymously) is an intrinsic reward that students work hard to achieve. Reading interesting passages to the class also sometimes generates discussions that probe the ideas further.

The papers that merely summarize the reading are less exciting to read, but they too inform my teaching. Sometimes I catch misunderstandings. More commonly, students mention confusions that need clarification. The written papers give me an opportunity to see how and to what extent they are absorbing the material.

Students' written questions are the backbone of my lesson planning. Some of their questions lend themselves to traditional teaching:

Explain what a logarithm is. I'm still a little confused.

or

Please explain the significance of $7 = 111_{two} = 1 + 2 + 4$. I've never seen this before.

I usually begin the week's exposition by answering such questions. One advantage of basing lessons on the students' questions is that at least one student is curious about the answer.

After I have taught the strictly mathematical topics, I turn to other questions where I believe I know the answer better than the student. For example:

When I think of a computer, I imagine an overpriced typewriter, and not a math-machine. I especially don't understand the system that a computer uses to translate the math information into word information. In other words, what is a bite (sic) and how does it work?

This is not a traditional mathematics question, but it does guide me in helping the students understand how mathematics "has significantly influenced our culture and society" (the course's aim). I answer such questions in the lecture format of a liberal arts or social science course, keeping the lecture short, of course.

For the final group of questions, I have the students arrange their chairs in a circle. I read a question and wait. My first class let hardly a moment pass before the ideas were flying. The second class taxed my ability to extend my "wait time." Sitting silent for long gaps was not one of my fortés before teaching this course, but developing it paid off. Student discussion would eventually happen, and it was almost always worth waiting for. Subsequent classes have been between these two extremes; none have been disappointing.

Sample discussion questions students contributed from the same lesson as above included:

> If we agree that "computerized polling procedures" give citizens a "new power over the government," is the reverse not also true?

> Page 30-bottom: "I construct in my own brain a pattern that has the same feel as your original thought." I disagree with this. Information obtained doesn't necessarily have the same "feel" to it as the transmitter's thought intended, for we decipher it according to our perception and, therefore, can and often do distort the intended meaning or end product of information.

This last question is related to a fascinating student story. A clerical worker from Newark, she wrote at the beginning of the semester in her introductory comments to me that she was scheduled to graduate; this was her last course. Then she added, "I have enormous math anxiety. I hope to be able to ease this somewhat and sweat in class only from the high temperature!" The student spoke not a word in class during its first month, but she was always there on time and made constant eye contact. Class reaction to her questions (such as the one above) was strong, however, and I remember her response when another student exclaimed to one of her questions, "What a genius must have asked that one!" I saw her sit a bit taller. After a few weeks, she began to participate, very cautiously at first. By early December, her term paper, "From Black Holes to the Big Bang," about Stephen Hawking and his theories, was so fascinating that I asked her if she would be willing to present it to the class. To my amazement, she hardly blinked, and a few minutes later gave a fine oral summary that mesmerized the class and stimulated an exciting discussion. She received a well-earned "A" at the end of the semester.

The best part of using essays and questions as the major assessment devices is that students take control over their learning. The essays reveal what they do know, while traditional tests find holes in what they don't know. Students rise to the challenge and put real effort into genuine learning.

Success factors

The sad fact is that the general public perceives traditional math tests as not objective because people can so easily "draw a blank" even though they knew the subject matter last evening — or even an hour ago. My student ratings on "grades objectively" have risen when using this method.

Students like to demonstrate what they know, instead of being caught at what they don't know. As the semester progresses, their efforts accelerate. Taking pressure off competitive grades results in remarkable leaps of understanding.

Combining Individual and Group Evaluations

Nancy L. Hagelgans
Ursinus College

One frequent concern of faculty members who have not yet tried cooperative learning is that giving the same grade to a whole group will be unfair both to the hard workers and the laggards. This article addresses this issue.

Background and Purpose

Assessment of student learning in a mathematics course with cooperative learning groups includes assessment of group work as well as assessment of individual achievement. The inclusion of a significant group component in the course grade shows the students that group work is to be seriously addressed. On the other hand, assessment of individual work is consistent with the goal of individual achievement. Such an assessment plan promotes learning through group interaction and also holds students accountable for their individual learning.

Ursinus College is an independent, coeducational, liberal arts college located in suburban Philadelphia. Over half the graduates earn degrees in science or mathematics, and biology is one of the largest majors. About 75% of the graduates enter graduate or professional schools. Mathematics classes are small, with a maximum of 20 students in calculus sections and even fewer students in upper level courses. In recent years, a majority of the mathematics majors has earned certification for secondary school teaching.

My first experiences with extensive group work occurred in the fall semester of 1989 when I took both my Calculus I sections to the microcomputer laboratory in order to introduce a computer algebra system. Throughout the semester, the students worked in pairs on weekly assignments, which counted about 20% of their course grade. I continued this arrangement for the next three semesters in my sections of Calculus I and Calculus II. Then, after the

summer of 1991, when I attended a Purdue Calculus workshop, I started using the cooperative learning aspects of that program in many of my courses. Subsequently I was a co-author of the book, *A Practical Guide to Cooperative Learning in Collegiate Mathematics* [2], which includes a full chapter on assessment. I now use cooperative learning and assess the group work in all the mathematics classes that I teach. The assessment method is applicable to any mathematics course, including first-year classes and majors' courses, in which cooperative learning groups are assigned for the semester. I have used similar methods in Calculus I, Calculus II, sophomore Discrete Mathematics, Topology, and Abstract Algebra.

Method

The assessment plan for classes with cooperative learning includes evaluation of most of the students' activities. I assess quizzes, tests, a final examination, group assignments and projects, and "participation." The participation grade reflects each student's involvement with all aspects of the course, including submission of journal entries each week, preparation for and contributions in class, and group interaction during class and outside class. A typical distribution of total points for the various activities assessed is shown Table 1.

With this scheme about 45% of a student's course grade is earned in the group since the first of the three hour exams given during the semester is a group test and about half the participation grade is related to group work. The quizzes,

Activity	Points	Percent of Course Grade
3 tests	100 points each	30%
4 quizzes	25 points each	10%
final exam	200 points	20%
group projects	300 points	30%
participation	100 points	10%

Table I

two tests, and the final examination all are individual tests. At times I give a few extra points on an individual test or quiz to all members of a group that has excelled or greatly improved. I give each student a midsemester report in which I allocate up to half the points for group projects and for participation. In this midsemester report, with one group test and two quizzes given, 275 out of a total of 350 points (about 79%) are related to group work. The final examination is a challenging comprehensive examination on the entire course.

The group test is given during a class hour around the fourth week of the semester. The goals of the group test include promoting group interaction early in the semester, reducing tension for the first test of the course, avoiding early withdrawals from the course (especially in Calculus I), and learning mathematics during a test. A group test is somewhat longer and harder than an individual test would be, and it certainly is too long for an individual to complete during the hour. A typical problem on an abstract algebra group test is: *Prove: If $x = x^{-1}$ for every element x of a group G, then G is abelian.* Seven of ten problems on one discrete mathematics group test asked students to determine whether a statement is true or false and to provide either a proof or counterexample with a full explanation; one such statement was: *The square of an integer has the form* $4k$ *or* $4k + 1$ *for some integer k.* The group test in any course is designed with completely independent problems so that students can start anywhere on the test. Each student receives a copy of the test questions and blank paper for the solutions. Each group may submit only one solution to each problem.

Preparation for a group test includes the usual individual studying. In addition, I frequently hold a review problem session in the class hour just before the test, and the groups continue to work on these problems outside class as they study for the test. Also, I instruct the groups to plan their test-taking strategies before the day of the test, and we discuss the possibilities in class. For example, a group of four students could work in pairs on different questions, one pair starting with the first question and the other with the last question; after all problems have been solved, students can check each others' work and discuss the solutions. I mention that individuals need time to think quietly before talking with other group members about a problem.

The group assignments, which vary with each course, comprise challenging problems that are submitted and graded

almost every week during the semester. The calculus classes solve problems using a computer algebra system; the laboratory book is chosen by the Department. The discrete mathematics students implement the fundamental concepts with some problems that I have written for the same computer algebra system [1]. The abstract algebra and topology students solve traditional problems without the use of technology, but with an emphasis on writing clearly.

The journal is a diary in which the students write about their own experiences in the class. On some weeks I ask them to concentrate on certain topics, such as their group meetings, the text book, the department's open tutoring service, or how they learn mathematics. I respond to the journals without determining a grade. Each week I give any reasonable submission a check to indicate credit toward the participation grade.

Findings

With the assessment described above, the students actively participate in their learning both in class and outside class every week of the semester. They interact with other students as they work together in long, problem-solving sessions, when they learn to discuss mathematics. They encourage each other to persist until the problems are solved. During the frequent group visits to my office for help, the students fluently talk about the mathematics, and they show that they have thought deeply about the problems. The students are motivated to complete the group assignments by the assignments' part in determining the course grades as well as by the responses I write on the work.

The test grades on the group test are quite high, and no group has ever failed a group test. The lowest grade ever was 68 in one Calculus I class, and one year all four groups in Discrete Mathematics earned grades above 90. This test forestalls a disastrous first testing experience for first-year students in Calculus I and for majors enrolled in a first abstract mathematics course. Few students have complained that the test grade lowered their individual grades, since, as a matter of fact, most students have a higher grade on the group test than on subsequent hour exams. On the other hand, some weaker students have written in their journals about their concern that they cannot contribute enough during the test, and no student has depended entirely on other members of the group. Since it is to the advantage of the group that all members understand the material, the group members usually are very willing to help each other study for the exam. The group exam does reduce tension and foster the team spirit. Many students claim that they learned some mathematics during the test, and they frequently request that the other tests be group tests. I see evidence that students have edited each others' work with additions and corrections written in a different handwriting from the main work on a problem.

Students in all the courses have time to become acclimated to the course before the individual work counts heavily, and no students have withdrawn from my Calculus I classes in recent years. This assessment plan is much more successful than a traditional plan that involves only periodic assessment of performance on timed tests. The plan promotes student behavior that supports the learning of mathematics, and it allows me to have a much fuller understanding of the current mathematical knowledge and experiences of each student.

Use of Findings

By reading the groups' work each week, I am able to follow closely the progress of the students throughout the semester and to immediately address any widespread weaknesses that become evident. For example, in one proof assigned to the abstract algebra class, the student groups had twice applied a theorem that stated the existence of an integer, and they had assumed that the same integer worked for both cases. During the next class I discussed this error, later in the class I had the groups work on a problem where it would be possible to make a similar mistake, and then we related the solution to the earlier problem.

I use the information in the weekly journals in much the same way to address developing problems, but here the problems usually are related to the cooperative learning groups. I offer options for solutions either in a group conference or in my response written in the journal. In addition, when students write about their illnesses or other personal problems that affect their participation in the course, we usually can find a way to overcome the difficulty.

Success Factors

I discuss the assessment plan and cooperative learning at the beginning of the course to motivate the students to fully participate and to avoid later misunderstandings. The course syllabus includes the dates of tests and relative weights of the assessed activities. The midterm report further illustrates how the grades are computed. At that time I point out that the individual work counts much more heavily in the remainder of the semester.

References

[1] Hagelgans, N.L. "Constructing the Concepts of Discrete Mathematics with DERIVE," *The International DERIVE Journal*, 2 (1), January 1995, pp. 115-136.

[2] Hagelgans, N.L., Reynolds, B.E., Schwingendorf, K.E., Vidakovic, D., Dubinsky, E., Shahin, M., and Wimbish, G.J., Jr. *A Practical Guide to Cooperative Learning in Collegiate Mathematics*, MAA Notes Number 37, The Mathematical Association of America, Washington, DC, 1995.

Group Activities to Evaluate Students in Mathematics

Catherine A. Roberts
Northern Arizona University

From reviewing before a test to various ways for students to take tests collaboratively, this article looks at ways that groups can be used to evaluate student learning while increasing that learning.

Background and Purpose

As a student, the opportunity to collaborate with other students was instrumental to my success in learning mathematics. As an instructor, I take these approaches one step further. If the structure of a class is largely group-based, why not use group activities during testing as well? Exams then become genuine learning experiences.

I have been teaching at large state universities for five years. I use these approaches in classes ranging from College Algebra to Advanced Calculus. The average class size is about thirty, but the techniques can be used in larger classes as well.

Below is a list of group activities that can be used to assess learning in the mathematics classroom. They are arranged from modest to radical, so if you are just starting group activities, you might find it helpful to move through the list sequentially. I discuss findings and use of findings with each activity, and make brief remarks that apply to all the methods subsequent the Method section.

Method

Group Exam Review

This approach allows you to cover a broad spectrum of problems in a limited amount of time. Hand out a worksheet with all the problems you would like students to review for an upcoming exam. Divide the class into groups, and have each group begin by solving only one problem and putting the solution on the board. Then, instruct groups to work on additional problems throughout the class period, always entering their solution on the board. When it is clear that several groups agree on a solution, mark that problem as "solved." When groups disagree, have the entire class work on the problem until agreement is reached. At the end of the class period, everyone has a completed solution set for the entire worksheet, even though as individuals they only wrestled with a few problems.

Findings. Students really like this approach for review. They spend the class period actively solving problems and feel that their time has been used wisely. They make contacts for forming study groups.

Use of Findings. This approach excites students to begin studying: they've gotten started and have a guide (the remaining problems on the worksheet) for continuing their studies. They have some sense of what they do and do not know, and thus how much studying will be necessary.

Think, Pair, Share

Write a problem on the board and ask students to think about how to solve this problem for a few minutes. Then, have them pair up with a neighbor to share solutions and require each pair to agree on one solution. Next, ask each group to report its result to the class.

Findings. Since some students are more prepared than others, this technique prevents the most prepared students from dominating an open class discussion on a new topic. It forces each student to spend some time thinking alone. They can then refine their ideas and build confidence by pairing before sharing their solutions with the entire class.

Use of Findings. Student confidence increases and they are more willing to participate in the class discussion.

Quiz for Two and Quiz Consulting

In *Quiz for Two*, students pair up on a quiz. They can catch each other's mistakes. In *Quiz Consulting*, students are allowed to go out in the hall empty-handed to discuss the quiz, or to talk to each other for one minute during a quiz. By leaving their work at their desks, students are forced to communicate ideas to each other verbally.

Findings. Students learn during the assessment itself. I've been surprised to see that a wide cross section of students (not just the weakest ones) choose to take advantage of opportunities to discuss quizzes with their peers. Performance on these quizzes rises while student anxiety declines

Use of Findings. I have found that I can make the problems a bit more challenging. One must take care, however, not to overdo this, as this may cause the class' anxiety level to rise, defeating the purpose of the technique.

Collaborative Exams

Option 1: I give an extremely challenging multiple choice exam to the class. I allow the class to discuss the exam, but require that they each turn in their own solution at the end of the class period. I've seen this technique used in a sociology course with hundreds of students at the University of Rhode Island. *Option 2*: Give a shorter exam, but have them work on it twice. For the first half of the testing period, students work alone and turn in their individual answer sheets. Then, allow them to work in small groups and turn in a group answer sheet at the end of the period.

Findings. One danger with collaborative exams is that students frequently defer to the person who is perceived as the "smartest"—they don't want to waste precious time learning the material but rather hope to get the best score possible by relying on the class genius. Option 2 gives students an opportunity to discuss the exam more deeply since they have already spent a good deal of time thinking about the problems. Discussion can deteriorate, however, if the students are more interested in trying to figure out what their grade is going to be than in discussing the problems themselves.

Use of Findings. With Option 2, the final score can be an average of their two scores.

Findings

While group work is a slower mechanism than a lecture for transmitting information, a well designed group activity can lead to deeper learning. Moreover, activities can combine several topics, thereby saving time. My main finding is that student anxiety about testing is reduced. Students think of quizzes as learning exercises instead of regurgitation exercises. Students are thinking more during testing and are learning more as a result.

I use *Group Exam Review* several times each semester. *Think, Pair, Share* is most useful when I am introducing a new concept. There is a definite penalty to pay in class time when using this technique, so I try to use it sparingly. I use *Quiz Consulting* frequently; it is one of my favorite techniques.

Use of Findings

I primarily use these methods as part of students' overall grade for the course. As with all teaching techniques, it is important to vary assessment methods so that any weaknesses are minimized.

Success Factors

I solicit regular feedback from my students and rearrange groups based on this feedback. I ask students individually, for example, how well they felt the group worked together, and whether they would enjoy working with the same group again or would prefer a new arrangement next time. Students who report that they do not want to work in groups at all are grouped together! This way, they can work individually (perhaps checking their answers with each other at the end) and don't spoil the experience for other students who are more invested in this collaborative approach.

A concern with allowing collaboration for assessment is that the strong students will resent "giving" answers away and carrying weaker students, or that weaker students will receive an unfair boost since they "scam" answers from others. This is a real worry, but there are ways to address it. It is helpful to rearrange the groupings frequently. I've found that students who do not contribute adequately feel embarrassed about their performance and consequently work harder to be well-prepared for the next time. If there are enough opportunities for students who are strong to demonstrate their individual understanding, there is not much grumbling.

Although many people use groups of four to five students, I've had more success with groups of three to four. Experiment with different grouping approaches — try putting students with similar scores together or pairing dominant and talkative people. For my initial group arrangements, I have students fill out information cards and then pair them according to similarities — commuters, parents, students who like animals, etc. Some instructors like to assign roles to each group member — one person is the record keeper, one person is the doubter/questioner, one person is the leader. I prefer to allow the leadership to emerge naturally. I am constantly rearranging groups until members unanimously report that they like their group. This can be a lot of work, however I feel it is worth the effort.

It helps to give specific objectives and timelines in order to keep the group on task. Group conversation can wander and it helps to forewarn them that you will be expecting a certain outcome in a certain amount of time.

Give your students guidelines, such as those found in [1]. These guidelines can, for example, encourage students to listen to each other's opinions openly, to support solutions that seem objective and logical, and to avoid conflict-reducing tricks such as majority vote or tossing a coin.

References

[1] Johnson D.W. and Johnson, F.P. *Joining Together; Group Theory and Group Skills*, 3ed., Prentice Hall, Englewood Cliffs, N.J., 1987

[2] Peters, C.B. (sociology), Datta. D. (mathematics), and LaSeure-Erickson, B. (instructional development), The University of Rhode Island, Kingston RI 02881, private communications.

[3] Roberts, C.A. "How to Get Started with Group Activities," *Creative Math Teaching*, 1 (1), April 1994.

[4] Toppins, A.D. "Teaching by Testing: A Group Consensus Approach," *College Teaching*, 37 (3), pp. 96–99.

Continuous Evaluation Using Cooperative Learning

Carolyn W. Rouviere
Lebanon Valley College

This article discusses several cooperative learning techniques, principally for formative assessment. These include ways to help students learn the importance of clear definitions and several review techniques (comment-along, teams-games-tournaments, and jigsaw).

Background and Purpose

From 1993–1995, I taught at Shippensburg University, PA, a conservative, rural, state university with about 6000 students. I used cooperative learning in a geometry course for students most of whom were intending to become teachers, but this method could be used with most students. The text was Greenberg [1].

The NCTM Standards [4] urge teachers to involve their students in doing and communicating mathematics. The responsibility for learning is shifted to the students as active participants and the instructor's role is changed from dispenser of facts to facilitator of learning.

I chose to use the cooperative learning format for two reasons: I felt my students would learn the content better in a more interactive atmosphere and they may need to know how to teach using these methods. I hoped experiencing these new methods would help them in their future teaching careers. Since most teachers teach the way they were taught, this would be one step in the right direction.

Method

I use heterogeneous groups (based on sex, ability, age, and personality) of four students, or five when necessary. I use a variety of cooperative activities, including team writing of definitions, Comment-Along, Teams-Games-Tournaments, and Jigsaw. I'll describe each briefly.

One of the first activities the students work on is writing definitions. Each group receives a geometric figure to define, and then exchanges definitions (but not the original figure). The new group tries to determine the shape from the definition given to them. In whole class follow-up discussion, we try to verify if more than one shape can be created using the definition, and in some cases they can. Students are thereby able to understand the difficulty and necessity of writing clear definitions.

To be successful in mathematics, students must be able to decide if their judgments are valid. Comment-along gives them practice with this. This activity begins as homework, with each student writing two questions and answers. In class, the teams decide on the correct answers and select two questions for to pass to another team. The new group considers the questions, adds their comments to the answers, and then passes the questions to a third team which also evaluates the questions and answers. Then the questions are returned to the original teams which consider all responses, and prepare and present the conclusions to the whole class. This activity is an assessment that students do on their own reasoning.

Teams-Games-Tournaments (TGT) is a Johnson & Johnson [3] cooperative learning activity which consists of teaching, team study, and tournament games. We use this at the conclusion of each chapter. The usual heterogeneous groups are split up temporarily. Students are put into homogeneous ability groups of three or four students for a competition, using the list of questions at the end of the chapter. Students randomly select a numbered card corresponding to the question they are to answer. Their answers can be challenged by the other students and the winner keeps the card. Students earn points (one point for each card won) to bring back to their regular teams, a team average is taken, and the teams' averages are announced and all congratulated. On occasion, I follow up with a quiz for a grade. The quiz takes a random selection of three or four of the questions just reviewed; students write the quiz individually.

Jigsaw is a cooperative learning activity [3] where each member on a team becomes an "expert" on a topic. After the teacher introduces the material, each team separates, with the members joining different groups who study one particular aspect of the topic. In effect, they become "experts" on that topic. The teacher's role is to move among the student groups, listening, advising, probing, and assuring that the groups make progress and correctly understand the concepts. When asked, the teacher should not try to "teach" the material, but rather pose questions which lead the students to form their own correct conclusions. When the students return to their original teams, they teach the other members what they have learned. The teacher is responsible for choosing topics and for monitoring the groups to assist and verify that what is being learned is accurate. Ultimately, all students are responsible for knowing all the information. I use this method to reinforce the introductory discussion of non-Euclidean geometry. Each expert team is responsible for the material on a subtopic, such as Bolyai, Gauss, Lobachevski, and similar triangles. When the students are divided into "expert" groups and each group is assigned a section to learn, I listen to, and comment on, what they are saying.

Findings

Because each team gives an oral analysis of the definitions it wrote, the whole class is able to share in the decisions and reasoning. Since students' understanding of the need for clear definitions is essential to their understanding and using an axiomatic system, this activity provides them with needed practice. This assessment provided me with immediate feedback verifying that this foundation was indeed in place.

One of the goals of TGT is to give weaker students an opportunity to shine and carry back the most points for their team. After one tournament, a student was ecstatic at her top score! She had never before been best at mathematics.

Once I tried using team TGT average grades in an attempt to promote accountability. Students howled, "It's not fair." So, I stopped counting these grades, and yet they still worked diligently at the game. (I then replaced this group quiz with an individual one.) Several students commented that they liked the game format for review.

During Jigsaw, the students seemed highly competent discussing their topics. During such activities, the classroom was vibrantly alive, as groups questioned and answered each other about the meaning of what they were reading. When the students finished teaching their teammates, one group assured me that the exercise was a great way to learn the material, that it had really helped their team. Students teach the content and assess their knowledge through questioning and answering each other.

One day, before class, a student told me that his teacher-wife was pleased to hear I was teaching using cooperative learning; it is where the schools are heading.

Use of Findings

Teacher self-assessment of effectiveness of the classroom assignment was done informally through the questions heard when traveling through the room. It let me know how the students were learning long before test time, while I could still effect improvements. Depending on what was learned, the lesson was revisited, augmented, or considered completed. For instance, students had a very difficult time "letting go" of the Euclidean parallel postulate. When they would incorrectly use a statement that was a consequence of the Euclidean parallel postulate in a proof, I would remind them of our first lesson on the three parallel postulates.

Some group assessment for student grades was done (one was a team project to write proofs of some theorems in Euclidean geometry), but most of the formal assessment of students for grades was done by conventional individual examination.

Daily and semester reflection on the results of my evaluations allowed me to refine my expectations, activities, and performance for the improvement of my teaching. The students were learning, and so was I.

Success Factors

I would offer some advice and caution to making the change to a cooperative learning format in your classroom. The cooperative classroom "feels" different. Some college students may feel anxious because the familiar framework is missing. The "visible" work of teaching is not done in the classroom. Good activities require time and considerable thought to develop. Consider working with a colleague for inspiration and support. Don't feel that it is an all or nothing situation; start slowly if that feels better. But do start.

I continued because of my desire to help all students learn and enjoy mathematics. But more, it was a joy to observe students deeply involved in their mathematics during class. There was an excitement, an energy, that was missing during lecture. It was addictive. Try it, you may like it, too.

References

[1] Greenberg, M.J. *Euclidean and Non-Euclidean Geometries* (3rd ed.). W. H. Freeman, New York, 1993.

[2] Hagelgans, N., et al., *A Practical Guide to Cooperative Learning in Collegiate Mathematics*, MAA Notes Number 37, Mathematical Association of America, 1995.

[3] Johnson, D. and R. *Cooperative Learning Series Facilitators Manual*. ASCD.

[4] National Council of Teachers of Mathematics. *Professional Standards for Teaching Mathematics*, National Council of Teachers of Mathematics, Reston, VA, 1991.

Collaborative Oral Take-Home Exams

Annalisa Crannell
Franklin & Marshall College

Giving Collaborative oral take-home examinations allows the instructor to assess how well students handle the kind of non-routine problems we would all like our students to be able to solve.

Background and Purpose

I began assigning Collaborative Oral Take-Home Exams (hereafter called COTHEs) because of two incidents which happened during my first year of teaching at Franklin & Marshall College. My classes have always had a large in-class collaborative component, but until these two incidents occurred, I gave very traditional exams and midterms.

The first incident occurred as I was expounding to my students in an integral calculus course upon the virtues of working collaboratively on homework and writing projects. One of the better students asked: if working together is so desirable, why didn't I let students work together on exams? What shook me up so much was the fact that I had no answer. I could think of no good reason other than tradition, and I had already parted from tradition in many other aspects of my teaching.

The second incident followed closely on the heels of the first. I visited the fifth *International Conference on Technology in Collegiate Mathematics*, where I heard Mercedes McGowen and Sharon Ross [3] share a talk on alternative assessment techniques. McGowen presented her own experiences with using COTHEs, and this—combined with my student's unanswered question still ringing in my ears—convinced me to change my ways. Although the idea behind COTHEs originated with McGowen, I will discuss in this article my own experiences—good and bad—with this form of assessment.

Method

I usually assign four problems on each COTHE, each of which is much more difficult than anything I could ask during a 50-minute period. I hand out the exam itself a week before it is due. Because of this, I answer no questions about "what will be on the midterm," and I hold no review sessions—they do the review while they have the exam.

These are the instructions I provide to my students:

Directions This is an open-note, open-book, open-group exam. Not only are you allowed to use other reference books and graphing calculators, but in fact you are encouraged to do so, as long as you properly cite your resources. However, please do **not** talk to anybody outside your group, with the possible exception of me; any such discussions will, for the purposes of this midterm, be considered **plagiarism** and will be grounds for failure for the midterm or the course.

You should work on this midterm in groups no larger than 3 people, unless you ask for permission beforehand. Everybody in the group will receive the same grade, so it is up to the group to make sure that all members understand the material.

You and your group will sign up for a half-hour time slot together. On the day of your exam, please bring your written work to hand in — the group may submit one joint copy or several individual ones. You may bring solutions, books, and calculators to use as reference. You should be prepared to show your graphs to me, and you may of course use your written work to aid you with the oral questions. However, you will not be graded directly on your written work; rather you will be graded on your understanding of the material as presented to me.

I begin each appointment by explaining the ground-rules for the interview. I will choose students one at a time to answer the questions. While that student is answering, all other group members must be silent. Once that student has finished, the other students will be permitted to add, amend, or concur with the answer. I accept no answer until the whole group has agreed. Then I move to a new student and a new problem. The questions I ask tend to go backwards: what did you find for an answer to this problem? How did you go about finding this answer? Did you try any other methods? What made this problem hard? Does it look like other problems that you have seen?

Although each exam appointment is 1/2 hour, I leave 15 minutes in between appointments for spill-over time, which I invariably end up using. The interviews themselves usually take a whole day for a class of 30 students — when I teach 2 calculus classes, I have interviews generously spread over a 3-day period, with lots of time left free. While this certainly seems like a lot of time, I spend no time in class on this exam: no review sessions, no questions about what will be covered, no in-class test, and no post-exam follow up. Better yet, I spend no time grading papers!

Since I assign the grades as the students take their exams, I have had to learn how to present the grades face-to-face. I've never had a problem telling students that they earned an 'A.' To students who earn 'B's, I explain that they are doing "solid" work, but that there is still room for improvement — and I explain just where (perhaps they need to be able to relate visual understandings back to the algebraic definitions). I tell students who do 'C' work that they're really not doing quite as well as I would like, for reasons X, Y, and Z. Notice that the emphasis here is not that I think that they are stupid or lazy, but rather that I have high expectations for them, and that I hope that this situation of weak work is temporary. For students who earn 'D's or worse, I explain that I am very worried about them (again, for reasons as broad as that their algebra skills are very weak and they seem to be confused about fundamental concepts in the course) and that they will need substantial effort to do better in the course.

Findings

The idea of an oral exam in mathematics strikes terror in the heart of the student, but knowing the questions beforehand and having their whole group present at the interview goes a long way to mitigating that fear. Despite the newness and scariness of taking oral exams, students wind up being fond of this form of exam. Indeed, by the end of the course what they are most worried about is taking the final exam as individuals—a turnabout I hadn't expected!

Use of Findings

From the instructor's perspective, the exam interviews become a mint of insight into the minds of the student. I gain a real appreciation of what was hard and what was easy about each problem. I get a chance to challenge misconceptions directly, and also to encourage flashes of insight. Moreover, because every student not only has to come to my office, but also talk directly to me, it becomes much easier to carry conversations about mathematics or study habits into out-of-exam time. The benefits of these brief obligatory encounters spill over into the rest of the semester.

Several of my students told me that for the first time they learned from their exam mistakes; that when they get traditional exams back covered with comments, they can not bear to read any of it, so painful is their remembrance of the midterm. Whereas in a COTHE, they get the feedback even as they present their results; and since they have debated the results with their teammates, they care about the answer. All of them appreciate the time they get to do the problems —a frequent comment I get is that they'd never be able to do these problems on a timed exam. (Of course, I would never assign such difficult problems on a timed exam, although that does not seem to occur to them.)

I now design my exams to be both forward and backward looking—in this way I can use the interviews as a way of preparing students for material to come, and myself for their reaction to it. (For example, Problem D at the end of this paper requires students to do something that we encourage them to do in the regular course of the class: to read a textbook and learn new mathematics. The COTHE allows me to test this skill and to build on this material.)

What my students tell me during their exam tells me a lot about their "trouble spots." One year several groups of students were concerned about whether a function could be concave up at a point of discontinuity (see Problem A at the end of this paper). Although this would not have been my own choice of an important topic to cover, I then spent a half day in class working with the students on this issue and its relative importance to the subject of Calculus, and the students seemed to appreciate the class.

Often I discover—and can immediately correct— difficulties with interpreting the displays on graphing calculators. Sometimes I discover, to my delight, that the students really understand a concept (such as horizontal asymptotes) and that I can safely use that concept to introduce new topics (such as limits of sequences).

I enjoy using "stupid mistakes" as an opportunities to see what students know about the problem. If a students makes an arithmetical or algebraic error at the beginning of a problem on a written test, it can change the nature of the rest of the problem. On an oral exam though, I might say, "At this step, you decided that the absolute value of $-x$ is x. Can you tell me why?" At that point, the students bang themselves on the head and then huddle together to work the problem through again. This opportunity to see the students in action is illuminating (plus it provides them the opportunity to get full credit on the problem). We discuss the correct solution before the students leave.

Success Factors

The most hard-won advice I have learned about assigning COTHEs has to do with the grades:

Do not assign grades by points. Providing so many points for getting so far, or taking off so many points for missing thus-and-such a fact, tends to be at odds with what the students really know. Also, because the students do so well on this kind of exam, the grades bunch up incredibly. I once was in the awkward position of explaining that '83–85 was an A' and '80-82 was a B.' Neither the students nor I appreciated this.

Do decide on larger criteria beforehand, and do share these criteria with the students. Pay attention to the larger, more important areas that make up mathematical reasoning: mechanics (how sound are their algebra skills; can they use the graphing calculator in an intelligent way?); concepts (do they understand the principles that underlie the notation? Can they make links between one concept and another?); problem solving (can they read mathematics, attempt various solutions, etc.?). I assign grades based on the strength of these areas, and try to provide at least cursory feedback on each.

There are occasional concerns from the students about getting group grades, and occasional concerns I have about losing track of how individuals are doing (particularly I worry about not catching the weak students early enough), and for that reason I'm moving in the direction of a two-part exam (similar to the "pyramid" exam described by Cohen and Henle in [1]). The first part of my two-part exam would be a COTHE, and the second would be a series of follow-up questions taken individually in class.

References

[1] Cohen, D. and Henle, J. "The Pyramid Exam," *UME Trends,* July, 1995, pp. 2 and 15.

[2] Levine, A. and Rosenstein, G. *Discovering Calculus*, McGraw-Hill, 1994.

[3] McGowen, M., and Ross, S., Contributed Talk, Fifth *International Conference on Technology in Collegiate Mathematics,* 1993.

[4] Ostebee, A., and Zorn, P. *Calculus from Graphical, Numerical, and Symbolic Points of View, Volume 1,* Saunders College Publishing, 1997.

Selected COTHE Questions

Problem A.

(a) Create a formula for a function that has all the following characteristics:

- exactly 1 root;

- at least 1 horizontal asymptote; and

- at least 3 vertical asymptotes.

(b) Using any method you like, draw a nice graph of the function. Your graph should clearly illustrate all the characteristics listed in part (a).

(c) You should be prepared to talk about direction and concavity of your function.

Problem B. Let $h'(x) = 3 - \left(\dfrac{x^2+1}{5}\right)^{(x^2-1)}$ defined on the domain $[-2, 3]$ — that is, $-2 \le x \le 3$. *Please notice* that this is , $h'(x)$, not $h(x)$!

(a) Sketch a graph of $h(x)$ on the interval. (Note: there is more than one such sketch.)

(b) Locate and classify the critical points of h. Which values of x between -2 and 3 maximize or minimize h?

(c) For which value(s) of x does h have an inflection point?

Problem C. Let $f(x) = \sqrt[5]{x^3 - Lx + 1}$ (where L is the average of the number of letters in your group's last names). The tangent line to the function $g(x)$ at $x = 2$ is $y = 7x - 4F$, where F is the number of foreign languages being taken by your group. For which of the following functions can we determine the tangent line from the above information? If it *is* possible to determine it, do so. If there is not enough information to determine it, explain why not.

(a) $A(x) = f(x) + g(x)$ at $x = 2$.

(b) $B(x) = f(x) \cdot g(x)$ at $x = 2$.

(c) $C(x) = f(x)/g(x)$ at $x = 2$.

(d) $D(x) = f(g(x))$ at $x = 2$.

(e) $E(x) = g(f(x))$ at $x = 2$.

Problem D. Attached is an excerpt from *Calculus from Graphical, Numerical, and Symbolic Points of View* [4]. Read this excerpt and then use Newton's Method to find an approximate root for the function $f(x) = x^3 - Lx + 1$ where L is the average of the number of letters in your group's last names. You should start with $x_0 = 0$, and should make a table which includes

$$x_0, f(x_0), f'(x_0), x_1, f(x_1), f'(x_1), \ldots, x_5, f(x_5), f'(x_5),$$

You might also refer to Project 3.5 in your own *Discovering Calculus* [2].

Assessment in a Problem-Centered College Mathematics Course

Sandra Davis Trowell and Grayson H. Wheatley
Florida State University

When using a problem-centered teaching approach, the instructor needs new methods of assessment. This article explores one such approach.

Background and Purpose

In this article, we describe an assessment procedure the second author used at Florida State University in an upper level mathematics course on problem solving. Analysis of this course formed part of the doctoral dissertation of the first author [3]. Florida State University is a comprehensive institution with a research mission. It is part of the state university system and has 30,000 students. The 16 colleges and schools offer 91 baccalaureate degrees spanning 190 fields. There is an extensive graduate program in each of the schools and colleges. Florida State University is located in Tallahassee and has strong research programs in both the sciences and humanities. The main goal of the course, taken by both undergraduate and graduate students, was to engender mathematics problem solving through the development of heuristics. A second goal was to provide opportunities for students to reconstruct their mathematics in an integrated way so that it could be utilized. The second author has taught this course several times using the methods described in this paper. We view mathematics as a personal activity — one in which each individual constructs his/her own mathematics. It is our belief and experience that a problem-centered instructional strategy is more effective than explain-practice which tends to emphasize procedures [4]. For assessment to be consistent with this view of mathematics, it should occur as students are engaged in problem solving/learning mathematics rather than as a separate activity.

Method

In order to promote reflection on their mathematical activity, it is important to negotiate a classroom environment which focuses on the students' problem solving and explanations of their methods rather than prescribed procedures. In this problem solving course, nonroutine mathematical problems such as the following constituted the curriculum.

> *Fraction of Singles* The fraction of men in a population who are married is 2/3. The fraction of women in the population who are married is 3/5. What fraction of the population is single?

> *Angle Bisector Problem* Write the equation of the bisector of one of the angles formed by the lines $3x + 4y = -1$ and $5x + 12y = 2$.

The assessment plan involves considering each students' problem solutions and providing written feedback in the form of comments, without points or grades. Overall, assessment is based on classwork, homework, a midterm examination, and a final examination. However, it differs from traditional policies in that the grade is not simply a weighted average of numbers for these components but reflects the students' progress in becoming competent mathematical problem solvers. The assessment plan and the goals of the course are discussed with the students on the first day of class and elaborated in subsequent sessions.

Reading the homework on a regular basis is the richest source of assessment data as the student's mathematical competence and reasoning become clear. For example, one student usually wrote very little on each problem, unlike other students who used numerous pages explaining their methods. Yet he frequently developed quite elegant solutions. Elegance of solutions came to be valued by all students as the course progressed. In contrast, another student, while getting correct answers, would often use less sophisticated methods, nongeneralizable methods such as trial and error or an exhaustive search. On another occasion, a graduate student who thought he knew mathematics turned in simplistic solutions which did not withstand scrutiny. He was sure his answers were correct and was surprised to see the methods that were explained by other students. He initially underestimated the course demands but adjusted and became a strong problem solver. A few students were not successful in the course because their mathematical knowledge was inadequate. They had been successful in procedural courses but had difficulty when required to construct solutions without prescribed methods. By carefully tracking homework, the instructor is able to chart progress and assess quality. Some students who have little success on problems during the first few weeks of the course come alive and become successful problem solvers by the end the course. In previous years, the second author tried assigning letter or number grades on homework, but found that this practice encouraged students to do homework for the grade rather than focusing on learning and doing mathematics. The less students are thinking about their grade the more they can be doing mathematics. An average of test scores will not show this progression as clearly. Thus homework papers are an essential source of assessment data.

While generalizable and elegant solutions are valued, students are free to use any solution methods they wish. They are encouraged to explain their solution process and good ideas are valued. Comments made on papers include "very nice," "how did you get this?" "I do not follow," "this does not follow from what you did before," or "how did you think to do this?" Quality work is expected and the instructor will write "unacceptable" on some of the first homework papers if necessary. Students are not expected to be expert problem solvers when they enter the course; so weaknesses on early assignments are not unexpected. Furthermore, a student is not penalized by poor work initially.

In class, students present *their* solutions to the class for validation. There is no one method expected by the instructor. Some students are frequently eager to present their solutions while others are reluctant. By making notes on class activities after each session, these data can be a significant component of the assessment plan. On some days the students work in groups of three or four solving assigned problems. During this time, the instructor is actively engaged in thinking about the group dynamics: who initiated ideas, who is just listening; and assessing each student's involvement. In an early group session, the instructor mentioned that during the small group sessions he had seen several different approaches. He followed this by saying that "we are always interested in other ways." He was pointing out that there was not necessarily one correct way to solve a problem and stressing the belief that problem solving is a personal activity.

The midterm and final examination questions are similar to the nonroutine problems assigned on homework and classwork. Each examination consists of a take-home part and an in-class part. Early in the course it is announced that an A on the final examination will result in an A in the course, since it would be evidence that students had become excellent problem solvers. The comprehensive final examination consists of approximately 10 challenging problems. The goal is a high level of competence and many students reach this by the end of the course. In determining the final grades for students in this course, all the information about the student is considered, not just numerical entries in a gradebook. Particular attention is given to the level of competence at the end of the course rather than using an average of grades throughout the semester. Early in the course, some students are not able to solve many of the problems assigned but as they come to make sense of mathematics previously studied and develop effective heuristics, their level of competence rises dramatically. The students are also encouraged to assess their own work, and near the end of the course they turn in to the instructor the grade they think they should receive. This information can serve as a basis for discussion, but in the end the instructor determines the course grade. The final grade reflects the knowledge at the end of the course without penalizing students for lack of success early in the course. Thus, students may receive a higher course grade than would result from an average of recorded numbers.

Findings

Students became more intellectually autonomous, more responsible for their own learning. This was particularly evident in the whole class discussions. Students began to initiate sharing solutions, suggestions, and ideas. By the third class session, some students would walk to the board without waiting for the instructor's prompt in order to elaborate upon another idea during the discussion of a problem. Their additions to the class discussion added to the richness of their mathematics. In addition, students would suggest problems that were challenging and also voluntarily hand in solutions to nonassigned challenging problems.

This assessment plan, which encourages self-assessment, reflects student competence at the end of the course rather than being an average of grades taken at various points throughout the course. This practice encourages students who might get discouraged and drop the course when they actually have the ability to succeed. Students tend to like this grading method because there is always hope. Four students of varying levels of mathematical sophistication were interviewed throughout this Problem Solving course by the first author. The students generally liked the assessment plan. One student said that this plan was very appropriate as mathematical problem solving should be evolving and that the comments on their problem solving were helpful. When each of these four students was asked during the last week of the course what grade they believed they would receive for this course, all four reported grades which were identical to those later given by the instructor, one A, two Bs, and one C. Even without numerical scores, these students were able to assess their own work and this assessment was consistent with the instructor's final assessment.

Use of Findings

The results of this study suggests that the holistic plan described in this paper has merit and serves education better than traditional systems [1, 2]. Furthermore, by a careful reading of the homework, the instructor is better able to choose and design mathematics tasks to challenge and enrich the students' mathematical problem solving. For example, if students are seen as being weak in a particular area, e.g. geometric problems or proportional reasoning, the instructor can focus on these areas rather than spending time on less challenging problems.

Success Factors

Students enter a course expecting practices similar to those they have experienced in other mathematics courses. Thus, in implementing an assessment method which encourages students to focus upon doing mathematics rather than getting points, it is essential to help students understand and appreciate this unconventional assessment plan.

References

[1] Mathematical Sciences Education Board. *Measuring what counts*. National Academy Press, Washington, 1993.

[2] National Council of Teachers of Mathematics. *Assessment standards for school mathematics*, NCTM, Reston, VA, 1995.

[3] Trowell, S. *The negotiation of social norms in a university mathematics problem solving class*. Unpublished doctoral dissertation, Florida State University, Tallahassee, FL, 1994.

[4] Wheatley, G.H. "Constructivist Perspectives on Science and Mathematics Learning," *Science Education* 75 (1), 1991, pp. 9–21.

Assessing Learning of Female Students

Regina Brunner
Cedar Crest College

The author discusses assessment techniques which she has found to be particularly effective with female students.

Background and Purpose

Some women students, including most of us who have acquired Ph.D.s in mathematics, excel at demonstrating what they've learned on traditional tests. However, most women students are less successful than men with comparable levels of understanding when only these instruments are used. It is thus important to look for assessment methods that allow women to demonstrate as well as men what they have learned. Cedar Crest College is a four year liberal arts women's college in Allentown, Pennsylvania. As a professor in a women's college, I have experimented with a variety of assessment techniques over the past 14 years. These include journal writing, group work, individual and group tests, and creative projects. My findings are that journals help form positive attitudes in females through encouragement and weekly guidance. Formative evaluation improves the teaching/learning process.

A research study of students enrolled in college remedial mathematics courses [4, p. 306] supported the assertion that beliefs such as self-confidence were more important influences on mathematics achievement and success for females than males. Research indicates that sex differences in mathematics self-efficacy expectations are correlated with sex differences in mathematical performance [2, p. 270]. Mathematics self-efficacy expectations refer to a person's beliefs concerning ability to succeed in mathematics. Such beliefs determine whether or not a person will attempt to do mathematics and how much effort and persistence will be applied to the task at hand and to obstacles encountered in the process. Suggestions for improving learning experiences of females include changing students' perceptions of the nature of mathematics by emphasizing creativity in mathematics, making connections to real life situations, permitting students to engage in problem posing, and encouraging debates, discussions, and critiques of mathematical works [1, p. 89–90].

I find that our students learn mathematics better with a hands-on teaching approach and with daily assessment activities to boost their self-esteem and to build their confidence as successful problem solvers. Making models helps our students visualize mathematical concepts. Since many women lack experiences with objects of physics which motivate much of mathematics, they require experience with such objects to continue studying mathematics successfully.

Creative projects help students visualize a mathematical concept. Also, projects require students to reflect on how all the concepts they learned fit together as a whole.

Women as nurturers thrive on cooperation and collaboration rather than competition. The first attribute of women leaders is collaboration [3, p. 26]. Women leaders elicit and offer support to group members while creating a synergistic environment for all and solving problems in a creative style.

Classes used in the following assessments include: Precalculus, Finite Math, Calculus I, II, III, and IV, Probability and Statistics, and Exploring Mathematics for Preservice K-8 teachers over the past fourteen years.

Method

1. *Group Work* Group work in class gives instantaneous assessment of each individual's daily progress. Students

thrive as active learners receiving constant feedback and positive reinforcement. Problems can be solved by groups working with desks drawn together or working in groups on the blackboard. When groups work on the blackboard, I can watch all the groups at the same time. It is easy for me to travel around the room from group to group giving encouragement to those who are on the right track and assisting those who are stuck. This works successfully with classes as large as thirty-two.

A natural outgrowth of working in teams during class is to have team members work together on an exam. All team members receive the same grade for their efforts if all contribute equally to the end result. Teams are given 7–10 days to complete the test, and hand in one joint solution. Each problem is to be signed by all team members that contribute to its solution. Teams are only allowed to consult each other and their texts.

A good group teaching technique is to play "Pass the Chalk." When the teacher says to pass the chalk, then the person with the chalk passes it to another group member who continues solving the problem. This technique brings everyone into the problem solving process. No spectators are allowed: all are doers.

Group work leads to working together outside of class. The class mathematics achievement as measured by hourly tests improves. They feel comfortable as a successful working unit and take tests together as a team. For more detailed discussion of group work, see articles by Crannell (p. 143), Hagelgans (p. 134), and Roberts (p. 137) in this volume.

2. *Journal* Daily journal writing provides a one-on-one conversation with the teacher about progress and stumbling blocks. I answer student inquiries about learning or about homework, and dispel negative feelings with encouragement. Students require about 5 minutes to complete a daily journal entry. Weekly reading of journal entries takes about 3–5 minutes per student. Scoring consists of a point for each entry. Total points are used as an additional test grade.

SAMPLE JOURNAL WRITING TEMPLATE

CLASS

1. In class, I felt...

2. In class, I learned...

3. The most positive result of class was...

4. The least positive result of class was...

5. Some additional comments related to class are...

I use the same questions (replacing "In class," by "While doing homework") for responding to homework. Questions 3 and 4 are both formed in terms of a positive response and not with the words, "most negative result," to emphasize a positive frame of mind. Question 5 allows individuality in response. Students freely share their progress and problems in these entries.

3. *Creative Projects* In Calculus II, for instance, I assigned students the task of making a model of an inverse function. One ingenious model was made from hospital-sized Q-tips. This student made a three dimensional cube using these over-sized Q-tips. Inside the cube she placed three axes. With wire, she constructed a physical representation of a function and its inverse. Another movable model was made from a pipe cleaner and ponytail beads. The function (a pipe cleaner attached to a piece of cardboard by the ponytail beads) was attached to a two-dimensional axes model. Then by lifting and rotating the pipe cleaner, a student could view a function and its inverse function by rotating the function (pipe cleaner) about the line $y = x$. One student brought in her lamp as an illustration since the contours of the lamp's shadow were representative of a function and its inverse function.

A possible assignment in precalculus is to draw a concept map, a scaffold, or a flow chart to classify conic sections. For additional information on concepts maps, see article by Dwight Atkins in this volume, p. 89.

I am always amazed at how creative projects put a spark of life into mathematics class and that spark sets off a chain reaction to a desire to succeed in the day-to-day mathematical activities in the rest of the course. These creative projects range from bringing in or constructing a model of a calculus concept, writing a poem or a song, or developing a numeration system and calendar for a planet in the solar system. In the latter case, students are required to explain their numeration system and why it developed as it did on this planet using relevant research from the Internet. Enthusiasm for mathematics increases as students research mathematical ideas and concepts, brainstorm, and engage in critical thinking and reasoning.

Findings

In the fall of 1996, I used journals, group work, a group test, and creative projects in Calculus I and Finite Math.

Final grades included two individual tests, one group test, and an individual comprehensive three-hour final. In comparing student grades on the first test of the semester to their final grades, I found that students performed substantially better on the final tests.

Written student final evaluations note that journal writing gives students time to reflect on how they learn best, a focus on class work, and a link between the professor and the student. In addition, journals provide the teacher an opportunity to view student thought processes and use this knowledge to teach more effectively. Group work increased student confidence in mathematics, replaced competition with cooperation, emphasized hands-on problem solving, and individual contact with the teacher. Projects enabled students to review and reflect on major concepts from the course.

Use of Findings

I find journal writing improves my students' attitudes towards mathematics, and thus their success in doing mathematics. As the course progresses, journal entries provide the stimulus for making changes in the learning environment. If many student responses indicate that a definite change is required, then the teacher may decide to reteach a concept or proceed at a slower pace. If just a handful require additional help, then the teacher may schedule an additional study session with these students.

Students realize that I care about their learning because I require and grade their journals each week. In turn, they work harder and harder to understand mathematics. I will solve original problems that they pose in the journal. So the journals become written, one-on-one, semester-long conversations and dialogues.

Group work helps the teacher view the interactions within a group, notice which group members need additional help in graphing or algebra skills, and provide help while the group is problem solving. I find group work invaluable. I do not enjoy giving long lectures anymore. I want to present a concept briefly and then solve problems in groups at the blackboard for the rest of the class period. I teach individual groups and assess learning effectiveness as they learn in small groups. My students always write in their journals of the value of the board work. It prepares them for the homework for that night. They are successful on the homework assignments because of the struggles in class that day.

Creative projects provide a needed respite to talk about concepts and to bring them to life in the classroom. I am always intrigued by student creativity and originality. A group in Finite Math shared a counting technique they found on the Internet for making sand patterns in India. The class went to the blackboard to draw these patterns because they were intrigued with this concept and wanted to try it also. Wanting to learn more than is required by the coursework and bringing mathematics into their lives to me makes our students mathematically literate and aware of the power, beauty, and mystique of mathematics.

Success Factors

1. Start on a small scale.
2. Be willing to fail and to try again.
3. Do not abandon lectures. Students cannot succeed unless you give them the background needed.
4. Answer everything asked in the journal. Try to be positive.
5. Shuffle group members during the semester. Working with a variety of partners encourages students to explore various ways to solve a problem, requires that our students be active learners, and increases communication within the class.

These techniques are especially helpful with female students yet they can also be beneficial for *all* students.

References

[1] Barnes, M. "Gender and Mathematics: Shifting the Focus," *FOCUS on Learning Problems in Mathematics*, Vol. 18, No. 1, 2, & 3, 1996, pp. 88–96.

[2] Hackett, G. and Betz, N.E. "An Exploration of the Mathematics Self-Efficacy/Mathematics Performance Correspondence," *Journal for Research in Mathematics Education*, Vol. 20, No. 3, 1989, pp. 261–273.

[3] Regan, H.B. and Brooks, G.H. *Out of Women's Experience: Creating Relational Experiences*. Thousand Oaks CA: Corwin Press, Inc., 1996.

[4] Stage, F. and Kloosterman, P. "Gender, Beliefs, and Achievement in Remedial College Level Mathematics." *Journal of Higher Education*, Vol. 66, No. 3, 1995, pp. 294–311.

Strategies to Assess the Adult Learner

Jacqueline Brannon Giles

Houston Community College System – Central College

Adult students are often better motivated than traditional-age students, but many have not taken an examination for many years. Thus, finding appropriate methods of assessment poses a challenge.

Background and Purpose

The Houston Community College System's Central College has about 27,000 full-time equivalent students who have various goals including changing careers, acquiring associate degrees and transferring to programs which will result in 4-year degrees. There are no residential facilities at the college which is located in downtown Houston, Texas. Some students are homeless, physically challenged, in rehabilitation programs; and others are traditional students. The average age of our students is 27 years old.

Older students present the faculty member with unusual challenges since they don't necessarily respond well to traditional methods of teaching and assessment. However, they are capable of succeeding in learning mathematics if given appropriate support and opportunities to show what they have learned. Many students have personal difficulties at the beginning of the semester. They are challenged with housing arrangements, parking and transportation problems. Day care is also a problem for many single or divorced women. Although the students have social problems and economic constraints, they are motivated to strive toward the academic goals of my class. They honestly communicate their difficulties and they share their successes. I design activities and assessment in my class to address these special needs of my students. Flexibility and compassion create an environment rich with opportunities to learn, while the expectations of excellence, mastery and skills acquisition are maintained.

These non-traditional students usually begin my course in college algebra with little confidence in their abilities.

They are reluctant to go to the board to present problems. Older students tend to do homework in isolation. Many of them do not linger on campus to benefit from interaction with their classmates. College Algebra students at Central College have no experience with mathematics software or computer algebra systems such as DERIVE, and are uncomfortable with computer technology generally. While older students may be uncomfortable with reform in college algebra, with time they do adjust and benefit from a pedagogical style which differs from the one they expected. In this article, I discuss some assessment techniques which I have found help these students grow, and allow them to show what they have learned.

Method

I usually start my college algebra classes with a 10 or 15 minute mini-lecture. I pose questions which encourage discussion among the students. These questions are sequenced and timed to inspire students to discover new ways of approaching a problem and to encourage them to persist toward solutions. Students are invited to make presentations of problems on the board.

Cooperative Learning

Cooperative Learning offers opportunities for sharing information and peer tutoring. Students are invited to form groups of four or five. I assign the work: for example, the section of Lial's *College Algebra* test on word problems.

Each group selects its group leader. Participants keep a journal of their work, and each member must turn in a copy of all solved problems in their own handwriting. Students are warned that each group member must be an active problem solver and no one is to be a "sponge." One class period is dedicated to forming the groups and to establishing the rules for completion of the assignment, and a second to work on the assignment. If the word problems are not completed in the designated class period, the assignment become a homework assignment to be completed by each group. Approximately 30 word problems requiring the use of equations are solved using this method.

As a result of this activity, students become aware of their strengths and weaknesses and become better at pacing themselves and at self assessment. Although I give a group grade for projects, I do not give a group grade for classwork or participation in discussion groups: each individual receives a grade for work completed.

Projects

Projects provide an opportunity for students to become familiar with the Mathematics Laboratory. The goal of the project is graphing and designing at least three images: a cat's face, a spirograph and a flower. Most designs are accomplished using lines and conics such as circles, ellipses and parabolas.

Some students do library or internet research for their projects: they look up trigonometric equations and discover the appropriate coefficients to produce the best (and prettiest) flowers, or use polar coordinates or parametric equations. An objective of the research is to identify role models and people of diverse backgrounds who persisted and succeeded in mathematics courses or professions that are mathematics dependent. A second objective is to inspire students to write about mathematics and to record their attitude toward mathematics and technology. I had the students in summer school find a web page on Mathematics and Nature, and write a short report. They also did research on biomimetics and composites to become aware of the usual connections in science, mathematics and nature. Connecting mathematics to the world around them helps motivate them to work harder learning the mathematics. The Mathematics and Nature web site that the students found is the one that I have linked to my page. I encouraged the students to use e-mail and to provide feedback to me by e-mail.

Examinations

Examinations usually contain 14-20 questions requiring the students to "show all work." This gives me an opportunity to see exactly what their needs are. Four examinations and a comprehensive final are administered. The student is given the option to drop the first examination, if the grade is below 70. This option provides a degree of flexibility and often relaxes tension.

Findings

Our most successful endeavor has been the establishment of my web page, using student research and feedback to improve the design. The web page can be accessed at http://198.64.21.135. I have an instructional page (with information for my students on syllabus, expectations, etc.) and a personal page that students viewed and provided corrective feedback. This activity inspired many students to do more research using World Wide Web. Encouraging students to use this technology helps some of those who are feel bypassed by technology overcome their fears. One student sent me an e-mail message: "This project to research your webpage was not only educational to students it gave them an opportunity to use current fast moving technology." Another commented, "I have been meaning to start on my own for quite some time, and now you have inspired me to do so." Attitudinal learning took place and students seemed more excited about the mathematics class and the use of technology.

The college algebra classes have developed into a stimulating and broad intellectual experience for my students. They gained the algebra skills and a new attitude about mathematics and careers in mathematics. The drop out rate in my college algebra classes is lower than most classes in my department. Approximately 89 percent of the students passed the course with grade C or above, while others realized that they did not have the time nor dedication to do a good job and they simply dropped the course. I believe students made wise judgments about their own ability and that they will probably re-enroll at a later date to complete the coursework.

With persistence on the instructor's part, older students become more confident learners. For example, toward the beginning of the semester, a 62-year-old businessman would often ask me to do more problems on the board for him. He would say, "Darling, would you work this problem for me?...I had a hell of a time with it last night." I didn't mind his style as long as he continued doing the work. Eventually, he would put forth more effort and go to the board to show the class how much progress he made. He would point to the area where he was challenged, and then a discussion would ensue. As I observed his change in behavior I noted that he and other students were taking more and more responsibility for their learning.

Participants in my classes seem more appreciative of the beauty of mathematics and mathematics in nature. Students who successfully completed their designs using DERIVE discovered the beautiful graphs resulting from various types of mathematics statements. They learned how to use scaling and shifting to design images that were symmetric with respect to a vertical axes. While designing, for example, the cat's face, they learned how to change the radius and to translate small circles to represent the eyes of the cat. Some students used a simple reflection with respect to the x-axis to obtain a portion of their design. As they pursue their work, they had conferences with me to gain more insight.

Many students needed to learn basic computer skills: how to read the main menu of a computer, search and find DERIVE, and author a statement. Tutors were available in the Mathematics Laboratory and I assisted students when questions arose. It was the first time many of my students had ever made an attempt to integrate the use of technology into college algebra.

More students are enthusiastic about the use of technology in mathematics. Three years ago only about 5 percent of the Central College students owned graphing calculators or used a computer algebra system. In 1997, approximately 35 percent of them are active participants in the Mathematics Laboratory or own graphing calculators.

Use of Findings

As a result of student input, via informal discussions and e-mail, I will lengthen my mini-lectures to 20–25 minutes. Furthermore, at the end of each class I will review concepts and provide closure for the entire group. More demonstrations using the TI-92 will help prepare the students for their project assignments and a bibliography on "Diversity in Mathematics" will assist the student research component. I usually document class attendance on the first and last day of class so that I have a photographic record of retention. The inspiration to design a more suitable attitudinal study using a pre-test and post-test to measure change in is my greatest gain as a result of these experiences.

Success Factors

The adult learner can be an exceptional student; adults learn very rapidly in part, because they don't need to spend time becoming adults, as traditional-aged students do. However in my experience, adults may be more terrified of tests, they may feel a great sense of failure at a low grade, they may be slower on tests, and they may have lost a lot of background mathematical knowledge. Also they have more logistical problems which make working outside of class in groups harder, thereby not benefitting from explaining and learning mathematics with others. The methods I have used are specifically designed to bring the adult learners together, to relieve anxiety, to encourage group work, the "OK-ness" of being wrong.

The Class Mission Statement

Ellen Clay
The Richard Stockton College of New Jersey

To set the tone of a course at the beginning, develop with the class a "class mission statement," which can be revisited as the course progresses to assess progress toward meeting course goals.

Background and Purpose

Several semesters of teaching expression manipulation-type courses tired me out. In order to re-establish my enthusiasm for mathematics in the classroom I remembered that mathematics is a science in which we look for patterns, generalize these patterns and then use them to better understand the world around us. Yet when I introduced these ideas into a traditional atmosphere, the students revolted because it wasn't what they expected or were accustomed to. It was difficult to acquire the new skills being asked of them: attacking a problem they had never seen before, experimenting with a few examples, choosing from the assortment of mathematical tools previously learned, sticking with a problem that had more than one answer, communicating their answers, and even asking the next question. During the transition period, frustration levels were high for everyone. I was disappointed with their work which showed little or no improvement throughout the semester. The students were angry and I found myself on the defensive. From their point of view, it was a desperate semester-long struggle to bring their grades up with very little idea of how to accomplish that. This led to the introduction of the Mission Statement into my classes during the first week of each semester. When I began to use Mission Statements I was at a small liberal arts college in upstate New York. I am currently at a state college in New Jersey. Although the atmospheres of the colleges are quite different, mathematics education at the K–12 level is fairly consistent. Students come out of the secondary system in the expression manipulation mode. For this reason I believe that Mission Statements can be useful at any institution and in every classroom. To date, I have used them successfully in service courses, general education or core distribution courses, and program courses. While the process takes a full class period, from the first day of class, students begin to take an active role in their own learning. Learning becomes a process over which students have control, something they can improve, not something to simply submit to. Once they have some control, once they see a purpose, their motivation to learn carries the class, rather than the more typical scenario where the professor carries the course dragging the students with her.

Method

I begin the first day with a preliminary process to determine students' expectations for the class. I read each of the following statements aloud and have the students complete them.

Each day I come to class I expect to (be doing) ...
Each night for homework I expect to (be doing) ...
Each day I come to class I expect the professor to (be doing) ...
I expect to do _____ hours of homework between each class (even on weekends??).
I expect my grades to be based on ...
If I could get anything out of this class I wanted to, I would make sure that I (got) ...

After each statement is completed, I pick up the responses, usually written on index cards, and react to them. This allows me to express my expectations for the class. It brings us a little closer together on what the course is all about and it opens up a dialogue between the students and me. Once this preliminary exercise is complete, I facilitate the creation of individual Mission Statements. Each student is handed 20 to 30 post-it notes or index cards and a large sheet of paper. Then I read through the following process.

1. To establish a PURPOSE for the course, ask yourself why you are taking this course. What's your purpose? What do you hope to achieve? You must write until I tell you to stop. Brainstorm everything that comes to your mind. Write each thought on a different card. (Have students write for five minutes.) Now organize your thoughts by prioritizing them, grouping them, or whatever works best for you. (Allow up to 2 minutes.) Then set these ideas aside for now.

2. To establish a PROCESS for the course, ask yourself how you expect to fulfill those purposes. What means and methods will you use? What skills and behaviors will you use, both inside and outside of the classroom? (Have students write for five minutes.) Organize in the same way. Then set these aside.

3. To establish your ACCOMPLISHMENTS, ask yourself, what results do I plan to achieve? What specific outcomes will be evidence that I've succeeded? What topics do I want to have learned? (Have students write for five minutes.) Organize in the same way.

4. Arrange your individual thoughts from the cards onto the large sheet of paper using the following format:

   I AM TAKING "COLLEGE ALGEBRA"
   (purpose statements)
   BY
   (process statements)
   SO THAT
   (accomplishment statements)

   At this point you may switch any statements between categories if you find they fit better elsewhere. The grammar necessary to make each statement follow the category headings will help you decide which statements go under which categories.

5. Take these home and type them up. (This ensures that they at least read them one more time and in the best case ponder over them again.) Also write a paragraph or so reacting to the entire process.

During the semester, we reflect on our Mission Statements several times, to make sure that those goals, processes, and accomplishments which we chose to the best of our ability at that time still work for us. For example, after returning a project I ask them to decide if the assignment pertained to any of their goals, if they actually used the processes on their Mission Statement. If so, did they work; if not are they willing to try them now? If the processes didn't work, I ask them either to think of other processes that might work, or to come to see me, because I have experiences which might be beneficial to them.

Findings

By the time the process is complete, I have assessed students' goals for the course, students' expectations of themselves, students' expectations of me, and students' definitions of success. Maybe even more importantly, students have taken the time to ponder these questions for themselves, many for the very first time.

Use of Findings

If the Mission Statements appear to be too diverse to work with, you can combine common elements from the individual Mission Statements, plus anything that was in at least one statement that you feel is essential, into a common Mission Statement for the course. Presenting this gives you another opportunity to get those students who chose to avoid the process, by writing shallow statements, to buy into it. If, during this process, you have had to eliminate goals that would lead to learning, you can have students create individual projects, related to the course material, which help them realize their individual goals. Keep in mind that each discussion about goals and how to achieve them, offers opportunities for students to learn about themselves, their learning styles, and their defense mechanisms.

This method of assessment has enabled me to make many changes in my classroom. One major change that I have incorporated into several of my general education courses is that I no longer use the textbook as the course syllabus. Our goal is no longer to complete sections 2.1–6.5 skipping as few sections as possible. Now we set our goals based on my own and my students' Mission Statements and use the textbook when and as needed. As an example, in my College Algebra class, students wanted to see the usefulness of mathematics. Therefore, we began the semester with a linear programming problem. As we solved it we found ourselves graphing lines and linear inequalities, finding points of intersection, and invoking major theorems. We turned to the textbook in order to remember how to accomplish some of these tasks. We began in the index and the table of contents, and often found ourselves in several different chapters in a single day. Many students are finally seeing that much of the mathematics they have previously learned is useful and that they can return to a textbook as a reference when necessary.

Another substantial change that I have incorporated into my classes is that of allowing revision on projects. Many of my students are being asked to make decisions based on quantitative data, and to communicate these decisions for the first times in their academic careers. This can be an onerous task for them. It is my experience that once a grade is assigned, learning stops. So instead of reading, commenting, and assigning grades to projects, I glance through them to find what is lacking and/or most misunderstood. Then I either make anonymous copies of the misconceptions or lackings on transparencies for the next class period or I create a peer review form in which students are asked to look for these particular misconceptions/lackings in their peers' work. Using transparencies on an overhead projector takes less time - the students look through their own paper for each suggestion I present. The peer review takes more time but better prepares them for critically looking at their own papers in the future. Peer review works better if each student is working with different data. In either case I offer the class the opportunity to work on projects one more time before I put a grade on them. Students are very grateful. This change took place as a result of our discussion of whether we were meeting our goals or not. Again the usefulness of the subject came up. We realized that in order for work to be useful, we need to be able to communicate it to others and the students had absolutely no training in this area. On a personal note, revision has been a life-saving device as far as time goes. Rather than write the same comments over and over, assignment after assignment, I can glance through a set of projects and decide if I should grade them or build in a revision. The second time around I have the same choice, request a second revision or grade. When I do decide to make individual comments and assign a grade, it takes very little time because I am reading the assignment for the second or third time and the comments are more often about how to extend the project, include deeper mathematics or simply suggest that next time the student take advantage of the revision process. A more impromptu change in the course took place when several of my students made statements about integrating mathematical ideas into their own lives. It developed rather like a challenge: I apparently claimed that mathematics is everywhere and as a result their Mission Statements challenged me to integrate any subject of their choice into the class. From that conversation the following project developed during a class session. Each of my thirty-two students had to think of a situation that could be modeled linearly. That night's assignment was to call two companies which provided the service they suggested in class. These ranged from mechanics' shops to pizza parlors to long-distance phone carriers. Students had to write a narrative description comparing and contrasting the two services to decide how one would choose between the two. The narrative had to be supported by appendices which included graphical, numerical and analytical representations of the data. Each representation had to point out all pertinent information used in their narrative argument. Both the creation of and reflection on our Mission Statements have us continually rethinking our intention for the course. For the first time I am treating general education students differently from mathematics majors. We spend significant time and effort thinking about how to communicate the mathematics we learn (as otherwise it is useless to us) and how to create assignments and processes that enhance this learning. One of the most exciting aspects of the new journey we are on is that for the first time we have student input at a time when students have a stake in the course rather than relying on student evaluations when often they are commenting on their success in the course not on improving or increasing the learning that takes place.

Success Factors

The more responsibility I give to the students, the more success we all have. However, you must be willing to relinquish the feeling of control! This has been difficult for me. Originally I asked the students to write their Mission Statements during week three or four so that they would have an idea of what the course was all about. Now I ask them to create them on the first day or two, before they feel that they are stuck with my choices. Then after three or four weeks we review the Mission Statements to decide whether they are working for us or not.

A second transition I have had to make to ensure success is accepting that this process takes time. You can't rush it! It takes an entire class period at the beginning of the course, and a few minutes of class time on occasion for reflection and discussion. I now know that the time is well worth it because when students and professor are working towards the same goal, progress is hastened dramatically. The time spent on the Mission Statement is made up in the first few weeks of class. I have had several students, reflecting on their Mission Statements, admit that they are not doing anything they claimed they would and this is affecting their grades. That is taking responsibility for one's own actions. If you feel as though your job is to educate students beyond covering content, you will have given them a useful, lifelong tool.

Early In-Course Assessment of Faculty by Students

William E. Bonnice

University of New Hampshire

Students learn to take responsibility for their learning by giving input on how the class is going and what needs to be changed. This works in classes of all sizes.

Background and Purpose

The University of New Hampshire (UNH) is predominantly an undergraduate public institution with an enrollment of approximately 14,000. The Department of Mathematics is in the College of Engineering and Physical Sciences. This encourages meaningful interaction between mathematicians and users of mathematics. The fact that the Mathematics Education Group (those at UNH who educate *mathematics* teachers for elementary and secondary schools) resides within the Department of Mathematics has fostered special concern about educational matters. Although I am a regular member of the Department of Mathematics and not part of the Mathematics Education Group, I am continually experimenting with new teaching methods. For example, I use student-centered methods in my standard sized classes of around thirty students and am experimenting with extending these methods to classes of up to 250 students.

In order to promote a student-centered approach in the classroom, several times during the course I formally gather student input. This input is then the basis for a class discussion about possible changes. Getting early feedback from students and implementing changes improve the course and my teaching. Students are also gratified to be able to give input and see their suggestions implemented. Much of what I do in the classroom has been inspired by the work of psychologist Carl Rogers. The three core conditions for successful psychotherapy that Rogers [1] formulated, in what is now called the Person-Centered Approach (PCA) apply to all human relationships, but, in particular, to student-teacher and student-student relationships.

These core conditions as I understand them are:

(1) Unconditional positive regard: The therapist is willing and able to value the client and trusts that the client has the ability to solve her/his own problems.
(2) Empathy: The therapist is willing and able to put her/his experience aside and really understand the experience and feelings of the client.
(3) Congruence: The expressions and communications of the therapist genuinely represent the true thoughts and feelings of the therapist.

In the classroom Rogers believed that "students can be trusted to learn and to enjoy learning, when a facilitative person can set up an attitudinal and concrete environment which encourages responsible participation in selection of goals and ways of reaching them." (from fly leaf, [2])

Method

Immediately after the first and second exams, at approximately the fourth and eighth weeks of the course respectively, I pass out questionnaires soliciting anonymous student feedback. Some questions seek open responses and others seek specific responses to matters of concern to me or my class. After collecting the questionnaires at the next class, I pass out a written summary of the results and we discuss possible changes. When there is general agreement, we adopt such changes. In cases of disagreement, we discuss possible actions and try to arrive at consensus. When someone has formulated what seems to be a good decision, we take a thumbs-up, thumbs-down, thumbs-to-the-side ("Will support but not my first choice") vote. We ask everyone who voted thumbs-down or thumbs-to-the-side to tell what would have to be done in order for them to change their vote. We try to incorporate these suggestions and come up with a better decision to try out.

Examples of issues and topics of discussion which have come up include: use of calculators on exams; use of "cheat sheets" on exams; the number of times cooperative groups should be changed during the semester and method of choosing the groups; grading criteria, methods of insuring individual responsibility in group work; use of too much time discussing class processes; etc.

Sometimes the class arrives at a decision that I don't like. Using Rogers' third condition, "congruence," I explain why I disagree with their decision, but if the students are not swayed to my point of view, if possible I go along with them. This is an application of the first condition, "unconditional positive regard."

The matters addressed on the questionnaire generally fall into three categories:

(1) *Attitudinal.* E.g.:

- I like working in cooperative groups as we have been doing in this class.
  Strongly Agree 10 9 8 7 6 5 4 3 2 1 Strongly Disagree
- I would suggest the following improvements:

(2) *Specific questions* address problematic areas. E.g.:

- What would better foster cooperation among team members; what would help teams function more effectively?
- When a team produces a single write-up for a project, what grading scheme will insure individual accountability?
- What would discourage talking and noise in the large lecture room? How can we move from active participation in team work to quiet attention to the lecture?

(3) *Open response items* seek general input. E.g.:

- Things I like/dislike about this class:
- Suggestions for improvement:
- The topic I found most difficult to understand:
- It would have been easier for me to understand this topic if:

Findings and Use of Findings

Attitudinal questions are especially useful after I have tried something innovative, and I want to know how the class feels about it.

Often students come up with better ways to deal with difficulties that arise in the classroom than methods I have

heard in professional conferences and workshops. Thus it seems that giving students "unconditional positive regard" is warranted. Examples of things implemented as a result of student feedback and discussion are:

- I let the students know the first day that the class will be unconventional, that probably most learning will take place in cooperative groups, that I will do little lecturing and that decisions about how the class will be run will be based on class discussions. I request that students be open to new ways of learning and that they be willing to work cooperatively to help one another learn. This early notice gives students the opportunity to withdraw if they prefer a conventional class.
- A grading scheme which factored into each individual team member's grade a small percentage of the team average was discarded. The intention had been to encourage group members to help one another. Instead it stirred up resentment in the better students whose grades were pulled down.
- Instead of keeping the same groups all semester, the composition of the groups was changed after the first exam and after the second exam. We made changing optional after the third exam.
- A flexible-weight grading system was initiated as described in my other article in this volume (p. 84).
- Several graphing calculators were put on reserve in the library.
- Each team developed a list of its own guidelines for group process.
- A grading scheme was initiated for projects which fosters individual accountability and virtually eliminates complaints about shirkers.
- Bonus questions were added to exams.
- New topics are motivated by illustrating their importance or usefulness.
- Students with identical calculators sit together whenever the class is working on programming an algorithm.

Success Factors

Of course students want their input taken seriously. Therefore, presenting a summary in writing at the class following the survey and using it as a basis for discussion of possible changes is very effective. If the discussion leads to a decision to make a change, it is important to implement it immediately.

As good as this all seems, not everything is rosy. When there is considerable disagreement on a matter, some students think that the resulting discussion is "wasting valuable class time." In fact, some students would like the teacher to make all the decisions and they object to using any class time to discuss the surveys. Other students resist change and object to trying different approaches in mid-semester. When a change is implemented and doesn't work, these students are particularly vehement. And certainly not everything works the way we think it will. For example, I have tried students suggestions about how to quiet a lecture hall after the students have been working in cooperative groups but none of them have worked. However if the entire class is invested in a decision, it has the greatest chance of success.

It takes extra time to make up survey questionnaires, and even more time to summarize the results while also grading exams. (Work-study students may help with the summariza-tion.) But the extra effort pays off because it makes the course more alive and interesting for both the students and the teacher. Students and teacher alike learn from carrying out early in-course assessments.

References

[1] Rogers, C. R. "The Necessary and Sufficient Conditions of Therapeutic Personality Change." *Journal of Consulting Psychology* 21(2), 1957, pp. 95–103.

[2] Rogers, C. R. *Freedom to Learn*. Merrill Publishing, 1969.

Student Feedback Teams in a Mathematics Classroom

David Lomen
University of Arizona

A student feedback team is a subset of the class which gives the instructor feedback on how the class is doing with new material.

Background and Purpose

The University of Arizona is a public institution with about 35,000 students. Each fall its Mathematics Department offers 15 sections of second semester calculus (Calc II), with about 35 students per section. Recently we noticed that students entering with credit on the AB Advanced Placement Calculus Examination do not seem to be well served by our Calc II. These incoming students are definitely not challenged by the standard integration material at the start of the semester, so instead of fully mastering the familiar material they relax. By the time new material is introduced they are well behind their classmates and many never catch up. Another possible reason is that in Calc II we emphasize written explanations along with numerical and graphical reasoning. Most incoming students are not very proficient with this. As a result, many of these students find their first mathematics course in college frustrating and unrewarding, even though they are often the most intelligent students in the class.

During the academic year 1996–1997, we made an effort to remedy this problem. We created a special year long course that uses differential equations to motivate topics from second semester calculus which are not covered in the AB exam. Enrollment was limited to incoming students who received a 4 or 5 on the AB Advanced Placement Examination. To determine the course content, I consulted several high school teachers, the Advanced Placement syllabus and past AP examinations.

The result was a course that starts with a very quick review of functions, limits, continuity, differentiation, and antidifferentiation, often within the context of some mathematical model. The emphasis here is on numerical and graphical interpretation. We then analyze simple ordinary differential equations using techniques of differential calculus. Standard integration topics are covered as they occur in finding explicit solutions of differential equations. Taylor series are introduced as a technique which allows solutions of difficult nonlinear differential equations. Here the ratio test for convergence of series is motivated by such series solutions. The textbooks for the course are Hughes-Hallett [1] and Lomen & Lovelock [2].

The course was limited to 30 students, as that was the number of chairs (and personal computers) in our classroom. With this agenda, a new type of mathematics classroom, and the nonuniform background of the students, it was evident that I needed some type of continuous feedback from the class.

Method

My first assessment effort solicited their response to the following two statements:

"One thing I understand clearly after today's class is ___ ."

"One thing I wish I had a better understanding of after today's class is ___ ."

The students were to complete their responses during the last two minutes of the class period and give them to me as they left. (This is a particular version of the One-Minute Paper—see David Bressoud's article in this volume, p. 87.)

It had seemed to me that this was a very appropriate way to have them evaluate the material I was covering, especially during the review phase at the start of the semester.

The second method entailed my meeting with a committee of individuals once a week to assess their response to our classroom activities, their learning from the class, and their homework assignments. From the 20 volunteers I chose four who had different majors and had used different calculus books in high school. (At our first class meeting I had all students fill out a one page fact sheet about themselves which greatly aided in this selection.) To facilitate communication among the class, I formed a listserve where I promised to answer all questions before going to bed each night. I also handed out a seating chart which contained each student's telephone number, e-mail address, and calculator type. Comments, complaints, suggestions, etc., were to be directed either to me or to this committee, either in person or via e-mail. I emphasized that this was an experimental class, and I really needed their help in determining the course content and the rate at which we would proceed.

During the second semester of the course, these meetings of the four students were disbanded, and then resurrected (see below) and made open to any four students who wished to attend, with preference given to those who had not participated previously. Also, during the second semester, I had the students work in groups for some of the more challenging homework exercises. Included with their write-ups was an assessment of how the group approached the exercises, how they interacted, and what they learned.

Findings

This class was the most responsive class I have ever experienced as far as classroom interactions were concerned. However, there were so many unusual things about this class, it is impossible to attribute this solely to the use of feedback teams. For example, almost all of the students used e-mail to ask questions that arose while they were doing their homework, especially group homework. For example, in one assignment they ended up with the need to solve a transcendental equation. A reminder that there were graphical and numerical ways to solve such equations allowed the assignment to be completed on time. Students could send messages either to me or to the listserve. When the latter was used, several times other students responded to the question, usually in a correct manner.

Use of Findings

It turned out that the students required no urging to express their opinions. At the inaugural meeting first semester, the committee of four very clearly stated that they KNEW my first means of assessment was not going to be effective. They reasoned that the Calc I review was being presented and

discussed in such an unusual manner that they had no idea what they understood poorly or well until they had some time to reflect on the material and worked some of the assigned homework. After much discussion, they agreed to a modification of this procedure where these two questions were answered AFTER they completed the written homework assignment. This was done, and provided very valuable feedback, including identifying their struggle in knowing how to read a mathematics textbook. To help with this reading problem, I had the students answer a series of True/False questions (available from http://www.calculus.net/CCH/) that were specific to each section of the calculus book. I would collect these as students came to class and use their answers as a guide to the day's discussion. While not all the students enjoyed these T/F questions, many said they were a big help in their understanding the material.

One other item this committee brought to my attention was the enormous amount of time it took them to work some of the word problems. To rectify this situation, I distributed the following seven point scheme, suggested by a colleague of mine, Stephanie Singer: 1) write the problem in English, 2) construct an "English-mathematics dictionary," 3) translate into equations, 4) include any hidden information, and information from pictures, 5) do the calculations, 6) translate the answer to English, 7) check your answer with initial information and common sense. This, together with some detailed examples, provided a remedy for that concern.

As the class continued on for the spring semester, I thought that the need for this committee had disappeared. However, after three weeks, three students (not on the original committee) wondered why the committee was disbanded. They had some specific suggestions they wanted discussed concerning the operation of the class. My reaction to this was to choose various times on Friday for a meeting with students. Whichever four students were interested and available at the prescribed time would meet with me. The first meeting under this format was spent discussing how to improve the group homework assignments which I instituted Spring semester. Because several homework groups had trouble arranging meeting times, they suggested these groups consist of three, rather than four students, and that any one group would not mix students who lived on campus with those living off campus. This was easily accomplished.

A meeting with a second group was spent discussing the "group homework" assignments which were always word problems, some of them challenging or open ended. These students noted that even though these assignments were very time consuming, they learned so much they wanted them to continue. However, they requested that the assignments be more uniform. Some assignments took them ten hours, some only two hours. The assignments given following this meeting were more uniform in difficulty.

At a another meeting the student's concern shifted to the number of exercises I had been assigning for inclusion in their notebook (only checked twice a semester to assess their effort). These exercises were routine, and for every section I had been assigning all that were included in a student's solution manual. They said there was not enough time to do all of these exercises and they did not know which they could safely skip. In response, I then selected a minimal set which covered all the possible situations, and let those students who like the "drill and practice" routine do the rest.

Success Factors

For larger classes, the student feedback teams could solicit comments from students before or after class, or have them respond to specific questionnaires. Occasionally they could hold a short discussion session of the entire class without the instructor present. In an upper division class for majors, this team could work with the class to determine what prerequisites needed to be reviewed for rapid progress in that class, see [3]. Another aid for this process are computer programs (available from http://math.arizona.edu/software/uasft.html) which help in identifying any background weakness.

Other instructors have used "student feedback teams" in different ways. Some meet with a different set of students each time, others meet after each class. A more formal process usually used once during a term involves "student focus groups" (see the article by Patricia Shure in this volume, p. 164). The essence of this process is for a focus group leader (neither the instructor nor one of the students) to take half a class period and have students, in small groups, respond to a set of questions about the course and instructor. The groups then report back and the leader strives for consensus.

Other suggestions for obtaining feedback from large classes may be found in [4].

References

[1] Hughes-Hallett, D. et al. *Calculus*, John Wiley & Sons Inc., 1994.

[2] Lomen, D. and Lovelock, D. *Exploring Differential Equations via Graphics and Data*, John Wiley & Sons Inc., 1996.

[3] Schwartz, R. "Improving Course Quality with Student Management Teams," *ASEE Prism*, January 1996, pp. 19–23.

[4] Silva, E.M. and Hom, C.L. "Personalized Teaching in Large Classes," *Primus* 6, 1996, pp. 325–336.

Early Student Feedback

Patricia Shure
University of Michigan

In introductory courses in mathematics at the University of Michigan, an instructional consultant visits the class one-third of the way into the semester. This observer holds a discussion with the class, in the absence of the instructor, about how the course is going, and provides feedback to the instructor.

Background and Purpose

Each fall at the University of Michigan, most of the incoming students who take mathematics enroll in one of the Department's three large introductory courses. These three courses, Data, Functions, and Graphs, Calculus I, and Calculus II, enroll a total of 3600 students each fall. Not only must the department design and run these courses effectively, but it is clearly crucial that we find out whether the courses are succeeding. Each course is systematically monitored by all the conventional methods such as uniform examinations, classroom visits, and student evaluations, but one assessment technique has proved especially helpful. This procedure, which we call Early Student Feedback, has the advantage of simultaneously giving both students and instructors an opportunity to review the goals of the course and providing a simple mechanism for improving instruction.

The courses have been planned around a set of specific educational objectives which extend beyond the topics we list in our syllabi. We are interested in encouraging the transfer of knowledge across the boundaries between disciplines, so our course goals reflect the needs of the many other disciplines whose students we teach. For example,

Course Goals for Introductory Calculus

1) Establish constructive student attitudes about the value of mathematics by highlighting its link to the real world.
2) Persuade more students to continue in subsequent mathematics and science courses.
3) Increase faculty commitment to the course by increasing the amount of student-faculty contact during each class period and having faculty grade students' homework themselves.
4) Develop a wide base of calculus knowledge including: basic skills, understanding of concepts, geometric visualization, and the thought processes of problem-solving, predicting, and generalizing.
5) Strengthen students' general academic skills such as: critical thinking, writing, giving clear verbal explanations, understanding and using technology, and working collaboratively.
6) Improve students' ability to give reasonable descriptions and to form valid judgments based on quantitative information.

All three of our large first-year courses are run in the same multi-section framework, a framework which makes achieving these objectives possible. Classes of 28–30 are taught by a mix of senior faculty, junior faculty who come to Michigan for a 3-year period, mathematics graduate students, and a few visitors to the department. Here is the general format.

Key Features of Introductory Courses

1) Course content: The course emphasizes the underlying concepts and incorporate challenging real-world problems.
2) Textbook: The textbook emphasizes the need to understand problems numerically, graphically, and through English descriptions as well as by the traditional algebraic approach.
3) Classroom atmosphere: The classroom environment uses cooperative learning and promotes experimentation by students.

4) Team homework assignments: A significant portion of each student's grade is based on solutions to interesting problems submitted jointly with a team of three other students and graded by the instructors.

5) Technology: Graphing calculators are used throughout the introductory courses.

6) Student responsibility: Students are required to read the textbook, discuss the problems with other students, and write full essay answers to most exercises.

Early Student Feedback was originally introduced to the Mathematics Department by Beverly Black of the Center for Research on Learning and Teaching (CRLT) at the University of Michigan. The procedure we currently use evolved from the Small Group Instructional Diagnosis (SGID), described by Redmond and Clark [1].

Method

A) *Overview*: Approximately one-third of the way into the term, each class is visited by an instructional consultant. The consultant observes the class until there are about 20 minutes left in the period. Then the instructor leaves the room and the consultant takes over to run a feedback session with the class. Soon afterward, the consultant and the instructor have a follow-up meeting to discuss the results and plan teaching adjustments. The results of each Early Student Feedback procedure are confidential.

B) *Consultants*: The consultants we use are the instructors who run our professional development program. They are upper-level mathematics graduate students and faculty members. Consultants are trained according to CRLT guidelines for observing classes. These guidelines stress the importance of having instructors reflect on their plans and goals for each class session in light of the overall course goals.

C) *Observing the class*: During the observation period, the consultant objectively records everything as it happens; the classroom setup, what the instructor says and does, student interactions with the instructor and with each other, etc.

D) *Running the feedback session*: First, the consultant explains to the class that the Mathematics Department routinely conducts these sessions at this point in the term and briefly explains the procedure. Then the class is divided into groups (of four or five students each) and the groups are given 7 minutes to discuss, reach a consensus, and record their responses on the Early Feedback Form:

EARLY FEEDBACK FORM

In your small group, please discuss the following categories and come to a consensus on what should be recorded for the instructor. Using detailed examples and specific suggestions will make your comments more useful to the instructor. Please have a recorder write down the comments that you all agree on.

List the major strengths in this course. What is helping you learn in the course? (Space is provided for five responses)

List changes that could be made in the course to assist your learning. (Space is provided for five responses.)

While the students are talking, the consultant writes two headings on the board:

Strengths   Changes

After getting a volunteer to copy what will be written on the board, the consultant calls on the groups in turn to read aloud one "strength." One by one, the consultant records the comments on the board in essentially the students' own words while continually checking for clarity and consensus. When all the "strengths" have been read, the consultant uses the same procedure to generate a list of "changes."

E) *Debriefing the instructor*: Before the meeting to debrief the instructor, the consultant makes a two page list of the comments from the board (strengths on one page and changes on the other). The beginning of this meeting is an opportunity for the consultant to get instructors to talk about how the class is going, what goals they had for the class, and what they see as their own possible strengths and weaknesses. The consultant then gives the instructor the list of students' comments (strengths first) and goes over it one point at a time. The consultant's own observation of the beginning of the class period provides valuable detail and backup for the students' interpretation of their experience. The final step is to plan with the instructor how to respond to the students and what adjustments should be made. For instance, if students suggest that the instructor spend more time going over homework problems, the answer may well be "OK." But, if they want to do less writing ("after all, this is a mathematics class, not an English class"), the instructor will need to spend more time explaining the connection between writing clear explanations and understanding the ideas.

Findings

During the feedback process, we find that everyone involved in the course, from the individual student, to the instructor, to the course director benefits from reflecting on the goals of the course. For example, students often complain (during the feedback session) that their instructor is not teaching them; that they have to teach themselves. They think that a "good" teacher should lead them through each problem step by step. In response, the consultant gets a chance to talk with the instructor about students' perception of mathematics as simply performing procedures, whereas what we want them to learn is how to become problem solvers. Commonly, both the students and their instructors respond well to the Early Student Feedback process. The students are pleased that the Mathematics Department cares what they think. We find that they will discuss the course freely in this setting. Similarly, there has been a uniformly positive response from instructors at all levels, from first-time graduate students to senior faculty. All instructors feel more confident about their teaching after hearing such comments as "he's always helpful and available" or "she knows when we're having trouble." In fact, since these feedback sessions have proved so useful in the introductory courses, the Department has begun to use them occasionally in more advanced courses.

Use of Findings

Each Early Student Feedback session is confidential and the results are never recorded on an individual instructor's record. However, the consultants talk over the results in general looking for patterns of responses. We ask questions about how the courses are going. Do students see the value of the homework teams? Is there lots of interaction in the classrooms? Do the students seem to be involved with the material? Are the instructors spending too much time on their teaching? The answers give a clear profile of each course. The timing of the procedure allows us to reinforce the course's educational objectives while the course is still in progress. Furthermore, it gives us the opportunity to improve the instruction in each individual classroom and make any necessary adjustments to the courses themselves.

Success Factors

Early Student Feedback has been used successfully in many courses throughout the University at the request of individual instructors. But in the Mathematics Department, we require it every term in every section. Making the process mandatory had the effect of making it seem like a routine (and hence less threatening) part of each term. Since almost everyone has had a consultation, there are always many experienced instructors who spontaneously tell newcomers that the sessions are very beneficial. It is easy to calculate the "cost" of using Early Student Feedback as an assessment technique.

Students – 20 minutes of class time
Instructors – 1/2 to 1 hour for debriefing
Consultants – (1 hour class visit) + (1 hour preparation time) + (1/2 to 1 hour debriefing)

The method is certainly cheap in comparison to the amount of information it generates about teaching and learning.

Reference

[1] Redmond, M.V. and Clark, D.J. "A practical approach to improving teaching," *AAHE Bulletin* 1(9–10), 1982.

Friendly Course Evaluations

Janet Heine Barnett
University of Southern Colorado

By having students write a letter to a friend about your course, you can get useful information both on what the students have learned and on what they thought of the course.

Dear Professor Math,

Thank you for your interest in my use of topic letters in lower division mathematics courses. I am sending you more information on these assignments in letter format so that you will have some sense of their style. Let me begin by outlining for you their **Background and Purpose**.

For a number of years, I have been using student writing both to develop and evaluate conceptual understanding in my lower division courses. The topic letters are one such assignment. In these letters, students are asked to write about mathematics to someone they know; for instance, I might ask them to describe the relation between various concepts (e.g., what is the relation between exponential and logarithmic functions?), the relation of a concept to previously studied concepts (e.g., how does knowledge of the logarithmic function extend our calculus repertoire?), or the relation of some concept to its motivating problem (e.g., how does statistics address the problem of prediction?). Because these types of questions are conceptual in nature, I have used topic letters primarily in my calculus and liberal arts courses, although there are ideas in more skills-related courses like college algebra for which a letter could also be used.

I am fortunate that class sizes at USC average 25–45, so that evaluating the letters is manageable. Initially, my objective with the topic letters was solely to evaluate student conceptual understanding. I have found, however, that they also provide an evaluation of how well I've done at designing course experiences to develop conceptual insight and understanding. Unlike traditional anonymous student questionnaires which focus almost exclusively on teaching mechanics like "punctuality," the topic letters tell me something about the *quality* of the learning environment I tried to create, my success in creating it, and its success in building conceptual understanding. An end-of-semester "Letter to a Friend" addressing the course as a whole was therefore a natural extension of the topic letter assignment.

My **Method** is illustrated in these partial instructions for an End-of-Semester letter in first semester calculus:

You have a friend who foolishly decided to attend another university in the east. (Like Harvard has anything on USC.) This friend has heard that calculus is a requirement for many disciplines, and is considering a course in calculus next semester to keep all major options open. Knowing that you are just completing this class, your friend has written to you asking for your insights.

Having never heard much about what calculus really is, this person would like to know what kinds of problems calculus can solve, what its most important ideas are, and what, if anything, is interesting or exciting about the subject.

Write a letter to your friend giving your answers and insights into these questions......

The audience for the assignment is set very deliberately. I emphasize to students that they must write to someone who has never studied the mathematics in question, but who has the same working knowledge that they had at the start of the topic, or the start of the course. Without such an audience, students are less inclined to describe their personal understanding of the mathematics, and more inclined to try to impress me with the use of formal terms that they may or may not understand...and I will probably let them.

I do assign a grade to the letters, based primarily on the insight and mathematical correctness displayed. I have found that it is important to stress that quality of insight (and not quantity of facts) is the key feature of a strong letter. Students are instructed to write no more or less than is needed to convey the ideas clearly. To allow them as much freedom as possible in how they do this, I do not set page limits or other formatting requirements. (Most letters fall into the 3–5 typed page range.)

Students who submit weak topic letters are encouraged to revise them, and I provide each student with specific written suggestions on areas that could be improved. Revisions of the end-of-semester letters are not permitted since they are collected during the last week of class. Since my main purpose with the end-of-semester letter is to get a more meaningful course evaluation, I simply give all students full credit as long as they seem to have made a serious effort with it.

Findings of note include the fact that the grading itself can take some time.....10–15 minutes per letter since I read each one twice (first for comments and a preliminary grade, then again to ensure consistency and to assign the actual grade). Most of this time goes to writing comments; in larger classes, this lead me to require fewer letters than I might otherwise, with four being the maximum I would consider for any class.

The time is worth it. First, the students receive an excellent opportunity to synthesize the material. Most students find this useful (although a few question the place of writing in a mathematics class), and quite a few students go out of their way to be creative. I've gotten interesting letters to a variety of real and imagined friends, family members, pets, plus the occasional politician and movie star, which makes the letters fun for me to read. I also enjoy the letters that take a more straight-forward approach to the assignment, since even those students use their own voices in the composition, something which rarely happens in other assignments.

The value of the topic letters for course evaluation became especially clear to me when I first began using topic letters in my liberal arts mathematics course about two years ago. I had been using them for some time before that in other courses, and knew that the assignment sheet needed to warn them (loudly) that working examples was *not* the point, and encourage them (strongly) to convey the "big picture" in a way meaningful to their "friend." Still, most students included numerous examples in their letters, and not much else. After reflecting on this, I realized that much of our class time focused on just that: working examples, and not much else. In fact, several of the letters came straight from

those class examples. Naturally, we looked at conceptual explanations in class (as did the book), but these were not included in the letters. Was this just because we spent more time on the examples? And why was I spending so much more time on examples? With further reflection, I realized that virtually all my homework, quiz and test questions were focusing on mechanics; the letters, I thought, would address the conceptual side of the course. But with such a heavy emphasis on mechanics in the course grade, I had (somewhat unwittingly) emphasized examples in class in order to prepare students to earn those points, and students responded to this in their letters.

My own **Use of Findings** came through various modifications to the course. After some initial experimentation, I replaced in-class quizzes based on mechanics with take-home quizzes that require students to discuss and apply concepts.. I've also developed a set of "course questions" which I use on the final exam, after giving them (verbatim) to the students at the beginning of the semester and returning to them during the semester as we cover the relevant material. Along with these changes in the course assessment structure, I modified instruction through the use of more discussions centering on the course questions and concepts, and less emphasis on examples of mechanics in lectures. These modifications are now having the desired effect; the topic letters are much better, as are students' responses to the conceptual questions on various tests and quizzes, and their mastery of mechanics has not suffered.

The information I gain from the end-of-semester letters also helps me to fine-tune course instruction and the overall syllabus, as well as instruction in specific topic areas. To get course evaluation information from these letters, I watch for comments that indicate the value of the various activities we did during the semester (such as advice to their friend concerning important things to do), an indication of the emphasis placed on concepts versus skills (such as a list of rules with no concepts mentioned), or an indication of a concept that may have been under (or over) emphasized (such as no mention of the Fundamental Theorem in a calculus class).

Some **Success Factors** to consider have already been mentioned; the time concern is one which I would reiterate. With the topic letters, there is also the difficulty of sorting course evaluation information from student assessment information; for example, when I get a letter emphasizing examples and algorithms instead of concepts (as I still do), it may not be due to the class experiences provided. The student may not have understood my expectations or the concepts, or may simply have taken the easy way out. This means that I must look at the collection of letters, as well as the individual letters, to get a "course" reading. All these factors make it essential to set the objectives of the letters carefully, so that both you and the students gain from the time investment.

Another concern I have is the fact that the letters are not anonymous, which may limit their validity as a course evaluation tool. On the other hand, some of the things students put in their letters suggest they are not overly-concerned with anonymity. In some of these, I suspect the student is going for shock value (with tales of drinking parties and other such escapades). As far as possible, I sort out the mathematical and course information in these, without reacting to their provocative side. A more distressing type of "no-holds barred" letter I have received comes from students who did not do well in the course, or whose confidence was shaken along the way, and who tell their friend very clearly how this has affected them. Although I've never read any anger in these, they are very difficult to read and comment on. Despite that (or perhaps because of that), this type of letter has given me good information about the course, and what kinds of experiences are discouraging to students.

I hope you too will find the letters valuable, and that you'll share your experiences with me. Best wishes...J.B.

The Course Portfolio in Mathematics: Capturing the Scholarship in Teaching

Steven R. Dunbar

University of Nebraska-Lincoln

The teaching portfolio is an alternative to the standard course questionnaire for summing up a course. The instructor collects data throughout the semester into a course portfolio, which can then be used by that instructor or passed on to others teaching the course.

Background and Purpose

The course portfolio documents course and classroom activity in a format open to reflection, peer review, standards evaluation, and discussion. The course portfolio can demonstrate the intellectual development in the students of the concepts in the course and documents that a transformation of students occurred because of the intellectual efforts of the instructor.

Bill Cerbin, at the University of Wisconsin - La Crosse was one of the first to form the idea of the course portfolio. The American Association for Higher Education (AAHE) Teaching Initiative Project "From Idea to Prototype: The Peer Review of Teaching" promoted the portfolio in a menu of strategies [2,3].

According to a report from the Joint Policy Board on Mathematics, a major obstacle to including evaluation of teaching in the reward system is the absence of evaluation methods. The course portfolio is one response to the need for better tools to document and evaluate teaching. It puts faculty in charge of monitoring, documenting and improving the quality of teaching and learning. It provides a vehicle for the faculty member to reflect on what's working and what's not, and for communicating that reflective wisdom and practice to colleagues.

Peer feedback on the course portfolio provides professional accountability, and improves the process of teaching. Both the presenter and the reader learn new pedagogical methods to improve their teaching.

I have created a course portfolio for a course on Principles of Operations Research. Our department teaches this course yearly in one or two sections to both senior mathematics majors and as a service course for senior students in actuarial science. From a prerequisite of linear algebra and probability, the course covers linear optimization, queuing theory, and decision analysis.

Method

The word "portfolio" conjures up a vision of a collection of papers, or a binder of whatever was available, perhaps assembled in haste. But this is not what I mean by the "course portfolio." The course portfolio starts with the course syllabus, with explicit mathematical goals for students in the course. The instructor measures progress toward the goals through a pre-course student background knowledge probe; informal course assessment by students; homework exercises, solutions, and scores; in-class active-learning exercises; tests; labs or projects; formal student course evaluations; and a post-course evaluation by the instructor.

A "course portfolio" is not a "teaching portfolio." A teaching portfolio is a larger, longer document that establishes over a period of time and a range of courses that you are an effective teacher.

Here are some questions to address in a portfolio:

- What were you trying to accomplish?

- Why were these goals selected?
- Did the course meet the goals? How do you know?
- What problems were encountered in meeting the goals?
- How did you conduct the course?
- How did you challenge students?
- What changes in topics, materials and assignments are necessary?

For the course portfolio on Principles of Operations Research, I began with 5 prerequisite skills for the course, including the ability to solve systems of equations by row reduction, and computing conditional probabilities. I listed 5 specific goals of the course, including formulating and solving optimization models, and formulating and analyzing queuing models. All prerequisite skills and goals of the course appeared on the course information sheet distributed to students on the first day of class. I selected mathematical modeling as a major focus of the course, since I believed that most senior mathematics majors were already adept in solving formulated problems, but needed practice in turning operations research situations into a mathematical model. The course information sheet became the first item in the course portfolio.

For the first day of class, I created a "background knowledge probe" testing student knowledge of both the prerequisite skills and the goals of the course. Problems on the background knowledge probe were of the sort that would typically be on the final exam for the course or its prerequisites. A copy of the background knowledge probe together with an item analysis of student scores went into the course portfolio. Comparing this with the scores on the final exam gives a measurement of student learning and progress.

I included homework assignments, copies of exams, and course projects in the course portfolio as evidence of the direction of the course toward the goals. Occasional minute papers and in-class assessments measured student progress through the term. I collated and summarized information from the minute papers and included this in the course portfolio as the "student voice." I also did an item analysis of the scores from tests as a further refinement on measuring student gains in knowledge. Angelo and Cross [1] have samples and ideas for constructing the background knowledge probe, lecture assessments, and other assessment instruments that become evidence in the course portfolio.

The final element of the course portfolio was a "course reflection memo." In this memo created at the end of the course, I reflected on the appropriateness of the goals, the student progress on the goals, and recommendations for the course. I included the course reflection as the second element in the course portfolio, directly after the course information sheet. There is no recipe, algorithm, or checklist for preparing a course portfolio. In practice, the course portfolio will be as varied as the faculty preparing it, or the courses taught.

Preparing a course portfolio requires time that is always in short supply. Nevertheless, a course portfolio is not such a daunting task since we already keep course notes from the last time we taught a course. My experience is that the course portfolio requires about an average of one hour per week through the term of the course. The time is not evenly distributed, however, since careful preparation of the goals and organization of the course takes place at the beginning of the term. Assessing student learning as a whole is a large part of the portfolio that occurs best at the end of the term. Thoughtful reflection on the course and its conduct and results naturally occurs at the end of the term. Fortunately these are the parts of the term when more time is available. During the term, collection of course materials and evidence of student learning is easy and automatic.

Findings

In the course on Operations Research that I documented with a course portfolio I found that students generally came to the course with good prerequisite skills. This factual evidence was encouraging counterpoint to "hallway talk" about student learning. I was able to document that students improved their homework solution writing through the course of the semester. The classroom assessments showed that the students generally understood the important points in my lectures, with some exceptions. The assessments also showed that the students generally were not reading the material before coming to class, so I will need to figure out some way to enforce text reading in the future. Analysis of the scores on the background knowledge probe, the exams, and the final exams documented that students learned the modeling and analysis of linear programming problems acceptably well, but had more slightly more trouble with queuing models and dynamic programming. The course portfolio was generally able to document that student learning had taken place in the course.

Use of Findings

One immediate benefit of the portfolio is a course record. A course portfolio would be especially useful to new faculty, or to faculty teaching a course for the first time. The portfolio would give that faculty member a sense of coverage, the level of sophistication in presenting the material, and standards of achievement. The course portfolio also gives a platform for successive development of a course by several faculty. This year, two other instructors taught the course on Operations Research. At the beginning of the term, I turned over my portfolio to them as a guide for constructing their course. The course portfolio informed their thinking about the structure and pace of the course, and helped them make adjustments in the syllabus.

Having a course portfolio presents the possibility of using the material for external evaluation. Peers are essential for feedback in teaching because they are the most familiar with the content on a substantive level.

Success Factors

The AAHE Peer Review Project [2] has some advice about course portfolios:

- Seek agreement at the outset about the purposes of portfolios, who will use the information, who owns it, what's at stake.

- Be selective. The power of the portfolio comes from sampling performance, not amassing every possible scrap of information.

- Think of the portfolio as an argument, a case, a thesis with relevant evidence and examples cited.

- Organize the portfolio around goals, your own goals or departmental goals.

- Include a variety of kinds of evidence from a variety of sources.

- Provide reflective commentary on evidence. Reflective commentary is useful in revealing the pedagogical thinking behind various kinds of evidence and helping readers to know what to look for.

- Experiment with various formats and structures, be deliberate about whether portfolios are working and how you can refine the structure.

It may be useful to include the *actual course* syllabus, to show what the instructor actually covered in the course in comparison with what the instructor intended to cover.

A variant is the "focused portfolio," which concentrates on one specific aspect or goal of a mathematics course and the progress made toward the goal. For example, a course portfolio might focus on the integration of graphing calculators or on the transition of students from "problem-solving" to "proof-creating." In any case, there should be a cover letter or preface to the reader, briefly explaining the purpose and methods of the portfolio.

An important addition to the portfolio is the "student voice," so the portfolio clearly represents not only what the instructor saw in the course, but also the students got out of it. Informal class assessments, formal student evaluations and samples of homework and student projects are ways of including the student voice.

References

[1] Angelo, T.A., and Cross, K.P. *Classroom Assessment Techniques*, second edition. Jossey-Bass, San Francisco, 1993.

[2] Hutchings, P. *From Idea to Prototype: The Peer Review of Teaching*. American Association of Higher Education, 1995.

[3] Hutchings, P. *Making Teaching Community Property: A Menu for Peer collaboration and Peer Review*. American Association for Higher Education, 1996.

Part III
Departmental Assessment Initiatives

Introduction

Sandra Z. Keith

St. Cloud State University

Mathematics departments are increasingly being asked, both internally by other departments and externally by accreditors, to reflect on their role on campus. At one time, a mathematics department was acting within the expectations of a college by providing research and teaching in a way that weeded out the poorer performers and challenged and pushed forward the elite. With the diversity of today's student clientele, changing expectations on the nature of education, the jobs crisis, and the downsizing of mathematics departments, we are having to examine the broader issue of how we can provide a positive, well-rounded learning environment for all students.

As more and more disciplines require mathematics, outside expectations of mathematics departments have risen; other departments are now competing for ownership of some of our courses. The issue of effective teaching in mathematics, for example, is a frequently discussed problem for students and faculty on campus. So is the content we are teaching, say, in the calculus sequence. And we shake our heads at what to do about general education courses in physics and chemistry in which students can't use mathematics because they can't do the simple, prerequisite algebra. However, assessment does not function at its best when aspects are pulled apart and examined under a microscope. The evaluation of mathematics teaching or the curriculum or the quantitative capabilities of our students cannot be separated from the evaluation of broader issues for which the department as a whole is responsible. Some of these issues involve finding ways to properly place and advise students, networking with client disciplines, examining new curricula and methods of teaching, and demanding and maintaining basic quantitative literacy requirements that will assure our students graduate even minimally competent.

All of these major departmental initiatives must operate within an assessment framework. Until recently, however, mathematics departments have not generally been at ease with the idea of going public. Even if we move beyond the idea that the word "assessment" is mere jargon, perhaps we feel we are being expected to compromise our high standards, which are naturally opaque to others, or that our freedom to teach in ways that intuitively feel best is being stripped from us. Many of us, overworked as we are, may be content to leave assessment issues to administrators. However it has become clear through a variety of publicized situations that if we neglect to deal with new expectations, then decisions about us will be made by others, with the result that we will be marginalized.

Some mathematics departments do not change; others may adopt changes (such as a change in a text or curriculum, a change in the use of technology, or a change in a policy of placement or admittance of transfer students) without ever paying explicit attention to the results of these changes. This unexamined approach to teaching will probably not be acceptable in the future.

In this section we have searched for efforts by departments which have created changes that feature assessment as an integral component. The method of assessment-as-process helps create an educational environment that is open and flexible in ways which benefit faculty, students and the institution as a whole. For example, everyone benefits when placement is running efficiently. We are not as concerned that the final results of the projects should be awesomely successful (although some are) as we are by the processes by which departments have examined and made public the results of their efforts. Good assessment is not a finished product but the input for more changes. In fact, most of the reform efforts of departments presented here are not at a stage of completion. Rather, the purpose of this section is to acknowledge that on campuses now there is some ferment on the issue of assessment, and that questions are being raised and brooded about for which we hope to have provided some plausible starting solutions.

Our contributions are highly diverse; we will see the efforts of small and large schools, research institutions and small inner city schools. Retention of majors is the issue in some cases, while elsewhere the focus is on developmental mathematics or general education mathematics. Some programs are personally focused on listening to students; others emphasize dealing collaboratively with faculty from other departments, and still others (not necessarily exclusively) have conducted massive analytical studies of student performance over the years for purposes of placement or for charting the success of a new course. We will learn about the efforts of one school with a bias against assessment to lift itself up by the bootstraps, and about some other schools' large-scale, pro-active assessment programs that operate with an assessment director/guide and encouraging, assessment-conscious administrators. We will also hear from some experts on quantitative literacy across campus, learning theory, and calculus reform so we may view concrete assessment practices against this theoretical backdrop.

In the diversity of methods exhibited here, it is interesting to observe how the rhetorical dimensions of assessment function; how the use of language in the interpretation of results has the power to influence opinion and persuade. Rhetorical interpretations of assessment procedures may indeed be intended to sell a program, but if this strikes us as an unnatural use of assessment, we would be equally unaccepting of the language of politics, law, business, and journalism. We would never recommend blanket adoption of these programs — this would not even be feasible. Rather we hope that the spectrum of efforts presented here, achieved through the obvious commitment and energy of the mathematics departments represented, will suggest ideas about what is possible and will serve to inspire the examined way.

Here is an overview of this section.

Placement and Advising

Judith Cederberg discusses a small college, nurturing a large calculus clientele in a flexible, varied calculus program, which recognizes the need for a careful placement procedure and studies the effectiveness of this effort with statistics that help predict success. Placement is also the issue at a large comprehensive school, in an essay by Donna Krawczyk and Elias Toubassi explaining, among other initiatives the department took to improve its accessibility, their Mathematics Readiness Testing program. Statistics about the reliability of the testing are studied, and the authors observe that their results have been found useful by other departments on their campus. At many institutions, advising operates at a minimal level — it may be the freshmen, or possibly even the seniors who are forgotten. Steve Doblin and Wallace Pye, from a large budding university, have concentrated on

the importance of advising students at all levels; from providing mentors for freshmen to surveying graduating students and alumni, this department operates with continuing feedback. Materials which carefully describe the program are distributed to students.

Quantitative Literacy

When our students graduate from four or more years of college, are they prepared for the quantitative understanding the world will require of them? This issue, which could be one of the most important facing mathematics departments today, has suffered some neglect from the mathematics community, in part because of the varied ways in which liberal arts courses are perceived and the general confusion about how to establish uniform standards and make them stick. Guidelines from the MAA for quantitative literacy have been spelled out in a document authored by Linda Sons who offers an assessment of how this document was actually used in a quantitative literacy program at her large university. For a variety of liberal arts mathematics courses and calculus, this school tested students and graded the results by a uniform standard, then compared results with grades in the courses. Results led to making curricular changes in some of the courses. Some of us may have tried student-sensitive teaching approaches in higher level courses, but when it comes to teaching basic skills, there seems to be a general taboo about deviating from a fixed syllabus. Cheryl Lubinski and Albert Otto ask, what is the point of teaching students something if they're not learning? They describe a general education course that operates in a learning-theoretic mold. For them, the content is less important than the process in which the students construct their learning. Elsewhere, Mark Michael describes a pre- and post-testing system in a liberal arts mathematics course, that among other things, raises some interesting questions about testing in general, such as how students may be reluctant to take such testing seriously, and why some students may actually appear to be going backwards in their learning. With an even broader perspective, how does one assure that all students graduate quantitatively literate at a large collective bargaining school with little coordination in its general education program and a tradition of diversity and disagreement among faculty? Is the mathematics requirement continually being watered down in other general education courses, in effect bowing to students' inability? What do students perceive and how can one interpret the broad findings? A non-mathematician, Philip Keith, discusses the ramifications of accommodating various departments' views of quantitative literacy and reports on using a simple survey to get an institution on track. How realistically can such broad data be interpreted? The value of this survey lay in its effect to prod the faculty to begin thinking and talking about the subject.

Developmental Mathematics

Developmental (remedial) mathematics is the subject of Eileen Poiani's report from an inner city school with a diverse, multi-cultural clientele. Deeply committed to raising students' ability in mathematics, this school has a richly expansive supportive program. They ask: does developmental mathematics help or is it a hopeless cause? On the receiving end of grants, an assessment plan was in order, and three major studies were completed over the course of ten years, with observations about retention and graduation data and performance in developmental and subsequent mathematics courses. As this study continues, it looks more in depth at student characteristics.

Mathematics in a Service Role

How is a mathematics department supposed to please its client disciplines? From a comprehensive university, Curt Chipman discusses a variety of ways in which an enterprising mathematics department created a web of responsible persons. Goals to students and instructors are up-front and across the board, course leaders are created, and committees and faculty liaisons are used to the fullest for discussions on placement and course content. In another paper, William Martin and Steven Bauman discuss how a university which services thousands of students begins by asking teachers in other disciplines — not for a "wish list" — but for a practical set of specific quantitative knowledge that will be essential in science courses. As determined by the mathematics department and client disciplines, pre-tests are designed, the questions on the tests asking about students' conceptual understanding relating to the determined skills. Results are used for discussions with students, the mathematics faculty, and faculty in client disciplines, creating a clear view for everyone of what needs to be learned and what has been missed. From a small university in Florida, Marilyn Repsher and J. Rody Borg, professors of mathematics and economics respectively, discuss their experiences in team teaching a course in business mathematics in which mathematical topics are encountered in the context of economic reasoning. Opening up a course to both mathematics and business faculty stimulates and makes public dialogue about the issues of a course that straddles two departments. It is hard not to notice that seniors graduating in mathematics and our client disciplines often take longer to graduate, and minority students, even longer. Martin Bonsangue explores this issue in his "Senior Bulge" study, commissioned by the Chancellor's Office of the California State University. The results directly point to the problems of transfer students, and Bonsangue suggest key areas for reform. At West Point, when changes were felt to be in order, Richard West conducted an in-depth assessment study of the mathematics program. With careful attention to the needs of client disciplines (creating a program more helpful to students), the department created a

brand new, successful curriculum—cohorts are studied and analyzed using a model by Fullan. Robert Olin and Lin Scruggs describe how a large technical institute has taken a "pro-active" stance on assessment: hiring Scruggs as an assessment coordinator, this institute has been able to create an assessment program that looks at the full spectrum of responsibilities of the department. Among other things, it includes statistical studies to improve curriculum, teaching, and relations with the rest of the university. These authors point out the benefits of using assessment to respond carefully to questions that are being asked and that will be asked more and more in the future. An unexpected benefit of this program proved to be increased student morale.

New Curriculum Approaches

In a discussion piece, Ed Dubinsky explains an approach to teaching based on learning theory. He asks how one can really measure learning in a calculus course. This question stimulates an approach in which continual, on-going assessment feeds back into the learning environment. The efforts of Dubinsky and RUMEC have often focused on the assessment of learning; one can observe in this volume how the methods he suggests here have influenced many of our authors.

Reform efforts in teaching (particularly in calculus) have been the subject of much discussion, and we could not begin to accommodate all the research currently conducted as to the success of such programs. Here we touch on a few that represent different schools and different approaches. Sue Ganter offers a preliminary report of a study at the National Science Foundation discussing what NSF has looked for and what it has found, with directions for future study. While it's true that the emphasis has shifted from teaching to student learning, she reports, we are not simply looking at what students learn, but how they learn, and what sort of environment promotes learning. At some departments, even where change is desirable, inertia takes over. Darien Lauten, Karen Graham, and Joan Ferrini-Mundy have created a practical survey that might be useful for departments to initiate discussion about goals and expectations about what a calculus course should achieve. Nancy Baxter Hastings discusses a program of assessment that stresses breadth, from pre- and post-testing, to journals, comparative test questions, interviews of students, end of semester attitudinal questionnaires, and more. Good assessment is an interaction between formative and summative. Larger programs demand more quantified responses. In their separate papers, Joel Silverberg and Keith Schwingendorf (both using C4L) and Jack Bookman (Project Calc) describe their efforts with calculus programs that have looked seriously at the data over time. These longitudinal studies are massive and broad, and not necessarily replicable. However, they indicate what is important to study, and suggest the ways in which gathering data can refine the questions we ask.

Administering a Placement Test: St. Olaf College[1]

Judith N. Cederberg

St. Olaf College

A small college nurturing a large calculus clientele in a flexible calculus program recognizes the need for careful placement, and studies the effectiveness of its efforts with statistics to derive a formula for placement.

Background and Purpose

St. Olaf is a selective, church related, liberal arts college located in Northfield, Minnesota. The majority of the 2900 students are from the North Central states and nearly all of them come to college immediately after high school. Over the past 5 years, the median PSAT mathematics score of incoming St. Olaf students has ranged from 58 to 62 and approximately 60% were in the top 20% of their high school graduating class. In recent years, 8–10% of the St. Olaf students have graduated with a major in mathematics.

The St. Olaf Mathematics Department has 24 faculty members filling 18.9 F.T.E. positions. In addition to the major, the department also offers concentrations (minors) in statistics and computer science (there are no majors in these areas). The entry level mathematics courses include a two semester calculus with algebra sequence (covering one semester of calculus), calculus (honors and nonhonors sections using the same text), gateways to mathematics (a non-calculus topics course designed for calculus ready students), finite mathematics, elementary statistics, and principles of mathematics (intended for non-science majors). Each of these courses (with the exception of the first semester of the calculus with algebra course) can be used to fulfill the one-course mathematics requirement. This requirement, one of a new set of general education requirements instituted after approval of the college faculty, took effect with the incoming students in 1994. Prior to this time there was no formal mathematics course required, but most St. Olaf students took at least one of these courses to fulfill a two course science/mathematics requirement. In the last three years, the average total of fall semester calculus enrollments of first-year students has been 498. In addition, approximately 10 first-year students enrolled in sophomore level courses each fall.

Fall Semester Calculus Enrollments

Calculus with Algebra	28
Calculus I (1st semester)	266
Math Analysis I (Honors Calculus I)	102
Calculus II (2nd semester)	9
Math Analysis II (Honors Calculus II)	93

The St. Olaf mathematics placement program began in the late 1960's in response to the observation that students needed guidance in order to place themselves properly. One of the major realizations in the development of the program was the inadequacy of a single test for accurate placement. So in addition to creating a test (later replaced by MAA tests[2]), regression equations were developed to use

---

[1] An earlier version of this article was published in the Fall 1994 issue of the Placement Test Newsletter, a publication of the MAA Committee on Testing.

[2] For detailed information on the MAA Placement Test program, contact the MAA headquarters in Washington, DC (phone: 800-741-9415).

admissions information and answers to subjective questions along with placement test scores to predict gradepoints as the one quantifiable measure of successful placement. Over the years, the one placement test evolved into three separate tests, the regression equations have been refined, and more sophisticated computer technology has been employed.

Method

All new students are required to take one of the three mathematics placement exams, with exceptions made for non-degree international students and for students who have received a score of 4 or 5 on the College Board Calculus BC exam. The placement examinations are administered early during Week One—a week of orientation, department information sessions, placement testing and registration (there is no early registration for new students) which immediately precedes the beginning of fall semester. All three exams contain an initial list of subjective questions asking students for information about (1) their motivation for taking mathematics; (2) the number of terms of mathematics they plan to take; (3) the area in which they expect to major; (4) what their last mathematics course was, and their grade in this course; and (5) how extensively they have used calculators. Questions on the regular and advanced exams also ask for (6) how much trigonometry and calculus they have had; and (7) the mathematics course in which they think they should enroll. This latter question provides helpful information, but also indicates the need for providing guidance to the student since many would not place themselves into the proper course.

The exams all are timed exams with a ninety minute limit. Students taking the Advanced and Regular Exams are allowed to use calculators without a QWERTY keyboard. Currently, students are not allowed to use calculators on the Basic Exam. The three exams and their audience are listed below.

1. *Advanced Exam*: Designed for students who have had at least one semester of calculus and want to be considered for placement beyond first semester calculus at St. Olaf. This test consists of a locally written exam covering topics in first semester calculus (25 questions) together with a modified version of an MAA trigonometry and elementary functions exam (15 questions each). Approximately 220 students take this exam.

2. *Regular Exam*: Designed for students with standard mathematics backgrounds who intend to take calculus sometime during their college career. This test consists of a trigonometry and functions section (identical to that on the Advanced Exam) together with an MAA calculator based algebra exam (32 questions). Approximately 425 students take this exam.

3. *Basic Exam*: Designed for students with weaker mathematics backgrounds who have no plans to take calculus and who are hesitant about taking any mathematics. This test consists of an MAA exam over arithmetic and basic skills (32 questions) and an MAA algebra exam (32 questions). Approximately 120 students take this exam.

Following the exams, the questionnaire and test data are scanned into the computer, which grades the tests, then merges the test scores with the admissions data and other relevant information to predict a grade. Finally, the computer assigns a recommendation to each student using a cutoff program that is refined from year to year. Borderline and special cases (e.g., low class rank from a selective high school) are considered separately. Individual student recommendations are printed on labels which are then pasted on letters and distributed to student mail boxes the next morning. In addition, each academic advisor is sent an electronic message containing the results for their advisees.

The current St. Olaf mathematics placement program is administered by a member of the mathematics faculty who is given a half course teaching credit (out of a six-course standard load) for serving as director. As the present Director of Mathematics Placement, I send each new student a letter early in the summer describing the placement process and handle questions and concerns that arise. I also manage the details of giving the exams and reporting the results. Following the exams, I counsel students who have questions about their placement recommendation; notify instructors of students in their classes who did not take an examination or who did not follow the placement recommendation; and assist students with changes among calculus sections.

Findings

The placement recommendations are computed using a large number of regression equations. Dr. Richard Kleber, a mathematics faculty member and our senior statistician, has spent numerous years developing and refining these equations. Each summer he has run a series of regression studies in order to "fine tune" the regression equations. There are regression equations for each of 24 cases, depending on the information available for a particular student. These equations are used in five different sets as follows:

- One set of equations is used to predict calculus grades for all students based only on information available from the admissions office in case the student never takes the exam.
- Two sets are used to distinguish between students who need to take Calculus with Algebra (or who need to do preliminary work before enrolling in this course) and those who are ready for Calculus I. One of these sets is used for students who took the Basic Exam and the other set for those who took the Regular Exam.
- Two more sets are used to recommend honors calculus versus regular calculus. One of these sets is used for

students who took the Regular Exam and the other set for those who took the Advanced Exam. Students who take the Advanced Exam and who achieve a score of 12 or higher on the calculus portion are placed into regular or honors Calculus II. Those with especially good scores on the calculus portion who indicate that they have had at least a year of high school calculus are asked to talk to the Director of Placement about the possibility of placing beyond first-year calculus.

Use of Findings

As an example of these regression equations, the equation below predicts a grade (gr) on a four point scale using a normalized high school rank (based on a 20 to 80 scale), PSAT and ACT math scores, a self-reported math grade from the ACT exam and the algebra, trigonometry and function scores from the placement exam.

$$gr = -2.662 + 0.0385nrank + 0.00792psatm$$
$$+ 0.0470actm + 0.218mgrade + 0.0237ascore$$
$$+ 0.0277tscore + 0.0244fscore$$

This particular equation, modified by information from the subjective questionnaire, is one of those used to make decisions about honors versus regular calculus placement for students who take the Regular Placement Exam. Typically, students with gr scores of 3.5 or above are given a recommendation for honors Calculus I; those with scores between 3.3 and 3.5 are told they may make their own choice between the honors and regular versions of Calculus I; and those with scores below 3.3 are given a recommendation for regular Calculus I.

Success Factors

The success and efficiency of this placement process is due to much "behind-the-scenes" effort by several people in the mathematics department and involves *everyone* in the department on the actual day of testing. This extensive effort is rewarded by the number of students who successfully complete their initial courses in mathematics. In the first semesters of the 1991–92 and 1992–93 academic years, 92% of the students who initially enrolled in a calculus course completed a semester of calculus; and of these, 92% received a grade of C– or above.

A Mathematics Placement and Advising Program

Donna Krawczyk and Elias Toubassi
University of Arizona

Placement was the issue at this large, comprehensive school. This article explains, among other initiatives
the department took to improve its accessibility to students, a Mathematics Readiness Testing Program.
Statistics measuring the reliability of the testing are included.

Background and Purpose

The University of Arizona is a large research institution with approximately 35,000 students. The mathematics department consists of 59 regular faculty, 8-12 visiting faculty, 20-25 adjunct lecturers, and 60-70 teaching and research assistants. Each semester the department offers between 250 and 300 courses serving about 10,000 undergraduates and graduate students. There are approximately 350 majors in mathematics including mathematics engineering and education students.

During the 1970s the University of Arizona experienced a large growth in enrollment in entry level mathematics courses. An advisory placement program was initiated in 1978. It did not work very well since anywhere from 25% to 50% of the students did not take the test and the majority of those who took it ignored the results. The placement test was given in a single session prior to the first day of classes. This, together with an inadequate allocation of resources, resulted in average attrition rates (failures and withdrawals) of almost 50%. In the mid 1980s the mathematics department, with support from the administration, took a number of steps to reverse the trend. They included the implementation of a mandatory Mathematics Readiness Testing Program (MRT), a restructuring of beginning courses, providing students a supportive learning environment, and an outreach program to schools and community colleges. The mathematics department appointed a full time faculty member as an MRT program coordinator with a charge to work with the student research office to analyze MRT scores. By 1988, the MRT program was fully implemented for all new students enrolling in courses from intermediate algebra to Calculus I.

Method

Over 5,000 students participate in the MRT program each year. Although placement tests are given throughout the year, most are taken during summer orientation sessions, which are mandatory for all new students. These two-day sessions accommodate 350-400 students at a time, and provide testing, advising, and registration for the following semester. During lunch on the first day, the MRT coordinator gives an explanation of the placement process and testing procedures. Students may ask questions during this hour before breaking into groups of 50 to be escorted to various testing sites. Administration of the tests, electronic scoring/coding, printing of results, and data analysis are handled by the University Testing Office.

Students choose between two timed multiple-choice tests adopted from the 1993 California Mathematics Diagnostic Testing project. The version currently being used allows a variety of calculators (including graphing calculators). For test securing reasons, those with QWERTY keyboards are not allowed at this time. Test A is a 50-minute, 45-question test covering intermediate algebra skills. This test is used to place students into one of three levels of algebra or a liberal arts mathematics course. Test B is a 90-minute, 60-question test covering college algebra and trigonometry skills. It is used to determine placement in finite mathematics, pre-calculus, or calculus. The students' choice of test depends primarily on their mathematical background and to some extent, on their major.

The tests are scored electronically, and results are printed that evening. Students pick up their results (in mathematics,

English, and languages) early the next morning prior to their academic advising sessions. Each student receives a profile sheet indicating their mathematics placement and a breakdown of their total score by topic. (Questions are grouped into 7–8 topic areas, such as linear functions, simplifying expressions, and solving equations.) Students may place into one of three levels of courses through Test A, the lowest level being a non-credit beginning algebra course offered by the local junior college (the community college course is offered on our campus), while the highest level is college algebra and trigonometry. Although Test B can also place a student into college algebra, it is primarily designed to place students into one of 5 levels above college algebra. The lowest of these includes finite mathematics and brief calculus, while the highest level is an honors version of Calculus I. Students may register for courses at their level or lower. The computerized registration process blocks students from registering for courses at a higher level.

Due to the complexity of the levels of courses and the wide variety of mathematics requirements by major, experienced mathematics advisors are available to answer student questions during the 90-minute period after profiles are distributed. Although placement is initially based on a test score, other factors are considered. One of these is the high school GPA which is automatically considered for freshmen whose scores fall near a cut-off. Appropriate GPA cut-off values have been determined by measuring success rates for courses at each level. A pilot placement program began this year in which freshmen can place into calculus based on high RSAT[1] or ACT mathematics scores without taking Test B. Transfer students are allowed to use prerequisite coursework in addition to their test results. The coordinator makes final decisions when overriding any initial placement based on test results.

Occasionally, students take the wrong test. In such cases, they may take the correct test on the second day of the session.

These tests are hand-scored. As a general rule, the same test cannot be taken more than once within any three-month period. This has lowered the number of appeals from students claiming, "I just forgot the material, but I know it'll come back to me once I'm in class."

Throughout the year, the coordinator monitors the number of students placing at each level, assists other academic advisors with placement issues, and distributes information to outside groups, such as high school counselors and teachers. At the end of each year, an analysis is made of the success of the various placement procedures. Adjustments may be made and monitored during the following semester.

Findings

Although our tests assess skills, we do not base placement on the correct performance of these skills. For example, we do not give questions which a student must answer entirely correctly. The minimum or cut-off scores, that were used initially were based on the data of students when the test was optional, in the early 80s. These cut-off scores are monitored periodically and adjusted if needed. We do not have a particular goal set for the percentage of students passing courses at each level. Our goal is to set a cut-off score that minimizes the errors on either side of the cut-off; i.e., we do not want to set an unreasonably high cut-off score that would deny registration to a student with a reasonable chance for success. The tables below give some information on our students.

Use of Findings

The data we gather, as well as the periodic studies provided us by the Office of Research on Undergraduate Curriculum, allows us to monitor the effectiveness of our program.

Percentages of new students placing at various levels in Fall 1996:

Placing above the college algebra/trigonometry level	33.9%
Placing at the college algebra/trigonometry level	46.4%
Placing at the non-credit, beginning algebra level	19.7%

Failure/withdrawal percentages for first time freshmen in Fall 1996:

Above college algebra/trigonometry	Failure 12%,	Withdrawal 9%
College algebra/trigonometry level	Failure 14.4%,	Withdrawal 4.8%
Calculus I*	Failure 5%,	Withdrawal 5%

(*Approximately half of the students participating in our pilot program for placement into calculus based on RSAT and ACT scores also took the MRT. Of those, 91% placed into calculus.)

---

[1] RSAT are recentered SAT scores. SAT did this recentering a few years ago. "RSAT" gives a way to distinguish between "new" SAT scores and "old" SAT scores.

Currently, we are generally pleased with the success rates above. However, we are about to embark on a significant change, due to higher university requirements and the consolidation of our two-semester algebra sequence into one. We are presently trying to determining the "best" cut-off for placement into our new college algebra course. During Fall 1996, we experimented with lower cut-off scores for entry into the new algebra course. The preliminary analysis of the data shows that these borderline students in fact were at significantly higher risk than students who met our regular cut-offs. More analysis and experimentation will be done before we settle on a cut-off score.

Another use of the data is to support the chemistry and biology departments as they determine the mathematics prerequisites for a few of their entry level courses. Both departments independently looked at their pass/fail rates for entry level courses. Chemistry studied their two-semester sequence for all science majors and minors. They found that students testing above the college algebra level or completing college algebra did significantly better. In fact, the students who placed lower, or only completed intermediate algebra received only D's and Failing E's. Biology studied their two-semester sequence for all science majors, minors, and health-related professions. The students who did not place above intermediate algebra had a 90% chance of receiving a D or E.

One important outcome of the data analysis is that there is no room for flexibility with regard to mandatory testing and forced placement. An optional test would be completely ineffective with the 5000 students in our entry-level courses. We have found that the majority of students taking the test are freshmen who have difficulty determining their own readiness for course work at a university.

Success Factors

Several features of the MRT program are essential to its success. Strong support by the department and university is crucial. The move from an optional program to the current one required the cooperation of administrators, faculty, and even students. Continuing communication is also important. Misinformation or lack of information can undermine the credibility of the program. It is imperative that this communication extend to the high schools and local colleges.

Funding and manpower are major obstacles. Until two years ago, the test was free. Now a $5.00 charge is automatically billed to the student's account. This offsets the costs of administering the test through the Testing Office, proctoring, reproduction, and scoring. The mathematics advisors and coordinator are funded through the department. The coordinator also receives a one-course reduction in teaching load during each semester. The commitment of the advisors and coordinator is essential.

Another significant factor for success is the ability to analyze the program and make adjustments. Here are some examples of the types of questions that come up: When the department phases out intermediate algebra next year, will we be able to place some of these students at the next level? Is it necessary to test all transfer students? When the new university entrance requirements of 4 years of mathematics begin in 1998, will Test A be necessary? Flexibility also extends to the appeals process. Although there are guidelines to follow, such as test scores, prerequisite courses, and GPA, a more subjective set of guidelines may be necessary to handle individual cases. In any case, continuity is important.

A Comprehensive Advisement Program

Stephen A. Doblin and Wallace C. Pye

University of Southern Mississippi

A large, budding university concentrates on the importance of advising students at all levels. From providing mentors for freshmen to surveying graduating students and alumni, this department operates with continuing feedback.

Background and Purpose

The University of Southern Mississippi (USM) is a relatively young, comprehensive research university with a student body of approximately 14,000 located on two campuses. The main campus in Hattiesburg typically enrolls 90% of these students, among them a large number of community college transfer students. In fact, in fall, 1996, the main campus enrollment of first-time students consisted of 1231 freshman and 1569 transfers. While no precise data are available, a large number of students in both groups are the first members of their families to attend an institution of higher learning. Mathematics is one of nine departments housed in USM's College of Science & Technology. In the fall of 1996, the Department of Mathematics had 123 prospective majors and taught approximately 3100 students in 80 course sections. Preservice secondary teachers are a major source of enrollment, as well as students pursuing preprofessional degrees (e.g., premedicine, predentistry, preveterinary medicine). The university core mathematics course is College Algebra, and the department offers an intermediate algebra course for the most mathematically underprepared university students. Mathematics faculty have kept the degree requirements at 128 semester hours, and the number of degrees in mathematics has remained relatively constant over the last few years (52 per year since 1991). Most serious students finish in the traditional four years.

Because of the enrollments cited above, our interest in attracting mathematics majors, and the fact that many of our students are transfers or naïve about college, advising for academic success and retention became a critical issue that aroused discussion in the 80s. In response to these concerns, the University Office of Admissions determined to involve faculty from the appropriate academic unit as early in the recruitment process as is feasible, often establishing the advisor-advisee relationship even before the student arrives on campus. The University instituted General Studies 101, a fairly typical "introduction to university life" course, as well as a summer orientation and counseling session for all new students. Even more significantly the computerized degree audit and advisement tool PACE (Programmed Academic Curriculum Evaluation) became available to the department and its college starting fall 1992.

Recognizing that retention and success in mathematics courses are nationwide concerns, at this time the mathematics department, in concert with our college and university, forged an advisement partnership with the college, which fostered a variety of advising, tracking and support strategies.

Method

The Advisement Program for mathematics majors is regarded as one of the primary responsibilities of the department (working in partnership with the college). In order to meet the needs of a student body with a mean ACT of 21.3 (19.9 in mathematics), reasonable high school mathematics preparation (at least two years of algebra and a year of geometry), and fuzzy expectations of college, our mathematics faculty believed that advising must be comprehensive, flexible, and afford a personal flavor.

Prospective majors are advised by a department faculty member in a summer orientation and counseling session. At this time, students are given the department's Guide for the Mathematics Major[1]. This guide has proven popular among both parents and potential majors, and serves as a resource for students after enrolling. It includes a summary of degree requirements, employment opportunities within the department, career opportunities in mathematics in general, job titles of recent graduates, course descriptions, faculty biosketches, and a five-year plan for the student regarding advanced course offerings. The students also select their fall classes in this counseling session. (For placement, the department relies on a combination of ACT scores, high school courses and grades, and self-selection based upon course catalog descriptions and discussions with the advisor.) The students are also encouraged at this time to participate with the department in its emphasis on undergraduate research.

The college also offers a Freshman Mentor Program; this program is discussed in another brochure which offers biosketches of the faculty volunteering. Mentors are volunteers drawn from the College of Science and Technology who provide freshmen support and advice. Students who do not apply for a mentor are contacted by phone so they can be personally encouraged to understand the benefits of the program. Although the mentor-mentee relationship is largely up to the individuals, students who have opted to participate in the program are matched up with their mentors in two planned group activities — usually the program reception and a Halloween pumpkin carving / pizza social. The reception is welcoming and the atmosphere cheerful — it includes a brief panel discussion led by a former program participant and a popular mentor, who gives some cautions about the pitfalls of freshman year.

As students progress through the years, it is important to keep information current. For this, we use the student's PACE form, which contains personal data and specific program requirements. The PACE form is automatically updated to include the results of courses taken, grade point averages in the various categories, credit completed at the various levels, remaining degree requirements, etc.

In the fall of the student's first year the student is assigned a permanent advisor and a mentor (if he/she opts for the latter); the advisor of a mathematics major comes from the mathematics department. During the first semester the student meets with the advisor to preregister for spring classes, discusses summer employment /undergraduate research opportunities, the department's undergraduate research program; and the student academic goals. Since the PACE

program frees the advisor from bookkeeping responsibilities, important issues about the students' immediate and long-range prospects can be discussed during these sessions.

During the year, students are kept on track by graduate students, undergraduate majors, and some faculty who provide tutorial assistance in a departmentally sponsored Mathematics Learning Center and new Calculus Lab.

Complementing these department initiatives, the College of Science & Technology Scholars Program recognizes the college's most outstanding sophomores and juniors by hosting a banquet at which attendees are provided certificates of achievement and information about local and national scholarships and awards, summer research opportunities, and requirements for admission to professional and graduate schools.

During the early fall of each year, the college also sponsors a Senior Workshop, primarily for students who plan to enter the world of work upon graduation. Conducted by USM's Director of Career Planning and Placement, the workshop provides tips on the job search — identifying career options, applying, interviewing, etc.

Advising of the Faculty: All department advisors are given a copy of the college's Academic Advisor's Handbook ([1], also available on our World Wide Web home page), and each fall personnel from the Dean's Office conduct an advisement workshop for new faculty.

Ongoing Assessment of the Program: Trying to incorporate all factors which impact on students' ability to achieve, we have opted for an approach which also assesses the effectiveness of our advisement initiatives. Hence, we:

- mail survey forms to mathematics majors 1, 5, 10, 15, and 20 years after graduation, soliciting each student views on department faculty's teaching and advisement effectiveness, perceptions of the quality of preparation for further education and/or employment, and opinions on ways in which the department could improve our educational offerings;

- require that each graduating senior in the college complete a questionnaire on items such as those mentioned above. This takes place at the time the application for the degree is filed, normally one semester before degree requirements have been satisfied;

- query faculty during the annual performance interview each spring concerning the advisement and support program (i.e., academic advisement, student mentoring, encouragement and direction of undergraduate research projects, etc.), their role in it, their perception of the

---

[1] Copies of the various survey forms and brochures, the college's Academic Advisor's Handbook, and the Guide for the Mathematics Major are available from the authors.

effectiveness of their contributions, and opportunities for improvement;

- obtain numerous student comments — many unsolicited — relative to these initiatives throughout the year;

- survey all mentor program participants — both students and faculty — to determine their perceptions of program effectiveness and value as well as suggestions for improvement; and

- conduct an exit interview with each student in the college who withdraws from the university during a semester, in an effort to identify the most common reasons for this action.

Findings

The University has seen increased success in earning national recognition for scholarship and service, improvement in academic records, and higher graduation rates for College of Science and Technology majors, and we feel that this is in part due to the success of our comprehensive approach to advisement and support and our continuing assessment of our efforts. For example, the information gleaned from these various mechanisms are discussed in appropriate forums (e.g., faculty meetings, chair's yearly performance reviews with faculty, dean's annual evaluation of departmental performance with chair), and has been used in modifying program components as well as in changing the recognition and reward structure for faculty at the department level and departments at the college level.

We believe that the advisement and support program which is currently in place effectively assists many of our students to persevere toward a degree and motivates our best students to establish enviable undergraduate records. The success of our program, we feel, is founded on some basic points:

- It is crucial to provide recognition and reward for those faculty who participate willingly and productively in such initiatives. The one-on-one interaction which we view as so important to the effectiveness of the advisement process is possible only if a substantial number of faculty believe in its value and cooperate with the process.

- It is important to remove most of the record-keeping tasks from the advisor so that he/she can devote interactions with the student to substantive issues (e.g., course performance, undergraduate research and employment opportunities, academic and career goals).

- To be effective, an advisement and support program must provide an assortment of initiatives that address our diverse student body and the variety of their needs and concerns.

- Surveys of current students and graduates can be quite valuable if one wishes to assess effectiveness and make improvements, but only if there is in place a formal mechanism for evaluating responses and developing and implementing new initiatives.

Use of Findings

With these observations in mind, here are some of the issues which we plan to address in the near future:

1. We are pleased that the vast majority of mentor program participants earn freshman-year grade point averages exceeding 2.0 and enroll for the sophomore year. However, many of the students who opt not to participate in the advice and support programs we offer are precisely those who need these programs most. We are seeking better ways of attracting the attention of these students.

2. Transfer students, who make up the largest portion of our new students each year, require extra effort in advising. While one could make the case that freshmen need the support far more than junior transfers, our observation is that there is little difference between these two groups, especially when the juniors transferred from a small, home-town community college.

3. Every senior who files an application for degree is required to complete the college's exit questionnaire. We are pleased that even though this form is submitted anonymously, the vast majority of students take the time to provide thoughtful and often provocative responses. On the other hand, the response rate to the department's mail-out survey to graduates is quite low (20–25%) (probably a common characteristic of mailing attempts, but below our expectations). A departmental advisory committee, consisting of current students and/or relatively recent graduates, could possibly provide perspective on this that the department now lacks.

Success Factors

In our experience, advisement is an important factor in student success and retention. University policy must be explicit; it is essential that faculty members from the academic unit be involved in an advisor-advisee relationship with their unit's majors. The university must be willing to pay for programs like PACE and for the increased staff who aid the faculty in their advisement and support role. The dean must actively cooperate in the student advisement and support process with broad-based imaginative initiatives that will complement and support the departments within the college. Most importantly, faculty must believe that their advisement and student support role does make a difference, and that it is duly valued by their department, college, and institution.

A Quantitative Literacy Program

Linda R. Sons
Northern Illinois University

The author helped create MAA Guidelines for Quantitative Literacy and here spells out how this document was used at her large university. This school tested students and graded results in a variety of courses. Results led to curricular changes.

Background and Purpose

In the mid-eighties the faculty at Northern Illinois University (De Kalb, Illinois) reviewed the requirements for their baccalaureate graduates and determined that each should be expected to be at least minimally competent in mathematical reasoning. Setting up an appropriate means for students to accomplish this expectation required the faculty to take a careful look at the diversity of student majors and intents for the undergraduate degree. The result of this examination was to establish multiple ways by which the competency requirement could be met, but that program carried with it the burden of showing that the various routes led to the desired level of competency in each instance. In this article the quantitative literacy program at NIU will be described, and some aspects of assessment of it will be discussed.

The author's involvement in this work stems from her service on the University's baccalaureate review committee, the subsequent establishment of a course at NIU related to competency, and the chairing of a national committee of the Mathematical Association of America (MAA) on quantitative literacy requirements for all college students. The latter resulted in the development of a report published by the MAA [1] which has also been available at the Web site: "MAA ONLINE" — http://www.maa.org.

Northern Illinois University is a comprehensive public university with roughly 17,000 undergraduate students who come largely from the top third of the State of Illinois. The University is located 65 miles straight west of Chicago's loop and accepts many transfer students from area commu-

nity colleges. For admission to the University students are to have three years of college preparatory high school mathematics/computer science (which means that most students come with at least two years of high school mathematics.) Special programs are offered for disadvantaged students and for honors students. The University consists of seven colleges and has 38 departments within those colleges, offering 61 distinct bachelor's degree programs within those departments. Designing a quantitative literacy program to fit a university of this type represents both a challenge and a commitment on the part of the university's faculty and administrators.

The term "quantitative literacy program" is defined in the 1995 report of the Mathematical Association of America's Committee on the Undergraduate Program in Mathematics (CUPM) regarding quantitative literacy requirements for all college students (See [1] Part III). The recommendations in that report are based on the view that a quantitatively literate college graduate should be able to:

1. Interpret mathematical models such as formulas, graphs, tables, and schematics, and draw inferences from them.

2. Represent mathematical information symbolically, visually, numerically, and verbally.

3. Use arithmetical, algebraic, geometric and statistical methods to solve problems.

4. Estimate and check answers to mathematical problems in order to determine reasonableness, identify alternatives, and select optimal results.

5. Recognize that mathematical and statistical methods have limits.

Such capabilities should be held at a level commensurate with the maturity of a college student's mind.

CUPM notes in Part III of the report that to establish literacy for most students a single college course in reasoning is not sufficient, because habits of thinking are established by means of an introduction to a habit followed by reinforcement of that habit over time as the habit is placed in perspective and sufficiently practiced. Thus, CUPM recommends colleges and universities establish quantitative literacy *programs* consisting of a foundations experience and continuation experiences. The foundations experience introduces the desired patterns of thought and the mathematical resources to support them, while the continuation experiences give the student the growing time essential to making new attitudes and habits an integral part of their intellectual machinery.

Method

At Northern Illinois University the foundations experience for the quantitative literacy program consists of taking a specific course as part of the general education program set for all students. However, the specific course may vary from student to student depending on placement data and the probable major of the student.

Many major programs at NIU require students to study specific mathematical content. For students in those major programs which do not require the taking of specific mathematics, the Department of Mathematical Sciences devised a foundations course called Core Competency. This course focuses on computational facility and facts, the interpretation of quantitative information, elementary mathematical reasoning, and problem solving. Topics in the course are taught in applications settings which may be part of the real life experience of a college graduate and include some probability and statistics, geometry, logical arguments, survey analysis, equations and inequalities, systems of equations and inequalities, and personal business applications. Elementary mathematics is reviewed indirectly in problem situations.

In contrast, the courses offered to meet the needs of students whose programs require specific mathematical content often demand a background of computational facility greater than that which is needed merely for admission to NIU and have objectives which relate to their service role. Thus while a student taking one of these latter courses may have attained the computational facility which the core competency necessitates, the student may not have acquired the broad mathematical skills the core competency entails. So in examining how the various service courses the Department taught met the competency objectives, the Department believed that the six courses (1) Elementary Functions; (2) Foundations of Elementary School Mathematics (a course

for future elementary school teachers and taken almost exclusively by students intending that career outlet); (3) Introductory Discrete Mathematics; (4) Finite Mathematics; (5) Calculus for Business and Social Science; and (6) Calculus I would satisfy the competency objectives provided a student completed such a course with a grade of at least a C.

The Calculus I course is always offered in small sections of about 25 students, and, in a given semester, some of the other courses are offered in small sections as well. However, some are offered as auditorium classes which have associated recitation sections of size 25–30. Since some students take more than one of the seven courses listed, it is not easy to know precisely which students are using the course at a given taking for the competency requirement, but approximately 13% meet the competency by taking the Core Competency course, 16% meet it by taking Elementary Functions; 8% meet it by taking Foundations of Elementary School Mathematics; 1% meet it by taking Introductory Discrete Mathematics; 25% meet it by taking Finite Mathematics; 25% meet it by taking Calculus for Business and Social Science; and 12% meet it by taking Calculus I.

In executing the competency offerings, two questions emerged which subsequently became the framework about which an assessment plan was set.

1. What hard evidence is there that these seven routes (including the Core Competency route) each lead to at least a minimal competency?

2. Can we obtain information from a plan for measurement of minimal competency which can be used to devise better placement tests for entering students? (The Department has historically devised its own placement tests to mesh with the multiple entry points into the mathematics offerings which entering students may face.) Could this same information be used to construct an exit examination for degree recipients regarding quantitative literacy?

Tests were prepared for use in the Fall of 1993 which could be administered to a sample which consisted of nearly all students in each of the courses taught that semester which were accepted as meeting the competency requirement. These tests were comparable in form and administered near the end of the semester so as to determine the extent to which students had the desirable skills. Because these tests had to be given on different dates in various classes, multiple forms had to be prepared. Each form included questions testing computational facility, questions involving problem solving, and questions requiring interpretation and mathematical reasoning. Each form consisted of ten multiple choice questions, three open-ended interpretive questions, and three problems to solve. To be sure that students took seriously their performance on these examinations, each course attached some value to performance on the examination which related to the student's grade in the course. For example, in the Core Competency course the test score could be substituted for poor performance on a previous hour examination, while in the Calculus for Business and Social Science course points were given towards the total points used to compute the final grade in the course.

A uniform grading standard was drawn up for the examinations, and a team of five graders evaluated all of the examinations except for the Calculus I papers which were graded by the course instructors according to the set guidelines. Examinations were scored for 2358 students. The course Introductory Discrete Mathematics was not offered in Fall 1993, so data from it needed to be gathered another semester. Clearly many hours were devoted to grading! The examinations were designed so that "passing" was determined to be the attainment of at least 50% of the possible 100% completion of the test for the Core Competency students. Test scores were divided according to the computational facility section (the multiple choice part) and the remainder, which was viewed as the problem solving component.

Findings

To interpret the scores on the graded examinations a comparison had to be made with the grades the students received in the course they took. These showed the pass rates in the table below for the foundations experience requirement on courses which are used to satisfy that requirement.

A comparison of the means of the various test forms suggested two of the seven were harder than the others (these forms were given in sections of more than one course and compared). Both of these forms were taken by students in the Elementary Functions course. For those students in the Elementary Functions course who did not take the harder forms, the pass rate was 71%. Looking at the total performance of all students in the Elementary Functions course who took an assessment test, it was seen that they scored high on the computational facility portion of the examination, but relatively low on the problem solving component. In contrast, the students in the Foundations of Elementary School Mathematics course showed weak scores on the computational facility portion of the test and relatively

high scores on the problem solving component. Students in the remaining courses averaged a respectable passing percentage on each examination component.

Use of Findings

Overall analysis of the assessment data led to the conclusions that:

- We should bolster the problem solving component of the Elementary Functions course and once again measure student performance on an assessment test;
- We should bolster the computational facility component of the Foundations of Elementary School Mathematics course and once again measure student performance on an assessment test;
- We should make no change in the requirement of a grade of C or better in the six courses cited earlier in order for a student to receive credit for meeting the competency requirement;
- We should use the item analysis of test responses to help construct an improved placement test which encompasses some measure of quantitative literacy;
- We should use the item analysis made of the test responses to construct a senior follow-up test intended to provide some measure of a degree recipient's literacy as the recipient leaves the University.

Other conclusions about the study were that we should periodically do an analysis across all the courses mentioned, but it need not be done in all courses during the same semester. Also devising many comparable tests is likely to result in some unevenness across test forms — a situation which needs to be carefully monitored. (Using test forms in more than one course and using statistics to compare the test results among the different forms should "flag" a bad form.) And although it was not desirable, multiple choice questions were used on the computational facility part of the test merely to save time in grading. (When fewer students are being examined in a short period of time, the use of multiple choice questions can be eliminated).

Course	Number of students taking test	Pass Percentage of those who received competency credit
Core Competency	278	84
Elementary Functions	358	44
Foundations of Elementary School Mathematics	122	64
Finite Mathematics	924	69
Calculus for Business and Social Science	398	80
Calculus I	278	84

Success Factors

This assessment process took considerable effort on the part of the author (a seasoned faculty member) and two supported graduate assistants. Two others were brought in to help with grading. However, the assessment process was organized so as to be minimally intrusive for the individual course and instructor.

This assessment process grew out of the Department of Mathematical Sciences' ongoing concern for the success of its programs and was furthered by the University's General Education Committee which then recommended the plan the Department proposed regarding assessment of the core competency requirement. Consequently the University's Assessment Office partially supported the two graduate assistants who assisted in the construction of the examinations, the administration of the examinations, and the grading of the examinations.

Now the Department needs to turn its attention to the follow-up uses noted above and to an assessment of the role of the continuation experiences. At present these continuation experiences vary from student to student also in accordance with the student's major. But there has been discussion within the University of having a culminating general education experience for all students which might involve a project having a quantitative component. In either case a current intent is to analyze student achievement based on the follow-up test noted above, where possible, and on the faculty judgment of student attainment of the five capabilities listed in the CUPM report (and listed above). The appendix of that CUPM report has some scoring guides to try.

Reference

[1] Committee on the Undergraduate Program in Mathematics. *Quantitative Literacy for College Literacy*, MAA Reports 1 (New Series), Mathematical Association of America, Washington, DC, 1996.

Creating a General Education Course: A Cognitive Approach

Albert D. Otto, Cheryl A. Lubinski, and Carol T. Benson
Illinois State University

What is the point of teaching students if they're not learning? Here a general education course operates from a learning-theoretic mold; mathematics education instructors become involved to help students construct their own learning.

Background and Purpose

Creating a general education course that is successful in developing quantitative reasoning is a reoccurring issue for most mathematics departments. A variety of approaches currently flourish, ranging from offering practical problems to viewing the course as linked to sociological, historical, or "awareness" issues. For many of us, such a course is a major source of a department's credit hours, yet the mathematics community has not been as aggressive about looking into new models for these courses as they have been, for example, about calculus reform. Moreover, general education courses in mathematics are usually not found to be enjoyable courses, either by students or by instructors. Students are often thrust into these courses because they need a single mathematics course and often have failed to place into a higher-level course. During a typical course of this nature, the instructor quickly proceeds through various skills, like Bartholomew shedding his 500 hats, providing exposure to a range of "interesting" topics rather than helping students develop in-depth understanding.

Illinois State is a midsize multipurpose university of 20,000 students, drawing primarily from the state. As members of the mathematics department who are doing research on undergraduate learning and who are interacting with RUMEC [1], we began to implement learning-theory ideas to design our general education course, "Dimensions of Mathematical Problem Solving." In order to do this, we asked ourselves: What does it really mean for students to "learn?"

What do we most want them to achieve from such a course? And, how can this course offer students an enriching, positive experience in mathematics? The professional literature [3, 5, 8] has stressed the need to promote active learning and teacher-student interaction. For example, the report *Reshaping College Mathematics* [8] speaks to "the necessity of doing mathematics to learn mathematics," "the verbalization and reasoning necessary to understand symbolism," and "a Socratic approach, in which the instructor works carefully to let the students develop their own reasoning" (pp. 111–112). This mode of instruction has been used by us and other mathematics education instructors who have worked with developing reasoning and problem-solving skills of students in our courses. Using this shared-knowledge base about teaching undergraduates, the pedagogy for this new education course developed.

Even though our students have had three or four years of high school mathematics, they still believe that doing mathematics means memorizing and identifying the appropriate equation or formula and then applying it to the situation. We wanted a course that would generate substantial improvement in their reasoning and problem-solving skills rather than exposing them to more topics to memorize and to not understand. The philosophy guiding the "Dimensions" course can best be summarized by the following paragraph from the syllabus of the course.

It is important that you realize that you cannot understand mathematics by observing others doing mathematics. You must participate mentally in the learning process. This participation includes studying the material; working with others; struggling with non-routine problems; using calculators for solving and exploring problems; conjecturing, justifying, and presenting conclusions; writing mathematics; listening to others; as well as the more typical tasks of solving problems and taking examinations. The emphasis in this course will be on ideas and on understanding and reasoning rather than memorizing and using equations or algorithms.

Mathematical topics are selected from number theory, discrete mathematics, algebra, and geometry. Special attention is given to topics from the 6-10 curriculum for which students have demonstrated a lack of 1) conceptual understanding and 2) reasoning and problem-solving skills. By focusing on material that students have presumably covered, such as proportional reasoning and algebraic generalizations, we can emphasize the development of reasoning and problem-solving skills as well as conceptual understanding. Guiding us are the following course principles.

- The course has no textbook because students tend to mimic the examples in the textbooks rather than to reason.

- The course is problem driven; i.e. we do not use the lecture format followed by assignments. Rather, problems are given first to drive the development of reasoning and problem solving and the understanding of mathematical concepts.

- The pedagogy of the class consists of students presenting their reasoning and strategies for solving problems. This is supplemented with questions from the instructor and other students to clarify the explanations of the students.

- The course focuses on quantitative reasoning [9, 10] and multiple representations.

- Class notes are used on examinations to reinforce that reasoning and problem solving are the focus and not the memorization of procedures and formulas.

- Homework used for assessment is completed in small group settings [7] outside of class. Students must submit their work with complete explanations of their reasoning.

- Students submit an individual portfolio at midterm and at the end of the semester in which they provide examples of how their problem solving and reasoning have improved.

- The faculty presently teaching as well as those who will be teaching this course meet weekly to share and discuss students' intellectual and mathematical progress.

- We use students' reasoning and explanations as we make instructional decisions for the next class period. However because of the focus on students explaining their reasoning

and problem solving, we are aware that students initially react negatively to this form of pedagogy.

Method

Many believe that proper assessment means collecting data for evaluations for grades. These data present only a snapshot or limited view of in-depth understanding. In "Dimensions," we have designed a course in which assessment is an ongoing, formative process. A guiding principle for our pedagogy is that what and how our students feel about learning mathematics affects what and how they learn mathematics.

To measure attitudes, at the beginning of the course we administer the "Learning Context Questionnaire." [4] This questionnaire is based on the works of William G. Perry Jr. [2, 6] and requires students to select a response from a six-point Likert scale (strongly agree to strongly disagree) on 50 items, some of which are "I like classes in which students have to do much of the research to answer the questions that come up in discussions," and "I am often not sure I have answered a question correctly until the teacher assures me it is correct."

We also administer a second survey which we developed, "Survey on Mathematics," that asks students more about their beliefs of what it means to learn mathematics, such as

1. How would you respond to someone in this class who asks, "What is mathematics?"
2. What does it mean to understand mathematics?
3. How do you best learn mathematics?
4. Explain the role of reasoning when doing a problem that uses mathematics.
5. Describe the role of a mathematics teachers AND describe the role of a mathematics student.
6. Describe a memorable mathematical experience you have had.

Findings

On the Learning Context Questionnaire, a low score (received by about 65% of our students) describes students who look to authority figures for correct answers or for making decisions. These students do not believe they have the ability to work out solutions by themselves, suggesting immature cognitive development. About 25% of our students fall into the second category. These students realize there may be different perspectives to situations; however they are not able to differentiate or to evaluate among perspectives. They still tend to look to authority figures for help in finding solutions and have little confidence in their own ability to find solutions. Only 10% of the students in these classes are at the upper end of the cognitive development scale. These students take a more critical look at situations and rely on their own ability to find solutions to problems.

The results of the Survey of Mathematics are analyzed both by noting clusters of similar responses for each question and by making connections among individual student's responses over all questions in order to better understand students' perspectives on learning mathematics. For example, responses to the first question typically indicate that students often think of mathematics as "the study of how to find an answer to a problem by using numbers and various theories." Some note that in mathematics there is only one correct solution. About 20% indicate that mathematics and understanding are related, such as "mathematics deals with understanding and computing numbers." However, question two provides additional insights into what students mean by understanding. Students feel that being able to explain their answer or being able to solve equations and perform procedures means to understand. For question #5, most students state that teachers need to go through procedures step by step, allow students time to practice, and show specific examples, "I learn the best when a teacher explains step by step the answer to a question." Few students state, as one did, "...to really learn you have to actually sit down and try to figure it out until you get it." Thus, students perceive that the role of the teacher is to show, tell, explain, and answer questions. Their perceived role of a mathematics student is to listen or to pay attention, ask questions, and some even state "to take notes." There is probably no better place to bring to student attention these fundamental misconceptions of our mutual roles than in a general education course.

Use of Findings

What we find from both the Learning Context Questionnaire (LCQ) and our own survey about how students anticipate learning mathematics forms our general approach as well as day-to-day decisions about our instruction. We do not lecture students on how to think about solving problems, but guide them to develop their own mathematical processes. Because the majority of our students are at the lower end of the scale on the LCQ, we recognize the need to help them realize that they are ultimately responsible for their own learning and "sense-making."

Class periods are spent analyzing varieties of solutions by having students use the board or transparencies to present their solutions. In addition, students must provide complete explanations about their reasoning and learn to include appropriate representations (pictorial, algebraic, graphical, etc.). Grading on written assessments is based on correctness of reasoning, understanding of the situation, and completeness and quality of explanations. However, we also informally assess by comparing students' mathematical performance with that suggested by their beliefs to evaluate how the students' problem-solving and reasoning skills are developing.

During a class session, we first determine if a strategy used by the student is comparable to a strategy used in class,

looking for whether the student has merely mimicked the class example or was able to generalize. An important component of our assessment is the degree to which students' explanations match the students' use of numbers and symbols: often the answer is correct, but the explanation does not match the symbols. In other words, we look for cognitive growth on the part of the student. Our ongoing assessment (both verbal and written) assists us to identify misconceptions that may have arisen as a result of classroom discourse.

Success Factors

The course as now developed and taught is labor-intensive. We continually search for content-rich problems. In addition, the instructor contributes a lot of effort to encourage students to volunteer and to respectfully respond to other students' reasonings and explanations. It is not surprising that with such a format we initially meet with resistance and frustration from students. It becomes important to resist the temptation to tell students how to do a problem. Likewise, we must reassure students that we are aware of their frustration but that making sense out of mathematics for themselves is a worthwhile and rewarding experience. Unfortunately, we find we cannot have all students understand every topic, so we focus our attention on those students who are willing to spend 6–10 hours each week struggling to understand and make sense of the mathematics; it is these students who influence our instructional pace.

As part of our ongoing course development, the weekly planning sessions for instructors provide an opportunity for instructors to share. These sessions become mini staff development sessions with all instructors learning from each other.

In spite of the many obstacles and demands this course creates, it is rewarding that we are able to collaborate with students in their learning. For many of our students, this course experience has allowed them for the first time to make sense out of mathematics. We see our course as a process course, and, as such, we feel we are able to offer students the best possible opportunity for improvement of their lifelong mathematical reasoning and understanding processes. Simultaneously, we put ourselves in an environment in which the drudgery of low-level teaching is replaced by the excitement of being able to dynamically reshape the course as our students develop their mathematical reasoning, understanding, and problem-solving skills.

References

[1] Asiala, M., Brown, A., Devries, D.J., Dubinsky, E., Mathews, D., and Thomas, K. "A Framework for Research and Curriculum Development in Undergraduate

Mathematics Education," *Research in Collegiate Mathematics Education II*, American Mathematical Society, Providence RI, 1996.

[2] Battaglini, D.J. and Schenkat, R.J. *Fostering Cognitive Development in College Students: The Perry and Toulmin Models*, ERIC Document Reproduction Service No. ED 284 272, 1987.

[3] Cohen, D., ed. *Crossroads in Mathematics: Standards for Introductory College Mathematics Before Calculus*, American Mathematical Association of Two-Year Colleges, 1995.

[4] Griffith, J.V. and Chapman, D.W. *LCQ: Learning Context Questionnaire*, Davidson College, Davidson, North Carolina, 1982.

[5] Mathematical Association of America, *Guidelines for Programs and Departments in Undergraduate Mathematical Sciences*. Mathematical Association of America, Washington, DC, 1993.

[6] Perry Jr., W.G. *Forms of Intellectual and Ethical Development in the College Years*, Holt, Rinehart and Winston, NY, 1970.

[7] Reynolds, B.E., Hagelgans, N.L., Schwingendorf, K.E., Vidakovic, D., Dubinsky, E., Shahin, M., and Wimbish, G.J., Jr. *A Practical Guide to Cooperative Learning in Collegiate Mathematics*, MAA Notes Number 37, The Mathematical Association of America, Washington, DC, 1995.

[8] Steen, L.A., ed. *Reshaping College Mathematics*, MAA Notes Number 13, Mathematical Association of America, Washington, DC, 1989.

[9] Thompson, A.G. and Thompson, P.W. "A Cognitive Perspective on the Mathematical Preparation of Teachers: The Case of Algebra," in Lacampange, C.B., Blair, W., and Kaput, J. eds., *The Algebra Learning Initiative Colloquium*, U.S. Department of Education, Washington, DC, 1995, pp. 95–116.

[10] Thompson, A.G., Philipp, R.A., Thompson, P.W., and Boyd, B.A. "Calculational and Conceptual Orientations in Teaching Mathematics," in Aichele, D.B. and Coxfords, A.F., eds., *Professional Development for Teachers of Mathematics: 1994 Yearbook*, NCTM, Reston, VA, 1994, pp. 79–92.

Using Pre- and Post-Testing in a Liberal Arts Mathematics Course to Improve Teaching and Learning

Mark Michael

King's College

This pre- and post-testing system in a liberal arts mathematics course raises interesting questions about testing in general, and asks why students may sometimes appear to go backwards in their learning.

Background and Purpose

King's College is a liberal arts college with about 1800 full-time undergraduate students. Its assessment program has several components. The program as a whole is described in [2] and [3], and [4] outlines the program briefly and details a component relating to mathematics majors.

The component of interest here is assessment in core (general education) courses. The assessment program was initially spearheaded by an academic dean, who built upon ideas from various faculty members. Assessment in each core course was to consist of administering and comparing results from a pretest and post-test. By the time the program had matured sufficiently for the College's Curriculum and Teaching Committee and Faculty Council to endorse a college-wide assessment policy, it became clear that not all disciplines would be well served by the pre/post-testing approach; the college policy was written in a more liberal fashion, and the use of a pretest became optional.

The use of both a pre- and post-test, however, proves to be very useful in the quantitative reasoning course, which is required of all students who do not take calculus, i.e., humanities and social science majors. It has no prerequisite, nor is there a lower, remedial course. The course covers a selection of standard topics — problem solving, set theory and logic, probability and statistics, and consumer math. Recent texts have been [1] and [5].

Method

As with every core course, quantitative reasoning is loosely defined by a set of learning goals and objectives for the student. These are initially formulated by the Project Team for the course. (The team typically consists of the instructors teaching the course; it may also have members from other departments to provide a broader perspective, though that is not currently the case.) The set of goals and objectives must ultimately be approved by the College's Curriculum and Teaching Committee, which represents the faculty's interest in shaping the liberal education our students take with them into society.

Learning objectives are distinguished from learning goals in that the former are more concrete and more easily assessed. For example, one of the eight objectives for quantitative reasoning is: "to be able to compute measures of central tendency and dispersion for data." By contrast, the six goals for the course are more abstract, as in: "to become more alert to the misuses of statistics and of graphical representations of data." All goals and objectives are phrased in terms of student (not teacher) performance.

The learning goals and objectives guide what a Project Team decides should be taught and how an assessment instrument should be designed. Teams differ in their approaches on the various sections of a course; for example, different instructors may be using different textbooks to teach the same core course. The Quantitative Reasoning Project Team has chosen to use a single textbook and a single assessment instrument. Atypical of most core courses, the pre- and post-tests for quantitative reasoning are identical, both being handed back at the same time. This approach limits how the pretest can be used as a learning tool during the course, but it provides the instructor the cleanest before-and-after comparisons. By not returning the pretest early, he/she does not have to worry about whether the second test is sufficiently similar to the pretest on one hand and whether students are unduly "prepped" for the post-test on the other.

While the quantitative reasoning course strives to provide the kind of sophistication we want each of our graduates to possess, the pre/post-test focuses even more intently on skills we might hope an alum would retain years after graduation — or an incoming freshman would already possess! While the test is sufficiently comprehensive to span the full range of areas covered by the course, it does not evaluate the entire collection of skills taught in the course. It fails intentionally to test for knowledge of the more complicated formulas (e.g., standard deviation). The pre/post-test also deliberately avoids the use of "jargon" that might be appropriate in the course, but which is unlikely to be heard elsewhere. For example, in the course we would discuss the "negation" of an English sentence; this term would be used in homework, quizzes, and tests. On the pre/post-test, however, a sentence would be given, followed by the query: "If this sentence is false, which of the following sentences must be true?"

The pre/post-test also makes a more concerted attempt to detect understanding of concepts than do traditional textbook homework problems. For example, given that the probability of rain is 50% on each of Saturday and Sunday, students are asked whether the probability of rain during the weekend is 50%, 100%, less than 50%, or in between 50% and 100%; a formula for conditional probability is not needed, but understanding is.

The pre/post-test differs from the course's quizzes and tests in other ways. It consists of 25 short questions, about one-third of which are multiple-choice. No partial credit is awarded. The problems are planned so that computations do not require the use of a calculator. This latter feature is imposed because the pretest is given on the first day of class, when some students come sans calculator. The pretest does not contribute to the course grade; even so, our students seem adequately self-motivated to do the best they can on the test. Each student should be able to answer several pretest questions correctly. The pretest thus serves as an early-warning system, since any student who gets only a few answers right invariably will need special help. Time pressure is not an issue on the pretest because the only other item of business on the first day of class is a discussion of the course syllabus. The post-test likewise is untimed, since it is given on the same day as the last quiz, and nothing else is done that day. The post-test counts slightly more than a quiz. It

becomes as an immediate learning tool, as it is used in reviewing for the last test and the final exam.

Findings

Since the pre- and post-tests are identical, and questions are graded either right or wrong, it is easy to collect question-by-question before-and-after data. During our six years of pre/post-testing, the precise balance of questions has changed slightly each year as we continually strive to create the most appropriate test. Still, consistent patterns emerge with regard to the learning exhibited in different subject areas. For instance, our students tend to be more receptive to probability and statistics than to logic.

Another finding is that the pretest is in fact, not a reliable predictor of success in the course except at the high and low ends of the scale; viz., students who score high on the pretest (better than 15/25) do not experience difficulties with the course while students who score very low (worse than 7/25), do. Similarly, post-test scores do not closely correspond to final exam scores. This is partly because some students make better use of the post-test in preparing for the final exam. Also contributing to the discrepancy is the difference in focus between the post-test and the final exam. This raises the legitimate question of whether the post-test or the more traditional final exam provides the "better" way of producing a grade for each of our students.

The most startling pattern to emerge is a good-news-bad-news story that is somewhat humbling. Most of our students arrive already knowing the multiplicative counting principle of combinatorics. By the end of the course, they have made good progress mastering permutations and combinations — but at a cost: consistently fewer students correctly solved a simple pants-and-shirts problem on the post-test than on the pretest!

Use of Findings

Even though students do not see the results of their pretests until late in the course, those results can be useful to the instructor. The instructor is forewarned as to the areas in which a class as a whole is strong or weak. Where a class is weak, additional activities can be prepared in advance. Where a class is strong, the instructor can encourage students by pointing out, in general terms, those skills which most of them already have. At the individual level, if a student has a very low pretest score, he/she may be advised to sign up for a tutor, enroll in a special section, etc. An obvious way to use before-and-after comparative data is for teachers and learners to see where they might have done their respective jobs better. For teachers, who typically get to do it all over again, comparing results from semester to semester can indicate whether a change in pedagogy has had the desired effect. Again, the news may be humbling.

A case in point: knowing that students were likely to "go backwards" in regard to the multiplicative counting principle, I attempted to forewarn students to be on guard against this combinatorial backsliding; pre/post-test comparisons revealed that my preaching had no impact! What was needed to overcome this tendency was a new type of in-class exercise designed to shake students' all-abiding faith in factorials.

Another potential use for pre/post-test data is to make comparisons among different sections of a course. Doing this to evaluate the effectiveness of faculty could be dangerous, as we all know how different sections of students can vary in talent, background, attitude, etc. But an important application of the data has been to compare two sections taught by the same professor. A special section of the quantitative reasoning course was set up for students self-identified as weak in math. The section required an extra class meeting per week and a variety of additional activities (e.g., journals). Pretest scores confirmed that, with one exception, the students who chose to be in the special section did so for good reason. Pre/post-test comparisons indicated that greater progress was made by the special section—not surprising since students who start with lower scores have more room to move upward. But, in addition, the post-test scores of the special section were nearly as high as in the regular section taught by the same instructor.

Success Factors

Comparing results from two tests is the key to this assessment method. Some particulars of our approach — using the same test twice, using simple questions, giving no partial credit — simplify a process which, even so, tells us much we would not have known otherwise about the learning that is or is not taking place; they are not essential to the success of testing students twice. However, pre/post-test comparative data only reveals what progress students have made. It does not reveal what brought about that progress or what should be done to bring about greater progress. In the case of the special section for weak students, the pre/post-test could not tell us to what extent the journal, the extra work at the board, and the extra homework assignments each contributed to the success of that section. Likewise, merely detecting negative progress in one area (combinatorics) was not enough to improve the teaching/learning process; a new pedagogical approach was needed.

Pre/post-testing does not provide the formula for improvement. It must be accompanied by a teacher's creativity and flexibility in devising new techniques. As with any powerful tool, it is only as good as its user! Ultimately the most important factor in the success of this assessment method is not how it is administered, but how it is used.

References

[1] Angel, A.R., and Porter, S.R. *Survey of Mathematics with Applications* (Fifth Edition), Addison-Wesley Publishing Company, 1997.

[2] Farmer, D.W. *Enhancing Student Learning: Emphasizing Essential Competencies in Academic Programs*, King's College Press, Wilkes-Barre, PA, 1988.

[3] Farmer, D.W. "Course-Embedded Assessment: A Teaching Strategy to Improve Student Learning," *Assessment Update*, 5 (1), 1993, pp. 8, 10–11.

[4] Michael, M. "Assessing Essential Academic Skills from the Perspective of the Mathematics Major," in this volume, p. 58.

[5] Miller, C.D., Heeren, V.E., and Hornsby, Jr., E.J. *Mathematical Ideas* (Sixth Edition), HarperCollins Publisher, 1990.

Coming to terms with Quantitative Literacy in General Education or, The Uses of Fuzzy Assessment

Philip Keith, General Education Assessment Coordinator
St. Cloud State University

This article presents an administrative look at the ramifications of accommodating various departments' views of how quantitative literacy is to be defined. The issue is: what are the students telling us—how do we interpret the answers they provide to the questions we've asked? The value of "fuzzy" assessment is discussed in the interpretation of a simple survey which helps move a collective bargaining institution on track.

Background and Purpose

St. Cloud State University is a medium-sized university of about 15,000 students, the largest of the state universities in Minnesota. As a public university within a statewide public university system, it operates with strained resources and constraints on its independence, and serves at the public whim to an extent that is at times unnerving. In 1992 a decision was made to merge the four year universities with the community college and technical college systems, and the state legislature has been demanding maximization of transferability across the public higher education system. This process has been developing with minimal attention to the issue of assessment: assessment functions have been basically assigned to individual universities so that institutions assess their own transfer curriculum for students moving to other institutions.

St. Cloud State has math-intensive major programs in science, engineering, education and business, but it also has many largely or totally "mathasceptic" major programs in arts, humanities, social science and education. The general education program has until now had no math requirement, but an optional mathematics "track" within the science distribution — students may count one of three non-major mathematics courses as part of the science distribution or, (under charitable dispensation!) use two advanced mathematics courses at the calculus level or higher to substitute for the math alternative. Until recently, admission standards permitted high school students with little in and less beyond elementary algebra to enroll. Looking at the national literature on the importance of mathematical "literacy" for career success and citizenship awareness, we felt this state of affairs was problematic.

The challenge of assessing quantitative literacy in a total general education curriculum lies in the amorphousness of the problem. Traditional testing trivializes the problem because restricted testable objectives in mathematics fails to take account of the variety of mathematical needs of the students in terms of their goals and expectations. Our purpose in doing assessment here is thus not to validate academic achievement, but to provide a rough overview of what is happening in the curriculum in order to identify general needs for future academic planning.

Rather than use expensive testing methods, we decided to settle for a cheap and simple way of getting some sense of what was going on with regard to general education learning objectives. General education courses can only be approved if they met three of the following five criteria: basic academic skills (including mathematical skills), awareness of interrelation among disciplines, critical thinking, values awareness, and multicultural/gender awareness. A 1987 accreditation review report had expressed concern that we had not been monitoring the student learning experience, but had settled merely for "supply side" guarantees that weren't guarantees of learning at all. To provide more of a guarantee, we established a five-year course revalidation cycle under which all general education courses needed to be reviewed in terms of demonstrations that they were each meeting at least three of the five objectives above.

Method

We developed a simple survey instrument that asked students in all general education classes whether they were aware of opportunities to develop their level of performance or knowledge in these criterion areas: 1) basic academic skills including mathematics, 2) interdisciplinary awareness, 3) critical thinking, 4) values awareness, and 5) multicultural and gender and minority class awareness. The survey asked the students to assess the opportunity for development in these areas using a 1–5 scale. The widespread surveying in conjunction with the revalidation process gave us a fuzzy snapshot of how the program worked, with which we could persuade ourselves that change was needed. Such a "rhetorical" approach to assessment may seem questionable to mathematicians, but I would argue that such techniques when used formatively have their value.

Institutional assessment at St. Cloud State has always been problematic. The unionized faculty have been extremely sensitive to the dangers of administration control, and to the need for faculty control over the learning process. This sensitivity has translated into a broad faculty ownership of such assessment processes. Thus, the changes in the general education program developing out of the assessment process have been very conservative, maximizing continuity. Also, the faculty has claimed the right to be involved in the design and approval of such a simple thing as the development of a student survey whose already determined purpose was to ask about whether already defined general education objectives were being emphasized in a particular class. The assessment committee and assessment coordinator have authority only to make recommendations for the faculty senate to approve. While this governance process makes development processes restrictive, it permits a fairly broad degree of faculty ownership of the process that provides a positive atmosphere for assessment operations.

We wanted to use the survey instrument in a way that would provide feedback to the instructor, to the department for course revalidation, and to the general education program to help trace the operationality of our program criteria. The surveying system has been working as follows. When we began the process in 1992, we received approval for a general survey of all general education classes being taught during the spring quarter. Because many lower level students are enrolled in various general education classes, this meant that we would need to be able to process something like 20,000 forms. We developed a form with questions based on the program criteria, and had the students fill out that form, and then coded their responses onto a computer scannable form. These forms were then processed by our administrative

computing center, and data reports returned to the appropriate departments for their information. Thereafter, the survey was available for use as departments prepared to have their general education courses revalidated for general education for the next five year cycle. A requirement for revalidation was that survey data be reported and reflected upon.

At the end of the first year, a data report of the whole program was used to identify criteria that were problematic. In 1996, as part of a program review for converting from a quarter calendar to a semester calendar, we reviewed and revised the program criteria, and created a new survey that reflected the new criteria. Again, we requested and gained approval for a broad surveying of all general education courses. This new survey has proved a more varied and useful instrument for assessing general education. However, it remains a student-perception instrument of fairly weak validity and reliability, not an instrument for measuring academic achievement. Even so, it has proven useful, as the following focus on the quantitative literacy criterion illustrates.

Findings

Figure 1 is a graph of the results from the first surveying from 1992 through 1995. It shows averages for classes in each of the program areas: communication, natural science distribution, social science distribution, arts and humanities distribution, and electives area (a two course area that allows students to explore areas for possible majors in more pre-professional areas). What the picture showed was that very little mathematics was happening anywhere in the general education program. In fact, we found only 3 classes in the whole program "scored" at a mean level over 3.5 out of a range of 1 to 5 (with 5 high). The courses in the science distribution showed the strongest mathematical content, the social science courses showed the weakest.

We found it easy, of course, to identify partial explanations — the serious mathematics courses with large enrollments were not in the general education program, the question was framed narrowly in terms of mathematical calculation rather than quantitative thinking, and so forth. But none of that got around the fact that the students were telling us that they were aware of having precious little mathematical experience in the general education program. The effect was to put quantitative literacy in a spotlight as we began monitoring the program.

Use of Findings

The information provided by the graph was widely discussed on campus. Our concern about mathematical awareness in general education became a major issue when the Minnesota legislature decided to require that all public universities and

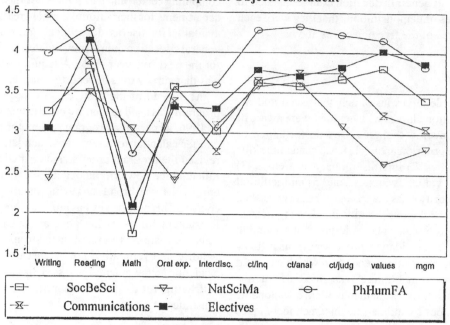

Figure 1. *Old GEd Survey*

colleges convert from a quarter to a semester calendar. This required a reshaping of the general education program, and in particular, reestablishing a quantitative literacy requirement as part of a general education core. In addition, the program criteria have been revised to provide a stronger basis for program assessment. The "mathematical component", rephrased as the "quantitative literacy" component, has been redefined in terms of quantitative and formal thinking, and additional basic skills criteria have been added relating to computer experience and science laboratory experience. Results coming from the new survey that started to be used in the winter of 1996 indicate that the students are seeing more mathematics

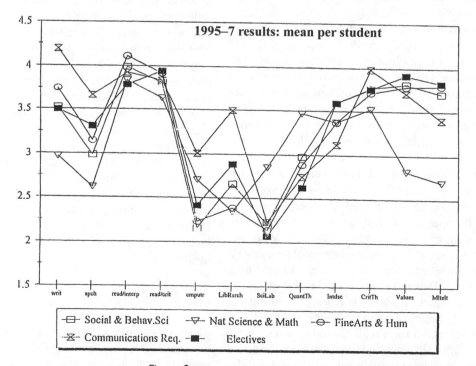

Figure 2. *New Gened Survey*

content everywhere (see the Quantitative Thinking results in Figure 2), but particularly in the social science distribution block as understanding data and numerical analyses become defining criteria. We have tried a small pilot of a Quantitative Literacy assessment of upper division classes across the curriculum using a test, admission data, mathematics course grades, and a survey. We found no significant correlations, probably indicating the roughness of the testing instruments, but did get a strong indication that the general education level math courses were not effectively communicating the importance of mathematics knowledge and skills to success in the workplace.

We will continue the surveying process, anticipating that as all students have to work with the new semester quantitative thinking requirement, they will be able to deal with more sophisticated numerical analyses in other general education courses. This improvement should be visible in individual courses, in the course revalidation review process, and in the program as a whole in the results from survey. Since our goal under the North Central Association assessment requirement is to be able to document the improvement of student learning in the program, we look forward to such results.

Success Factors

The first success factor as been the way a simple survey has generated meaningful discussion of definitions and learning expectations of general education criteria. Whereas course evaluation has tended to generate anxiety and obfuscation in reporting, the survey has provided a vehicle for productive collegial discussion within the faculty at different levels. For example, the revalidation process requires that faculty rationalize the continuation of a particular social science course using survey results. Suppose the math result for that course was 3.8 out of 5, with a large variance and 25% of the students indicating that this goal seemed not applicable to the course. That might reflect the extent to which understanding data analyses and measuring social change is basic to that course, but it definitely leads to some thinking about just how quantitative-thinking expectations for a course are to be defined, and just where the numbers are coming from. This then would be discussed in the application for revalidation, and some decisions would probably be made to include some alternative reading and perhaps some work with a statistical analysis package. Admittedly the definition

of quantitative literacy is loosely defined by non-mathematicians, but once it is defined, however loosely, it becomes discussible at the program level. Part of the five-year program review involves reviewing overall survey results and course descriptions to see what is meant in various departments by "quantitative thinking" and how it is taught across the program curriculum.

A second success factor is the current framework of institutional and system expectations concerning assessment. Our faculty have been aware that assessment reporting will be required in the long term to justify program expenditures and that, to meet accreditation expectations, assessment reporting needs to include data on academic achievement. Student learning data is thus required in the second revalidation cycle, and survey results have been providing some student input that help to define critical assessment questions. In the case of the hypothetical sociology course, the survey score is something of an anomaly: in spite of the emphasis in the class on the use of data to describe and evaluate social realities, half of the 75% who even thought mathematical thinking was relevant to the course scored the emphasis from moderate to none. However "soft" that result from a validity perspective, it provides something of a shock that would motivate faculty interest in emphasizing mathematical thinking and tracking student ability and awareness of mathematical reasoning in tests, papers, and other activities in the class. In this sense, the survey has served as a crude engine for moving us along a track we are already on.

Editor's Note

When asked if the numbers could be used to present more specific results, Philip Keith replies, no, because, "We are dealing with a huge program that has been flying in the dark with pillows, and the point is to give questions like 'why did the students give this score to this item and that to that' some valid interpretation. For example, students in Geography XXX feel that writing is more important than other students do, in English Composition. How is this possible? Well, this Geography class serves a small band of upperclassmen who work continually on research reports which count toward their major in chemistry and meteorology. In other words, the numbers don't determine or judge anything, they point! When they are counterintuitive, then you look for explanations. And that's a useful way to get into assessment in murky areas."

Does Developmental Mathematics Work?

Eileen L. Poiani
Saint Peter's College

An inner city school with a diverse, multicultural clientele is deeply committed to raising students' mathematical abilities. The school has operated with grants that are now drying up, but that help authored some assessment studies over a ten-year period. They ask: does developmental mathematics help, or is it a hopeless cause?

Background and Purpose

The June 1997 issue of Network News [1] states:

> Remediation at the postsecondary level has long been a controversial topic. Those in favor argue that postsecondary remediation provides a second chance for underprepared students, while those opposed maintain that it is duplicative and costly, and may not be effective.

Saint Peter's College has tried to address both aspects of this issue. As a medium-sized (about 3800 students) Catholic, Jesuit liberal arts college in an urban setting with a richly diverse student body, Saint Peter's has long applied resources to assist underprepared students in mathematics. The roots of the developmental mathematics courses go back to the Educational Opportunity Fund's Summer Mathematics Program in 1968. The department then initiated College Algebra (now MA 021, 3 credits, for the day session) in 1975, and added Introductory Algebra (now MA 001, no credit) in 1980. All students must fulfill a six-credit core mathematics requirement, consisting of a Calculus or Finite Mathematics sequence. (Students normally progress from MA 001 to the Finite Mathematics sequence or from MA 001 to MA 021 and then to the Calculus Sequence.) College advisors use placement test results and student past academic performance to assign students to appropriate developmental courses.

How well students are being "mainstreamed" from the developmental into regular college courses was the subject of three major studies. The first Mainstreaming Study was prompted by a request from the Middle States Association of Colleges and Schools, our regional accrediting agency, for a follow-up report on developmental programs.

Method

Over the course of years, we have conducted several investigations[1]:

I. The 1990 Mainstreaming Study completed in May, 1990, analyzed freshmen who entered in Fall 1984, by considering three entrance categories of students: (1) regular admits, (2) students in the Educational Opportunity Fund (EOF) program — a state supported program for selected students with income and academic disadvantage — and (3) students in the College's Entering Student Support Program (ESSP), who were identified at admission as academically

---

[1] To produce these reports, we have had the able assistance of Institutional Research Directors Thomas H. Donnelly, Brother James Dixon, Kathy A. Russavage; Developmental Mathematics Director and current department chair, Gerard P. Protomastro, former Mathematics chair, Larry E. Thomas, and Director of the Institute for the Advancement of Urban Education, David S. Surrey, as well as the entire mathematics department.

underprepared in one or more subjects. (ESSP began under a Title III grant in 1981.)

This study conducted by Dr. Thomas H. Donnelly had two components:

(1) retention/graduation data for students in these categories. Here we looked at entrance categories, racial/ethnic background, gender, verbal and mathematics SAT scores, high school quintile ranking, college GPA, status and credits earned by non-graduates, and consistency in entering and final major for graduates and non-graduates.

(2) student success in developmental mathematics courses and subsequent mainstream college courses. We studied enrollment and grades among developmental (ESSP and EOF) students and non-developmental students in selected developmental courses and subsequent core courses, and enrollment and grades for these groups in other core courses.

II. A 1992 study conducted by Brother James Dixon produced performance data on freshmen entering from Fall 1984 -1991. These data examined outcomes in a core mathematics sequence (Finite or Calculus) relative to whether or not a developmental course had preceded it.

III. A 1996 study conducted by Dr. Kathy Russavage replicated and expanded the earlier Mainstreaming Study, examining outcomes of freshmen (day session) who entered college from Fall 1990-1995. These data were considered in preparing the Five-Year Review of the Mathematics Department and the results continue to be used to shape our progress. In this study, the 1990 study was expanded to examine enrollment patterns in developmental and core courses, to identify repeaters of developmental courses, and to examine retention and graduation relative to various student characteristics.

Findings

Later studies echoed trends of the original 1990 study. The results in 1990 showed that students participating in EOF and ESSP developmental programs did move, in sizeable proportions, into mainstream courses and did persist to complete bachelor's degrees. Since 1990, the performance and persistence of ESSP students has declined somewhat, due to diminished services possible with available funding. In the 1997-1998 academic year, we have worked to restore some of the positive aspects of the ESSP program based on the results of our studies. As expected, developmental mathematics works less successfully for students in the Calculus sequence than in the Finite Mathematics sequence. Findings of the most recent, 1996, study which Dr. Russave identified are as follows:

- An average of approximately 27% of incoming freshmen from Fall 1990–1995 were required to take a developmental mathematics course.

- An average of about 23% of the freshmen enrolled in MA 001 needed to repeat it to achieve successful completion (16% twice, 7% more than twice). Meanwhile, 6% of MA 021 students took it twice and 1% more than twice.

- Performance in developmental courses strongly correlates with the successful completion of core curriculum mathematics courses, as the following table shows:

Developmental Course Grade	% Freshmen Completing Core
4.0 (highest)	53%
3.0 or 3.5	44%
2.0 or 2.5	36%
1.0 or 1.5	26%
0.0	3%
WA/WD/IC*	2%
No developmental course	60%

(* Withdrawal for Absence/Withdrawal/ Incomplete)

Approximately 55% of entering freshmen completed the core mathematics requirement within two years of entrance.

- For the cohorts studied, 86% of those who completed the core mathematics requirement are continuing or have graduated by the end of the Fall 1995 term.

- EOF students, a small group of 50-55 students, who receive remedial mathematics in the summer before college, and continue to receive special tutoring thereafter, normally have better retention and graduation rates than other special admits.

Use of Findings

The Mathematics Department has long been attentive to the need to continually evaluate its program outcomes, and results such as these have helped us pinpoint our weaknesses and address them. We have introduced, over time, computer supported learning modules, personalized instruction, and greater use of graphing calculators in developmental work. The CALL program (Center for Advancement of Language and Learning) also provides student tutors free of charge. Furthermore, several core mathematics sections are taught by the "Writing to Learn" method, whereby students maintain journals and enhance problem-solving skills through writing and frequent communication with the instructor. Faculty members undergo intensive training to become part of the "Writing to Learn" faculty. Saint Peter's has also used the results of these studies in its Institute for the Advancement of Urban Education, which reaches out to promising high school juniors and seniors in need of mathematics and other remediations by offering special after school and Saturday programs.

A concerted effort is currently underway to study the factors contributing to the repetition of developmental courses. We are considering running special sections for repeaters. Whether or not learning disabilities play a role also needs to be explored. The most important feature of these studies is our goal to identify "at risk" students early so we may impress on them the need for regular attendance, good study habits, and persistence. To this end, a three-week Summer Academy was created for August, 1997, seeking to replicate within limited internal resources, some of the features of the EOF program. Open to any incoming freshman (except EOF students, who have their own six-week program), the Academy focuses on a successful transition to college, emphasizes academic and life skills, effective communication, and expectations for achieving success in college.

Success Factors

Saint Peter's College is aware of the fact that every walk of life requires mathematics literacy. But to help developmental students understand its importance is no mean feat. Student anxieties and frustration need to be replaced by confidence and persistence. Students need to cultivate good study habits and understand the importance of regular attendance. To this end, personal attention and support of students is important. Peer tutoring, cooperative learning, and encouragement from faculty and family, where possible, are ingredients in a successful developmental program.

Overall, we are encouraged by our assessment data, to say that our developmental program has succeeded in assisting many under-prepared and overanxious students to achieve satisfactory performance in developmental courses and related core courses, and ultimately reach their career goals. However, in the Saint Peter's tradition, we are continuing to examine the data to find areas for improvement.

References

[1] National Center for Education Statistics (Project of the State Higher Education Executive Officers). *Network News, Bulletin of the SHEEO/NCES Communication Network*, 16 (2), June 1997.

"Let them know what you're up to, listen to what they say." Effective Relations Between a Department of Mathematical Sciences and the Rest of the Institution

J. Curtis Chipman
Oakland University

How can a mathematics department please its client disciplines? This department finds a solution in establishing a web of responsible persons to establish goals for students and instructors, course leaders, committees, and faculty liaisons, for placement of students into courses and for the content of those courses.

Background and Purpose

It's one thing to say that effective departmental relations with the rest of the institution depend upon good communications, it's quite another to determine who should be talking to whom and what they should be discussing. This article illustrates how common assessment issues form a natural structure for such a dialogue — one that can actually produce results.

Oakland University is a comprehensive state university located on the northern edge of the Detroit metropolitan area. Its total enrollment is approximately 13,500, 20% of which is graduate and primarily at the master's level. Most students commute, many are financing their education through part-time employment, and pre-professional programs are the most popular. The average ACT score of matriculating students is in the low 20s.

The Department of Mathematical Sciences has 26 tenure track faculty positions. A mixture of part-time instructors and graduate teaching assistants brings the total instructional staff to a total of 35 full-time equivalents. The department accounts for over 10% of the institution's total credit delivery. The bulk of these credits are at the freshmen and sophomore level in courses required for major standing in the various professional schools or elected to satisfy the university's general education requirements. Average course success rates (measured in terms of the percentage of students enrolled in a course who complete the course with a grade of 2.0 or higher) have varied widely over the last 15 years, from a high of 70% down to 33%. By the end of the 1980's, these low rates had led to very negative perceptions of the department's instructional efforts. There was general departmental consensus that these perceptions threatened institutional support for the department's priority aspirations in the areas of research and emerging industrial collaborations. An effective response was clearly necessary. The one developed had two phases: the first took place solely within the department and the second, which continues to the present, involved the department and major units of the university.

Method

To improve its instructional image (and hopefully its instructional effectiveness) the department developed a unified policy for the delivery of courses at the freshman and sophomore levels. This policy included a general statement of departmental goals and objectives for all of the courses, detailed policies specific to each course, and a process for continuing course development with responsibilities allocated between teaching faculty and the department's curriculum committee.

For example, the general policy set the goal of "an academically sound curriculum in which most conscientious students could expect to be successful." It committed to "provide the skills and understandings necessary for later courses," to insure "consistent course policies in all sections of a course during a given semester," and "as consistent with other course goals, to adopt changes likely to increase the number of students being successful in each class."

Specific course policies would be described in Student and Instructor Information Sheets distributed at the beginning of each course. Issues such as prerequisites, grading, calculator usage, syllabus, and suggestions for successful study habits were addressed in the Student Information sheets. The Instruction Information sheets addressed issues such as typical student clientele, template processes for common test construction, current course issues, and student success rates in the course for the preceding eight semesters. The initial approval of these sheets would be made by the department. Future revisions in these sheets would be at the initiative of the designated faculty Course Leader with the approval by the department's committee on undergraduate programs, who would seek departmental approval for major course changes.

Further details are given in Flashman's panel article [1], but for this account, the key fact is that with the adoption of this policy, typical major assessment elements were in place. The department had fully considered, debated, and decided what it was trying to do, how it would try to do it, how it would measure how well it was doing, and how it would make changes that could assist in doing it better.

Findings

Since the unified policy was approved, there has been and continues to be a series of interactions with the rest of the university in the context of this general policy. The department found these to be a natural consequence of the implementation demands and communication needs of the new policy.

Three interaction examples are discussed in the next section. What is illustrative about these examples is not so much the items under consideration or the actual participants, but the manner in which the implementation of a specific assessment mechanism naturally leads to a process for effective interaction with the rest of the institution.

Use of Findings

Interaction 1. Calculus Reform and Relations with other Science Departments

The first example concerns the issue of calculus reform; the external units were the School of Computer Science and Engineering along with natural science departments in the College of Arts and Sciences. The policy impetus for this interaction was the initial departmental approval of the information sheets for the mainstream calculus course. Here's what happened.

The department's undergraduate committee decided this was the time to grasp the nettle of calculus reform and determine a departmental reaction to the various national efforts underway. The Chair wrote to the Dean of Computer Science and Engineering and to his counterparts in the natural sciences informing them of this effort and requesting faculty in their units to be identified as liaisons for consultative purposes. These colleagues were initially interviewed by the department committee concerning the state of the current course and later invited to review texts under consideration and drafts of the materials to be submitted for departmental approval. The process resulted in the departmental approval of new materials for the calculus sequence. A university forum was held for the science liaisons and other interested colleagues to describe the coming course changes. All of this was widely publicized across the university and covered in the student newspaper.

The department's currency with, and willingness to actively consider, issues of national curricular change was demonstrated. Its concern for the views of its major clients in the sciences was emphasized. A positive precedent for external consultation was established (which was later to be reinforced as described in the third example that follows). Subsequently, there was little surprise and general support at the most recent departmental meeting when the current Chair announced his intention to ask that 1) a permanent engineering faculty liaison be appointed and 2) a rotating (nonvoting) seat be created on the undergraduate committee for the chief academic advisors from the various professional schools.

Interaction 2. Student Support and Relations with the Division of Student Affairs

The second example concerns the issue of support for student work outside of class; the external unit was the Academic Skills Center in the Division of Student Affairs. While there had been a number of positive faculty interactions with the Center prior to the policy, the policy context for this interaction was the initial departmental approval of the student information sheets which addressed issues such as homework, office hours, and other help outside of class. Here's what happened.

Some of the support services offered by the Academic Skills Center include free peer tutoring, supplemental instruction, and luncheon seminars on various study skills. Many of these are routinely promoted in the student information sheets. In addition, departmental faculty participate in the Center's training sessions for supplemental instruction, hold review briefings for tutors, and lead study skills seminars. It has also become a common practice for many faculty to allocate some of their office hours directly to the Center, meeting with walk-in students at the Center itself. The thank you notes for this departmental commitment to student success reached all the way up to university's president.

There has also been an economic component to this active departmental support of another unit's efforts. The vast majority of the tutors hired and Supplemental Instruction sessions offered by the Center are for the direct support of courses in the mathematical sciences. Indeed, the dollars spent far exceed those which the department could possibly allocate from its College budget.

Interaction 3. Placement, Course Content, and Relations with the School of Business Administration

The third and final example concerns the issues of student placement and course content; the external unit was the School of Business Administration. The impetus for this effort was a task force appointed by the dean of that school to review and make recommendations concerning the two course sequence in mathematics, Linear Programming/ Elementary Functions and Calculus for the Social Sciences, required in the pre-major program in business. Since in its policy governing these courses, the department had already committed itself to the goal of increasing the number of students successful in such courses and had also assigned responsibilities and processes for considering and instituting course improvements, it was well positioned to respond positively to this external initiative. Here's (some of) what happened.

The task force included mathematical sciences faculty among its membership, met for a full academic year, commissioned a number of studies, and issued its recommendations. During the academic year which followed the filing of its final report, the department developed formal responses to all of the task forces recommendations, assisted by a number of formal studies and pilot projects supported by joint funding from the deans of Business Administration, the College of Arts and Sciences, and the Vice President of Student Affairs. Faculty liaisons from the department and the school were appointed to oversee this process. For the purposes of this article, the focus will be upon two specific items which illustrate well the assessment issues of careful information gathering, data-based decision making, and resulting change. As described at the beginning of this example, the two issues are student placement and course content.

Course Placement

The task force had recommended that the department review its method of student placement, given the low success rates in the two courses required of its pre-majors (which typically ranged in the low 50% range.) To assist in the formulation of its response, the department's undergraduate committee accepted the invitation of the business school's liaison to conduct, with another colleague trained in industrial/ organizational psychology, a formal validation study of the department's existing placement test. The committee also undertook a survey of other departments' placement practices throughout the state. As a result of this external study and its

own survey, the department determined to change its placement process. The existing test had been based upon a version distributed by the MAA more than ten years ago which the department had never updated. The statistical results from the validation study revealed low correlation between test scores and later results in some courses. It also identified many test questions as invalid since their unit scores deviated strongly from total test scores or otherwise failed to differentiate among students taking the test. The implementation of a placement system based upon ACT scores for all beginning courses outside the mainstream calculus sequence begins in this current academic year. For the mainstream calculus course, a process of revising the current test is underway with continuing advice and consultation of the business faculty who conducted the original study.

Course Content

In the area of course content, the task force had conducted a careful survey of course topics used in advanced business courses, surveyed the content of corresponding courses across the state, interviewed instructors of the courses, and developed a statement written by a working group of business faculty describing their goals for the course in a business major's curriculum. In formulating their response, the department's undergraduate committee asked the course leaders of these two courses to draft new syllabi and commissioned a question by question analysis of the departmental final examinations in these courses. Through a series of revisions and consultations between the committee and the working group, new syllabi for both courses were finally approved and implemented. The final result was a 20% reduction in the topic coverage for each course.

In the years since these studies were completed, both the department and the school have continued to appoint faculty liaisons who meet monthly to discuss the implementation of these and other changes that resulted from the process, as well as developing other means for potential course improvements. Their current efforts were recently the subject for a major article in the university's newspaper.

Success Factors

The actual adoption of a specific assessment policy by a real department and its subsequent implementation by real people in a real university set in motion an entire chain of interactions whose ramifications would have been difficult to predict. In addition to the constructive context which the department's policy provided for its relations with the rest of the university, there were four lessons learned from the initial round of interactions. They appear to be particularly relevant for any department contemplating such a process and can be summarized as follows.

- Faculty and administrative professional colleagues outside the department are much more willing and able to support your efforts if they know what you are trying to do and how you are trying to do it.

- Inviting and receiving recommendations from others does not obligate you to accept them, only to seriously (and actually) respond to them. Indeed, many faculty outside the department have professional skills and interests much more suited to the development of relevant data than you do. Even though your own colleagues may be initially quite leery of such external involvement, positive precedents can allow (mutual) trust and confidence to develop quickly.

- Just as in a class, important messages have to be continually repeated, not just to the participants, but to interested observers as well. Publicity is not a dirty word, it is an important way of letting people know what you're up to.

- Effective university relations require a great deal of time and energy. This means someone's real time and someone's real energy. If a department wishes to make this investment, it needs to carefully consider who will be involved and how their efforts will be assessed for the purposes of both salary and promotion.

Reference

[1] Roberts, A.W., Ed. *Calculus: The Dynamics of Change*, MAA Notes, Number 39, The Mathematical Association of America, Washington, DC, 1996.

Have Our Students with Other Majors Learned the Skills They Need?

William O. Martin and Steven F. Bauman

University of Wisconsin-Madison and North Dakota State University

A large university begins by asking teachers in other disciplines not for a "wish list" but for a practical analysis of the mathematical knowledge required in their courses. Pretests for students reflect these expectations, and discussion of results encourages networking.

Background and Purpose

Quantitative assessment at Madison began for a most familiar reason: it, along with verbal assessment, was externally mandated by the Governor of Wisconsin and the Board of Regents. Amid increasing pressure for accountability in higher education ([3]), all UW system institutions were directed to develop programs to assess the quantitative and verbal capabilities of emerging juniors by 1991. Although the impetus and some support for the process of assessment were external, the implementation was left up to the individual institutions.

The University of Wisconsin at Madison has been using a novel assessment process since 1990, to find whether emerging juniors have the quantitative skills needed for success in their chosen upper-division courses; a similar program began at North Dakota State University in 1995. A unique characteristic of both programs is a focus on faculty expectations and student capabilities across the campuses, rather than on specific mathematics or statistics courses.

The important undergraduate service role of most mathematics departments is illustrated by some enrollment data for the UW-Madison Department of Mathematics: in Fall 1994, the department had about 200 undergraduate majors and enrollments of about 6500 in courses at the level of linear algebra, differential equations, and below. Some of these students go on to major in a mathematical science; most are studying mathematics for majors in other departments. Mathematics faculty must perform a delicate balancing act as they design lower-division course work that must meet diverse expectations of "client faculties" across the campus.

Method

In a program of sampling from departments across the campus, we have gathered information about quantitative skills used in specific courses and the extent to which students can show these important skills at the start of the semester. Instructors play a key role in helping to design free-response tests reflecting capabilities expected of incoming students and essential for success in the course. Two important characteristics of this form of assessment are direct faculty involvement and close ties to student goals and backgrounds. We have found that the reflection, contacts, and dialogues promoted by this form of assessment are at least as important as the test results.

The purpose of assessment is to determine whether instructional goals, or expectations for student learning, are being met. We have found that explicit goals statements, such as in course descriptions, focus on subject content rather than on the capabilities that students will develop. Such statements either are closely tied to individual courses or are too broad and content focused to guide assessment of student learning. Complicating the situation, we encounter diverse goals, among both students and faculty. In response to these difficulties, we sample in junior-level courses from a wide range of departments across the campus. (e.g.

Principles of Advertising, Biophysical Chemistry, and Circuit Analysis). Instructors are asked to identify the quantitative capabilities students will need to succeed in their course. With their help, we design a test of those skills that are essential for success in their course. We emphasize that the test should not reflect a "wish list," but the skills and knowledge that instructors realistically expect students to bring to their course.

By design, our tests reflect only material that faculty articulate students will use during the semester — content that the instructor does not plan to teach and assumes students already know. This task of "picking the instructor's brain" is not easy, but the attempt to identify specific, necessary capabilities, as opposed to a more general "wish list," is one of the most valuable parts of the assessment exercise.

A significant problem with assessment outside the context of a specific course is getting students (and faculty!) to participate seriously. We emphasize to participating faculty members the importance of the way they portray the test to students and to inform students that

- the test does not count toward their grade, *but*
- test results will inform students about their quantitative readiness for the course
- the instructor is very interested in how they do, so it is crucial that students try their best
- results of the test may lead to course modifications to better match content to student capabilities.

On scantron sheets, each problem is graded on a five-point scale (from "completely correct" to "blank/irrelevant"); information is also coded about the steps students take toward a solution (for example, by responding yes or no to statements such as "differentiated correctly" or "devised an appropriate representation"). Within a week (early in the term) the corrected test papers are returned to students along with solutions and references to textbooks that could be used for review.

Although we compute scores individually, our main focus is on the proportion of the class that could do each problem. Across a series of courses there are patterns in the results that are useful for departments and the institution. Over time, test results provide insight to the service roles of the calculus sequence. We also use university records to find the mathematics and statistics courses that students have taken. Without identifying individuals, we report this information, along with the assessment test score, to course instructors.

Findings

During the first five years of operation at UW-Madison nearly 3700 students enrolled in 48 courses took assessment project tests of quantitative skills. We have found that instructors often want students to be able to reason indepen-

dently, to make interpretations and to draw on basic quantitative concepts in their courses; they seem less concerned about student recall of specific techniques. Students, on the other hand, are more successful with routine, standard computational tasks and often show less ability to use conceptual knowledge or insight to solve less standard problems ([1]), such as:

Here are the graphs of a function, f, and its first and second derivatives, f' and f''. (Graph omitted.) Label each curve as the function or its first or second derivative. Explain your answers.

(In one NDSU engineering class 74% of the students correctly labeled the graphs; 52% of them gave correct support. In another engineering class, which also required the three-semester calculus sequence, 43% of the students supported a correct labeling of the graphs.)

Here is the graph of a function $y = f(x)$. Use the graph to answer these questions:
(a) Estimate $f'(4)$. (*On the graph, 4 is a local minimum. 84% correct*)
(b) Estimate $f'(2)$. (*On the graph, 2 is an inflection point with a negative slope. 44% correct*)
(c) On which interval(s), if any, does it appear that $f'(4) < 0$? (*65% correct*)

(Percentages are the proportion of students in a UW engineering course who answered the question correctly — a course prerequisite was three semesters of calculus)

To illustrate common expectations, these two problems have been chosen by instructors for use in many courses. Our experience suggests that many instructors want students to understand what a derivative represents; they have less interest in student recall of special differentiation or integration techniques. Few students with only one semester of calculus have answered either problem correctly. Even in classes where students have completed the regular three-semester calculus sequence, success rates are surprisingly low. Most students had reasonable mathematics backgrounds, although more than half of the 87 students mentioned here had a B or better in their previous mathematics course, which was either third semester calculus or linear algebra. Problem success rates often are higher if we just ask students to differentiate or integrate a function. For example, over three-quarters of the students in the same class correctly evaluated the definite integral $\int_0^{-2} te^{-t} dt$. (See [5] for a discussion of student retention of learned materials.)

Indicative of the complex service role played by the lower division mathematics sequence we noted the differing balance of content required by faculty in the three main subject areas: (a) Mathematics (four distinct courses); (b) Physical Sciences (five courses); and (c) Engineering (six courses), and we structured our problems to lie in four main groups: (A) non-calculus, (B) differential calculus, (C) integral calculus, and (D) differential equations. In mathematics courses, for example, 60% of the problems used were non calculus; physical science drew heavily from differential calculus (56% of the problems), while engineering courses had a comparatively even balance of problems from the four main groups.

Use of Findings

Important advantages of this assessment method include:

- Faculty members must focus on specific course expectations to prepare an appropriate test.

- Student needs and backgrounds are reflected in this process because the test is tied to a course the student has chosen, usually at the start of their studies in the major.

- Faculty from mathematics, statistics, and client departments talk about faculty expectations, student needs, and student performance in relation to specific courses and programs.

- The conversations are tightly focused on the reality of existing course content and written evidence from students about their quantitative capabilities.

- Everyone is involved; students and faculty gain useful information that has immediate significance apart from its broader, long-term institutional meaning.

Instructors have mostly reacted very favorably to the assessment process. Some report no need to make changes while others, recognizing difficulties, have modified their courses, sometimes through curriculum, or by omitting reviews or including additional work. Students report less influence, partly because many mistakenly see it as a pretest of material that they will study in the course. Some fail to see the connection between a mathematical problem on the test and the way the idea is used in the course. In technical courses, typically around half the class may report studying both before and after the assessment test and claim that the review is useful. Most students, when questioned at the end of the semester, recognize that the skills were important in their course but still chose not to use assessment information to help prepare.

We report annually to the entire mathematics faculty, but we have probably had greater curricular influence by targeting our findings at individuals and committees responsible for specific levels or groups of courses, particularly precalculus and calculus. Findings from many assessed courses have shown, for instance, that faculty members want students to interpret graphical representations. This had not always been emphasized in mathematics courses.

- After finding that many students in an introductory course were unable to handle calculus material, one department increased their prerequisite from first semester business calculus to two semesters of regular calculus.

- In another department, many students with poor records in mathematics apparently did not realize that material from a prerequisite calculus course would be expected in later work. This illustrated the importance of advising , especially regarding the purpose of general education requirements.

- Faculty in other departments typically welcome the interest of our committee. One nontechnical department restructured their undergraduate program to incorporate more quantitative reasoning in their own lower level courses.

- In another department, following a planning session, the coordinator for a large introductory science course remarked that he "couldn't remember having spent even five minutes discussing these issues with mathematics faculty."

An early, striking finding was that some students were avoiding any courses with quantitative expectations. These students were unable to use percentages and extract information from tables and bar graphs. A university curriculum committee at UW-Madison viewing these results recommended that all baccalaureate degree programs include a six-credit quantitative requirement. The Faculty Senate adopted the recommendation, a clear indication that our focus on individual courses can produce information useful at the broadest institutional levels. Result of assessment not only led to the policy, but aided in designing new courses to meet these requirements. We are now refining our assessment model on the Madison campus to help assess this new general education part of our baccalaureate program.

How do faculty respond when many students do not have necessary skills, quantitative or otherwise? Sometimes, we have found a "watering down" of expectations. This is a disturbing finding, and one that individuals cannot easily address since students can "opt out" of courses. Our assessment can help to stem this trend by exposing the institutional impact of such individual decisions to faculty members and departments.

Success Factors

Angelo and Cross, in their practical classroom assessment guide for college faculty [1], suggest that assessment is a cyclic process with three main stages: (a) planning, (b) implementing, and (c) responding (p. 34). Although we have cited several positive responses to our assessment work, there have also been instances where assessment revealed problems but no action was taken, breaking our assessment cycle after the second stage. We expect this to be an enduring problem for several reasons. First, our approach operates on a voluntary basis. Interpretation of and response to our findings is left to those affected. And the problems do not have simple solutions; some of them rest with mathematics departments, but others carry institutional responsibility.

Some of our projects' findings are reported elsewhere ([2], [4]). While they may not generalize beyond specific courses or perhaps our own institutions, the significance of this work lies in our methodology. Because each assessment is closely tied to a specific course, the assessment's impact can vary from offering particular focus on the mathematics department (actually, a major strength), to having a campus-wide effect on the undergraduate curriculum.

Assessment has always had a prominent, if narrow, role in the study of mathematics in colleges and universities. Except for graduate qualifying examinations, most of this attention has been at the level of individual courses, with assessment used to monitor student learning during and at the end of a particular class. The natural focus of a mathematics faculty is on their majors and graduate students. Still, their role in a college or university is much larger because of the service they provide by training students for the quantitative demands of other client departments. It is important that mathematicians monitor the impact of this service role along with their programs for majors.

References

[1] Angelo, T.A., & Cross, K.P. *Classroom assessment Techniques* (second edition), Jossey-Bass, San Francisco, 1993.

[2] Bauman, S.F., & Martin, W.O. "Assessing the Quantitative Skills of College Juniors," *The College Mathematics Journal*, 26 (3), 1995, pp. 214–220.

[3] Ewell, P.T. "To capture the ineffable: New forms of assessment in higher education," Review of Research in Education, 17, 1991, pp. 75–125.

[4] Martin, W. O. "Assessment of students' quantitative needs and proficiencies," in Banta, T.W., Lund, J.P., Black, K.E., and Oblander, F.W., eds., *Assessment in Practice: Putting Principles to Work on College Campuses*, Jossey-Bass, San Francisco, 1996.

[5] Selden A. and Selden, J. "Collegiate mathematics education research: What would that be like?" *College Mathematics Journal*, 24, 1993, pp. 431–445.

Copies of a more detailed version of this paper are available from the first author at North Dakota State University, Department of Mathematics, 300 Minard, PO Box 5075, Fargo, ND 58105-5075 (email: wimartin@ plains.nodak.edu).

A TEAM Teaching Experience in Mathematics/Economics

Marilyn L. Repsher, Professor of Mathematics
J. Rody Borg, Professor of Economics
Jacksonville University

Opening a course to both mathematics and business faculty teaching as a team creates public dialogue about problems that straddle two departments.

Background and Purpose

Networking with client disciplines is a role that mathematics departments must be prepared to take seriously. Business disciplines have large enrollments of students in mathematics, and many mathematics departments offer one or two courses for business majors, generally including such topics as word problems in the mathematics of finance, functions and graphing, solution of systems of equations, some matrix methods, basic linear programming, introduction to calculus, and elementary statistics. This is a dazzling array of ideas, even for the mathematically mature. As a result, the required courses often have high drop and failure rates, much to the frustration of students and faculty. Nevertheless, at joint conferences, business faculty consistently urge the mathematics department to keep the plethora of topics, assuring the mathematicians that business has a strong interest in their students' development of a working knowledge of mathematics.

This article discusses one attempt to incorporate economic concepts with mathematics. What if students encountered mathematical topics in the context of economic reasoning? Would understanding of both disciplines be increased? Would student satisfaction improve? Would students stop asking, "What is this good for?" An integrated mathematics and economics course ensued: TEAM — Technology, Economics, Active learning, and Mathematics. The plan was to obtain a dual perspective on the problems of first year instruction in both economics and mathematics, and to use these assessments to feed back to both disciplines for further course development and more cooperation among the departments.

Jacksonville University is a small, private, liberal arts college which draws students from the Northeast and from Florida. Students tend to be career-oriented, fairly well-prepared for college, but lacking self-motivation. They view education as an accumulation of facts, and synthesis of these facts is largely a foreign notion. But if instructors make no synthesis of course content, it is unrealistic to expect freshmen to do so. The following is an account of success and failure of a totally integrated program* in which mathematical concepts are developed as needed, within the framework of a two-semester course in the principles of economics.

Method

We decided to develop mathematical concepts as needed in the study of the principles of economics. For example, slope would be couched within the topic of demand curves, and derivatives would emerge in an investigation of marginal cost and marginal revenue. The year-long syllabus included the standard topics of business calculus and elementary

* Supported in part by DUE grant 9551340.

213

statistics, both required by the College of Business. The classes were taught in a two-hour block with little distinction between topics in economics and mathematics. Both the economics and the mathematics instructors were present for all class meetings, and each pair of students had a computer equipped with appropriate software. The subject matter involved a series of problem-solving exercises which enabled the students to construct their knowledge via active, cooperative learning techniques. The instructors served as facilitators and coaches, and lecturing was kept at a minimum. Some of the exercises were done by the student pairs, but most were completed by two pairs working together so that four students could have the experience of setting priorities and assigning tasks. Reports were group efforts.

Apart from a desire to increase conceptual understanding in both disciplines, the instructors' intentions, we found later, were initially too vaguely formulated. But as the course evolved over the two-year period, more definite goals emerged. Ultimately, we assessed a number of fundamentally important criteria: (1) student achievement, (2) student retention, (3) development of reasoning skills, (4) attitude of students, and (5) attitude of other faculty toward the required courses.

Findings

Student achievement. The Test of Understanding in College Economics (TUCE) was administered at the beginning and the end of each semester both to TEAM members and to students in the standard two-semester Principles of Economics course. Slightly greater increases were indicated among TEAM members, but the difference was not statistically significant. Student achievement in mathematics was measured by professor-devised examinations. Similar questions were included in the TEAM final examination and in examinations in the separate, traditional mathematics courses. No significant differences were noted, but students in the TEAM course performed on the traditional items at least as well as the others. Students in the TEAM course performed better on non-traditional questions; for example, open-ended items on marginal analysis within the context of a test on derivatives. Since confidence and a sense of overview are ingredients of successful achievement, this anecdotal evidence supports the hypothesis that integrated understanding of the subjects took place.

Student retention. In the TEAM course, the dropout rate was substantially reduced. In comparable mathematics courses it is not unusual for ten percent to fail to finish the course, but nearly every one of the TEAM members completed the courses. Numbers were small, just twenty each semester for four semesters, so it is possible that this improvement in retention would not be replicated in a larger setting. The students evidently enjoyed their work and expressed pride in their accomplishments.

Reasoning skills. TEAM students were more proficient than others in using mathematics to reason about economics. For example, class discussions about maximizing profit showed TEAM students had an easy familiarity with derivatives. Students on their own seldom make such connections, and this may be the most important contribution of the TEAM approach. When students link mathematics to applications, they become more sure about the mathematics.

Student attitudes. TEAM members demonstrated increased confidence in attacking "what-if" questions, and they freely used a computer to test their conjectures. For example, in a project about changes in the prices of coffee, tea, sugar, and lemons, the students first reasoned that increasing the price of lemons would have no effect on the demand for coffee, but after some arguing and much plotting of graphs the teams concluded that the price of lemons could affect the demand for coffee since coffee is what economists call "a substitute good" for tea.

TEAM students uniformly reported satisfaction with their improved computer skills. On the other hand, they showed some resentment, especially at the beginning of the course, that the material was not spoon-fed as they had come to expect. Even though some economic topics are highly mathematical, the resentment was more pronounced in what the students perceived to be "just math." Indeed, mathematical concepts often became more acceptable when discussed by the economist.

The emphasis on writing added some frustration. One student asked, "How can I do all the computations and get the right answers but still get a C?" Not every student was entirely mollified by the explanation that a future employer will want, not only the correct answers, but a clear report of the results.

Faculty attitudes. The authors were surprised that they did not meet the kind of opposition that has sometimes attended calculus reform efforts, but there was some reluctance among colleagues to consider expanding the program to a larger audience. Even people who are favorably inclined to participate in an integrated course are made nervous by the fact that some topics must of necessity be curtailed or eliminated. In hindsight, it would have been well to involve more faculty from both departments at the early stages.

Use of Findings

Although the TEAM course is no longer offered, this does not mean that the experiment failed, for much of what was learned is being incorporated into the existing courses. The lab assignments have been revised and expanded for use with a larger audience both in principles of economics classes and in the business calculus and elementary statistics sessions. Teachers are finding, often based on discussions with the two experimenters, the many advantages of cooperative learning together with an interdisciplinary approach to mathematical content. More importantly, plans are being laid to offer a new integrated course in the 1999-2000 academic year.

Success Factors

A small institution with computerized classrooms will find the experiment worthwhile, but the labor-intensive delivery system is probably too expensive for widespread application. Desirable as it is to have both economics and math-ematics instructors present for all class meetings, it may not be feasible on a routine basis. A revision of the economics curriculum is under discussion. It has become desirable to offer an introduction to economics aimed at students with a strong mathematics background with a syllabus which would easily incorporate the experimental materials. Finally, and even more importantly, the dialogue between mathematicians and members of their client disciplines will continue, not just with business. Our work has set a model for networking and assessing across the curriculum, with physics, engineering, and others. We must decide what concepts students should carry with them and we must work for the development of those concepts in all related disciplines. A team approach teaches some lessons about how this can be done.

Factors Affecting the Completion of Undergraduate Degrees in Science, Engineering, and Mathematics for Underrepresented Minority Students: The Senior Bulge Study

Martin Vern Bonsangue
California State University, Fullerton

Commissioned by the California State University's Chancellor's Office, this study looks at transfer students and suggests key areas for reform.

Background and Purpose

The Alliance for Minority Participation program is a nationally-based effort designed to support underrepresented minority students enrolled in science, engineering, and mathematics programs at four-year colleges and universities. The primary goal of the Alliance for Minority Participation program is to increase the number of minority students graduating in a science, engineering, or mathematics (SEM) major. While an increasing number of minority students have enrolled in SEM programs in this decade, not all of these students are completing their degree in a timely way (see, e.g., [6]). Indeed, anecdotal comments from mathematics, science, and engineering departments indicate that it seems that a large number of students either seem to "hang around" a long time, or are behind schedule in their programs (e.g., senior enrolled in lower-division mathematics courses). Concern about what appears to be a bottleneck, or "bulge," for many minority seniors enrolled in SEM programs prompted the Chancellor's Office of the California State University (CSU) to commission a study relative to this issue. Thus, the purpose of this study was to identify possible factors affecting the completion of degrees in mathematics-based disciplines and how departments and institutions might help to streamline the path to graduation for their students.

Method

The research was limited to students with senior status, and included two components: transcript analyses and student interviews. Transcript analyses of student records were done at six participating CSU campuses by local administrative offices and academic departments, gathering information on three criteria:

- fulfillment of university general education requirements
- fulfillment of upper-division major requirements
- fulfillment of mathematics requirements

Follow-up interviews, by phone or writing, or in person, were used to help understand student perceptions of their own experiences.

The sample for transcript analysis was comprised of 813 students currently enrolled as seniors at one of six of the 22 campuses of the California State University. About three-fourths (74%) of these were transfer students from a community college coming in as third-year students. More than half of the students were majoring in engineering (55%) or the natural or physical sciences (41%), with about 4% majoring in mathematics.

Findings

Only one-half (52%) of the students had completed their general education requirements, while fewer than that (40%) had completed the upper division requirements in their major. Moreover, nearly half (45%) of these SEM students had not yet completed their mathematics requirement (first and/or second year calculus) even though the students were seniors. While many students owned more than one of these three deficits, relatively few seniors were qualified to take senior-level courses in their major, that is, were deficit-free. Typically, a student was at least two semesters from having completed all prerequisites for senior-level work, and in some cases, was essentially a freshman in the major.

There were notable differences in the trajectories towards successful completion of graduation requirements between transfer and non-transfer students. Virtually all of the transfer students had at least one of the three deficits listed above. By comparison, non-transfer students who had attended the CSU as freshmen were much more timely in their completion of courses, with more than half of them having completed all three requirements by the end of the junior year. Thus, students in mathematics-based majors who began their careers as freshmen in the CSU had a reasonable expectation of graduating in a timely way (for this study, within 6 years for engineering majors, and within 5 years for math and science majors). By comparison, SEM students who had transferred seemed to have no chance to finish on time. Graduation checks showed that while non-transfer students comprised only one-fourth of the sample, these students comprised more than 90% of the minority graduates in SEM majors. Moreover, this trend was true for all disciplines, including mathematics.

While the data here suggest that transferability (or lack of) is the real culprit, problems associated with changing schools are more severely felt by the minority community. Recall that three fourths of the SEM minority students were transfer students, compared with typical ratios of around 30-40% for non-minority students. Since the non-transfer students are more timely to graduation than are transfer students, it becomes a minority issue.

Follow-up interviews seemed to confirm that transferring creates problems, both obvious and subtle. Three hundred forty students were interviewed by telephone, in person, or in writing. The interviews were not sympathetic, but informational, in nature. While the interview format and questions asked varied somewhat by campus, student responses centered on the following issues:

- course availability
- course repeating
- inaccurate or unhelpful advising
- problems associated with isolation
- financial and personal issues

While financial and personal issues were mentioned by virtually all students as a factor affecting their academic progress, transfer students raised the other four issues as being stumbling blocks much more frequently (more than 3:1) than did non-transfer students. Typically, transfer students had not yet completed their required mathematics courses, and so had to accommodate these courses in their schedules. Once in the courses, they failed at a rate more than double that of their non-transfer counterparts. Some students reported that they had been advised to take unnecessary courses, but had not been advised to take courses that were really needed. Transfer students found that the academic level and expectation were much higher than they had experienced at the community college, and had often felt "on the outside" compared to students who had been in the department for all four years. Specifically, transfer students were much less likely to be involved with formative undergraduate activities such as conducting student-faculty research, attending departmental functions, or participating in social gatherings.

Use of Findings

This study showed the presence of a significant bottleneck for many California State University minority senior students currently enrolled in mathematics-based programs. The problems seemed to be triggered by issues relating to transferring from another institution, typically a community college. To what extent can the university take responsibility for these problems, or create changes that are genuinely effective? In California, the "Senior Bulge" study did result in helping to convince the Chancellor's office to initiate voluntary programs for interested campuses. Each campus was invited to devise a plan to address the issues associated with untimely graduation, with funding available (between $20-45 K) to help implement the plan. While each participating campus (12 of the 22 CSU campuses are now involved) customized its plan, there were at least three common elements shared by all:

1. "Catch" transfer students early. It is easy to assume that since transfer students have already attended college, they do not need guidance from the university. This study found that transfer students are an at-risk group in terms of adjusting to the academic rigors of a university, enrolling in the right classes, and forming early connections to their academic department. Academic departments identifying, contacting, and meeting with transfer students early in their university career may eliminate some of the problems later as seniors. (As Uri Treisman once remarked, "Care for your own wounded.")

2. Provide accurate academic advising within the department. Interview data showed that most students, especially transfer students, felt varying degrees of isolation

in their quest to gather accurate information about specific requirements and prerequisites, scheduling, and academic support programs and services. Having the academic department individually advise students throughout their enrollment at the university may be extremely helpful in streamlining their paths to graduation.

3. Provide effective academic support for key courses. Providing (and perhaps requiring participation in) academic support, such as Treisman [8] workshop-style groups, together with scheduling key classes to accommodate student needs, may be a significant way that the department can not only facilitate the success of its students but increase student involvement as well.

Success Factors

Studies by Treisman [8], Bonsangue [2, 3], and Bonsangue and Drew [4] have suggested that the academic department is the key element in facilitating changes that will make a real difference for students. Academic departments may be the key link in addressing each of the specific needs identified above, especially in providing academic advising and in creative scheduling to accommodate "off-semester" transfer students. Institutions whose mathematics courses are supported by academic programs such as the Academic Excellence Workshop Program have reported significantly higher rates of on-time course completion and subsequent graduation than they observed before instigating such programs [4, 5, 8]. While the majority of students in this study were not necessarily mathematics majors, the successful and timely completion of mathematics courses seemed to play a crucial role both in the students' time to graduation as well as in their attitudes about school [1, 7].

The programs described to address the Senior Bulge phenomenon are works in progress, with most still in their first or second semester at the time of this writing. Each program has its own character, with most programs run by an SEM faculty member and a person working in student support department. Interested persons should feel free to contact me to discuss gains (as well as mistakes) that we have made.

References

[1] Academic Excellence Workshops. *A handbook for AcademicExcellence Workshops,* Minority Engineering Program and Science Educational Enhancement Services, Pomona, CA, 1992.

[2] Bonsangue, M.*The effects of calculus workshop groups on minority achievement and persistence in mathematics, science, and engineering,* unpublished doctoral dissertation, Claremont, CA, 1992.

[3] Bonsangue, M. "An efficacy study of the calculus workshop model," *CBMS Issues in Collegiate Mathematics Education,* 4, American Mathematical Society, Providence, RI, 1994, pp. 117–137.

[4] Bonsangue, M., and Drew, D. "Mathematics: Opening the gates—Increasing minority students' success in calculus," in Gainen, J. and Willemsen, E., eds., *Fostering Student Success in Quantitative Gateway Courses,* Jossey-Bass, New Directions for Teaching and Learning, Number 61, San Francisco, 1995, pp. 23–33.

[5] Fullilove, R.E., & Treisman, P.U. "Mathematics achievement among African American undergraduates at the University of California, Berkeley: An evaluation of the mathematics workshop program," *Journal of Negro Education,* 59 (3), 1990, pp. 463–478.

[6] *Science* (entire issue). "Minorities in science: The pipeline problem," 258, November 13, 1992.

[7] Selvin, P. "Math education: Multiplying the meager numbers," *Science,* 258, 1992, pp. 1200–1201.

[8] Treisman, P.U. *A study of the mathematics performance of black students at the University of California, Berkeley,* unpublished doctoral dissertation, Berkeley, CA, 1985.

Evaluating the Effects of Reform

Richard West

United States Military Academy at West Point

West Point turned an entire department around. Using an in-depth assessment study with careful attention to the needs of client disciplines, the department created a brand new curriculum, and continues to study it with the "Fullan model" which the author investigated in his dissertation.

Background and Purpose

In 1990, the US Military Academy at West Point changed to a bold new core mathematics curriculum that addressed seven topics in four semester-long courses. The needs for change were both internal and national in scope. Internally, math, science and engineering faculty were disappointed in the math abilities of the junior and senior students. Externally, the national reform movement was providing support in the form of interest, initiative and discussion. Initially, my research evaluated this curriculum change from three perspectives: how the new curriculum fit the national recommendations for reform, how the change was implemented, and what the effects were on student achievement and attitudes toward mathematics. This paper will report on the resulting longitudinal comparison of two cohorts of about 1000 in size, on the "steady-state" assessment of subsequent cohorts, and on the changes as a result of these assessments. This study informs the undergraduate mathematics and mathematics education community about the effects of mathematics reform on student performance, about the implementation and value of department-level reform and evaluation, and the implications and prospects of research of this type.

a) Description of Students: Admission to West Point is extremely competitive. Current levels of admission are greater than ten applicants for each acceptance. The goal of the admissions process is to accept students who are as "well-rounded" as possible, including both physical and leadership aspects. Average Math SAT scores are around 650. Education

at West Point is tuition-free. In general, West Point cadets are very good students from across the nation with diverse cultural backgrounds.

The comparison cohort entered West Point in July 1989 and began the old core mathematics curriculum in August 1989. There were approximately 1000 students who finished the four core courses together. Most of these students graduated in May 1993. The reform cohort entered West Point in July 1990 and was the first group to take the new curriculum, starting in August 1990. There were approximately 1000 students who finished the four core courses together. Most of these students graduated in May 1994.

b) Descriptions of Old and New Curricula: West Point's core curriculum comprises 31 out of the 40 courses required for graduation. Throughout the first two years, all students follow the same curriculum of five academic courses each semester in a two-semester year. Approximately 85% of any freshman or sophomore class are studying the same syllabus on the same day. Over the first two years, every student must take four math courses as well as year-long courses in chemistry and physics. Approximately 85% choose a major toward the end of the third semester. During their last two years, all students take one of seven five-course engineering sequences. Thus, the core mathematics program provides the basis for much of the student's education, whether he or she becomes an English, philosophy, or math, science, and engineering major.

The old core math curriculum was traditional in context and had no multivariable calculus, linear algebra, or discrete math. The courses were Calculus I, Calculus II, Differential Equations, and Probability and Statistics. For those who majored in most engineering fields there was another required course called Engineering Mathematics that covered some multivariable calculus, some linear algebra, and some systems of differential equations.

The current core curriculum (initiated in August 1990) covers seven topics in four semesters over the first two years. The topics are discrete mathematics, linear algebra, differential, integral and multivariable calculus, differential equations, and probability and statistics. A discrete mathematics course focused on dynamical systems (or difference equations) and on the transition to calculus (or continuous mathematics) starts the two-year sequence. The mathematics needed for this course is new to the majority of high school graduates but is also intuitive and practical. Linear algebra is embedded in a significant way, as systems of difference equations are covered in depth. So, two of the seven-into-four topics are addressed in this first course. Further, this course by design provides a means to facilitate the accomplishment of many other reform goals and the goals of the curriculum change (see appendix), such as integrating technology, transitioning from high school to collegiate mathematics, and modeling "lively" application problems.

The current Calculus I course finishes differential calculus and covers integral calculus and differential equations through systems, thus addressing three of the seven-into-four topics. Calculus II is a multivariable calculus course. In addition to the above, these new calculus courses differ from their predecessors by integrating more technology, utilizing more interactive instruction, and including more group projects that require mathematical modeling, writing for synthesis, and peer-group interaction. The probability and statistics course, the last of the four core courses, has been taught for over thirty years to all second-year students and is gaining in importance to our engineering curricula.

The use of technology and the integration of the content in this curriculum into one program provide the opportunities to fit the topics from seven courses into the four semesters. In short, the gains are coverage of linear algebra, discrete math, and multivariable calculus, while totally integrating modeling and technology use. The losses are relatively minor: reduced emphasis on analytic geometry, series, and integration techniques and movement of Laplace transforms to an Engineering Mathematics elective specifically for those majoring in engineering.

c) Framework for Evaluation: Utilizing the three perspectives of (1) reform, (2) implementation process, and (3) comparison of two student cohorts, I conducted three different analyses of the curriculum change. Although this change appears to have come from needs internal to West Point, during the time period much was being said nationally about mathematics education at all levels from kindergarten through college. While the NCTM *Standards* [3] showed the way for K–12, the colleges have had their own voices for reform particularly the Committee on the Undergraduate Program in Mathematics (CUPM) *Recommendations* [5] in 1981 and the whole Calculus Reform movement which appears to have its beginnings around 1986. In my studies the recommendations of this national reform movement were best synthesized in *Reshaping College Mathematics* [4]. I used these recommendations and others to analyze whether West Point's curriculum reform had core characteristics similar to those of the national reform movement.

Large-scale educational change is difficult to initiate, to implement and to maintain. There are many obstacles to overcome in starting up and maintaining a new curriculum, not the least of which is a resistance to change itself. Consequently, the documentation of a major curriculum reform is extremely valuable to all those who wish to attempt such an innovation. One model whose designer addresses the procedures and factors that make up a successful educational change is posed by Michael Fullan in [1]. Fullan says that educational change has two main aspects: what to change and how to change. The national reform movement has provided a consensus of what to change and Fullan provides a theoretical model to compare the West Point innovation against. According to Fullan educational change has three phases: initiation, implementation, and continuation, all leading to outcomes. Each of these phases interacts with its sequential neighbor. In this study, Fullan's factors for each of these phases were analyzed for relevance and impact to the change process. In short, the Fullan model provides a cogent framework to evaluate the change process.

The outcomes for this evaluation were student achievement and attitudes. Mathematics reform at the college level, as with most other educational reforms, seeks to improve student learning and attitudes in the hope that this improvement will in turn motivate students to further study and application of the mathematics they have learned. Further, these outcomes are addressed to determine if the goals for the curricular reform are being accomplished.

Method

My focus was to evaluate the impact of reform on student performance and attitudes toward mathematics by comparing the achievement and attitudes of the two student cohorts described above. To conduct the evaluation, I formulated twelve guiding questions, nine of which describe the context of my study and provide input for the reform and implementation perspectives. The remaining three questions focus on the comparison of the two cohorts.

Data for the two contextual perspectives were obtained through an extensive literature search of recent reform and educational change references, review of historical documents, and interviews of students and faculty. Data for the comparison of the two cohorts were obtained from quizzes, exams, questionnaires, interviews and grades. Except for the interviews, most data were collected and stored for later analysis. There was almost no a priori experimental design.

Findings

The details of this analysis are contained in my dissertation [6], which was completed while I was an associate professor in mathematics at West Point. In addressing the three perspectives I have outlined above, I found for the reform perspective that the revised curriculum at West Point used for the core mathematics curriculum was consistent in most ways with the national call for reform in mathematics curriculum at the college level. For the implementation perspective, I reported that the processes for implementation of the curriculum change involve many factors, but that the change studied was successful in accomplishing its articulated goals (see appendix). The informed and empowering leadership of the department head and the involvement and consensus-building style of the Department of Mathematical Sciences senior faculty in the change process were the key factors that motivated the implementation of the revised curriculum. Over the two years prior to the August 1990 implementation, the senior faculty planned and built institution-wide consensus for the initiation of the revised core mathematics curriculum. The planning and implementation continued through January 1992, when the last of the four courses began. Improvements to this original revised curriculum have continued through the present. Articulated goals

for this curriculum for the most part appear to have been accomplished, and having continued with the curriculum for seven years, the change appears to be institutionalized.

Finally, for the comparison perspective I evaluated the effects of the change in curriculum in terms of student mathematics achievement and attitudes toward mathematics. Students in the reform cohort under the current core curriculum were compared with students under a traditional curriculum. The comparison of the two cohorts in terms of student achievement and attitudes was difficult. My planned data collection included results of math tests, common quizzes, questionnaires, and interviews. I found these data very informative about a certain cohort. Yet, a direct comparison was like comparing apples and oranges, or the results were inconclusive. The lesson learned is that experimental design is needed before-the-fact for these instruments to be compared. At the same time, this needed before-the-fact planning may not be feasible.

In contrast, my analysis of grades was intended purely for informational purposes, but produced the most compelling results. The comparison of grades in follow-on courses such as physics and engineering science, which pride themselves in standardization from one year to the next, showed significant improvements between cohorts. The tables below from my dissertation show the results of the comparison of grades for the two-semester physics sequence. Similarly, I looked at eight engineering science courses taken by a total of 85% of each of the cohorts. Four of these eight courses showed significant results (p-value < 0.05) and a similar shift in grades to the physics courses. The reform group performed better in these courses.

Table 1. *Percentage of Grade Category and Median for PH201 Physics I*

$N_C = 1000$ and $N_R = 1030$

PH201	F	D	C	B	A	Median
Comparison	2.6	16.3	43.5	27.0	10.6	C
Reform	1.8	12.3	39.7	33.7	12.5	C+

Note. $\chi^2 = 17.60$ with $p < 0.002$, $t = 3.82$ with p < 0.001.

Table 2. *Percentage of Grade Category and Median for PH202 Physics II*

$N_C = 970$ and $N_R = 1003$

PH202	F	D	C	B	A	Median
Comparison	2.9	18.9	48.4	22.1	7.8	C
Reform	0.3	5.9	46.4	38.2	9.3	C+

Note. $\chi^2 = 132.75$ with $p < 0.00001$, $t = 9.90$ with p < 0.001.

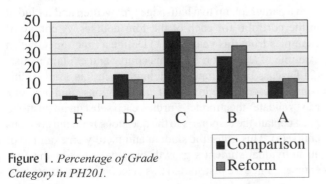

Figure 1. *Percentage of Grade Category in PH201.*

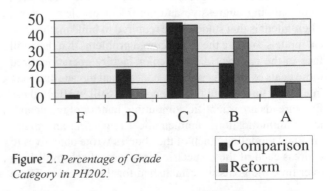

Figure 2. *Percentage of Grade Category in PH202.*

I was able to respond to all my guiding questions except the comparison of attitudes of the two groups. Attitude data for the comparison group were either not available, of such a small scale, or not of similar form to make a reasonable comparison feasible. I found myself comparing different questions and having to draw conclusions from retrospective interviews of students and faculty. If questionnaires are to be used, some prior planning is needed to standardize questions. However, the data from the student and faculty interviews indicate some improvement by the reform group in the areas desired to be affected by the revised curriculum. Attitudes had not been measured until 1992, after the revised curriculum had been implemented.

Use of Findings

While I concluded that the reform curriculum was successfully implemented, the change process in the Mathematical Sciences Department at West Point is still ongoing. This current year saw the adoption of a new calculus text. Further, for varying reasons of dissatisfaction, availability, cost, and adjusting to other changes, four different differential equations texts have been used over the six years of the new curriculum. In addition, the text for the discrete dynamical systems course will change this fall. All courses are using interdisciplinary small group projects designed with other departments and disciplines. The student growth model (more below) developed in 1991 and used to shape the four-course program is being updated continuously. At the same time, the faculty development program that supports improvements to the curriculum and the way we teach has been significantly enhanced in the last two years. Finally, assessment instruments are currently under great scrutiny to ensure that they mirror the goals of the student growth throughout the four-course program and attitude questionnaires have been administered each semester since spring 1992.

The most significant framework for change over the past eight years has been the department's focus on improving student growth over time. The senior faculty started by establishing program goals in the spring of 1990 (see appendix). Basically, these goals are difficult, as they are geared to developing aggressive and confident problem solvers. Their intent is that students are required to build mathematical models to solve the unstructured problems that they will face in the real world. The senior faculty operationalized these goals by establishing five educational threads that were integrated throughout the four courses. Still in effect, these five threads are scientific computing, history of mathematics, communications, mathematical reasoning and mathematical modeling. Each of the courses wrote objectives to address each of these specific ideas toward accomplishing over time the goals we established for the program. These threads and the course objectives are called the "student growth model." As a result, all assessment instruments are designed to address these objectives and thereby measure student growth.

The department turns over about one third of its faculty each year. As a result, over time the student growth model becomes unclear and open to interpretation. As students and faculty become less familiar with the student growth objectives they become unfocused. Therefore, approximately two years ago the senior faculty found it necessary to articulate the student growth model in terms of content threads. The result was nine: vectors, limits, approximation, visualization, models (discrete and continuous, linear and nonlinear, single- and multi-variable), functions, rates of change, accumulation, and representations of solutions (numerical, graphical, symbolic, descriptive). The intent was that both students and faculty could more readily identify growth if it was articulated in mathematical terms. Further, these content threads provide avenues for better streamlining of the curriculum to enhance depth on those topics essential to growth in the program. Finally, they facilitate the design of assessment instruments by having objectives that are content-specific. Further articulation of the goals and objectives for these content threads are forthcoming as well as a requisite assessment scheme.

Evaluation of the program continues. As a result of this initial study, since 1992 I have created a database for each cohort of grades of all mathematics-based core courses. Further, I have used common attitude questions on entry and at the end of each mathematics course. Study of these data are ongoing and are used to inform senior faculty about specific mid-program and mid-course corrections. These instruments actually tell more about the growth of the cohort over time rather than serve to compare one group to the other. However, looking at the same course over time does inform about trends in that course.

A recent additional evaluation tool uses portfolios to measure conceptual growth over time. We have been using student portfolios since 1993, mostly as self-evaluation instruments. This past year we instituted five common questions following the themes of our five educational threads that all students must respond to in each of their core mathematics courses. Each is supposed to be answered with a paragraph up to a half-page typewritten and included in the portfolio for each of the four courses. A couple of examples from this year are: (1) Define a function. Give an example of a function from this course and explain its use. (2) Discuss how a math modeling process is used in this course. Describe the impact of "assumptions" and how one can "validate" their model. Further, each subsequent portfolio must contain the responses to the questions from the previous course(s). This gives the student and faculty an example of an individual student's growth over the span of the four courses. Since we just started this year, we do not know how this will work. But we have hopes that this snapshot will be valuable to both the student and the faculty.

Success Factors

The references below proved excellent in shaping a consensus interpretation of the national reform movement. Further, the West Point Math Sciences Department was very good about articulating goals for the new curriculum (see appendix). The fact that they were written down allowed me to understand very quickly the stated focus of what I was trying to evaluate. It further allowed me to conclude that their stated goals had been accomplished. At the same time, the Fullan model provided a cogent model for evaluating educational change at the undergraduate level.

I understand that West Point is not the typical college, and for my dissertation [6] I devoted an entire appendix to the issue of generalizability. In this appendix I enclosed letters from prominent faculty familiar with our curricula from large research universities to small liberal arts colleges that supported the generalizability of the results. I believe that most of what I have related here is generalizable to other schools and other programs. The evaluation I have conducted is an example of the use of the model in [2] The bottom line is that the MAA model works and the assessment process can be very useful in informing senior faculty who must make curricular decisions.

In closing, assessment at the department level is a process that can involve the entire faculty, build consensus, inform decisions about improving curricular programs, and evaluate student learning over time. My experience with the evaluation of the curriculum reform at West Point is an example of this. I hope that the ideas posed here will encourage others to proactively design assessment programs with the goal of improving student learning.

References

[1] Fullan, M.G. *The New Meaning of Educational Change* (2nd ed.), Teachers College Press, New York, 1991.

[2] Mathematical Association of America. "Assessment of Student Learning for Improving the Undergraduate Major in Mathematics," *Focus*, 15 (3), 1995, pp. 24-28.

[3] National Council of Teachers of Mathematics. *Curriculum and Evaluation Standards for School Mathematics*, NCTM, Reston, VA, 1989.

[4] Steen, L.A., ed. *Reshaping College Mathematics*, MAA Notes Number 13, Mathematical Association of America, Washington DC, 1989.

[5] Tucker, A. ed. *Recommendations for a General Mathematical Sciences Program: A Report of the Committee on the Undergraduate Program in Mathematics*, Mathematical Association of America Washington, DC, 1981.

[6] West, R.D. *Evaluating the Effects of Changing an Undergraduate Mathematics Core Curriculum which Supports Mathematics-Based Programs*, UMI, Ann Arbor, MI, 1996.

Appendix

West Point Goals for Student Learning in the Revised Curriculum

1. Learn to use mathematics as a medium of communication that integrates numeric, graphic, and symbolic representations, structures ideas, and facilitates synthesis.

2. Understand the deductive character of mathematics, where a few principles are internalized and most notions are deduced therewith.

3. Learn that curiosity and an experimental disposition are essential, and that universal truths are established through proof.

4. Understand that learning mathematics is an individual responsibility, and that texts and instructors facilitate the process, but that concepts are stable and skills are transient and pertain only to particular applications.

5. Learn that mathematics is useful.

6. Encourage aggressive problem solving skills by providing ample opportunities throughout the core curriculum to solve meaningful practical problems requiring the integration of fundamental ideas encompassing one or more blocks of lessons.

7. Develop the ability to think mathematically through the introduction of the fundamental thought processes of discrete, continuous, and probabilistic mathematics.

8. Develop good scholarly habits promoting student independence and life-long learning ability.

9. Provide an orderly transition from the environment of the high school curriculum to the environment of an upper divisional college classroom.

10. Integrate computer technology throughout the four-semester curriculum.

11. Integrate mathematical modeling throughout the curriculum to access the rich application problems.

A Comprehensive, Proactive Assessment Program

Robert Olin and Lin Scruggs

Virginia Polytechnic Institute

A large technical institute using the author as assessment coordinator, creates a broad new assessment program, looking at all aspects of the department's role. Statistical studies guide improvements in curriculum, teaching and relations with the rest of the university.

Background and Purpose

Virginia Polytechnic Institute ("Virginia Tech") is a land grant, state, Type I Research University, with an overall enrollment of 26,000 students. Since 1993, the semester enrollment of the mathematics department averages 8,000 - 12,000 students. Of this group, approximately 90% of the students are registered for introductory and service courses (1000 and 2000 level). These first and second-year mathematics courses not only meet university core course requirements, but are also pre- or co-requisites for a number of engineering, science and business curricula.

A comprehensive assessment program, encompassing the activities of the mathematics department, began in the spring of 1995 at Virginia Tech. In the preceding fall of 1994, a new department head had been named. Within days of assuming the position, he received opinions, concerns, and questions from various institutional constituencies and alumni regarding the success and future direction of Calculus reform. The department head entered the position proactively: finding mechanisms which could provide faculty with information regarding student performance and learning; developing a faculty consensus regarding core course content and measurable objectives; recognizing and identifying the differences in faculty teaching styles and learning styles for students; synthesizing and using this information to help improve student learning and academic outcomes. Concurrently, the mathematics department was in the midst of university restructuring, and "selective and differential budget adjustments" were the menu of the day.

From the departmental perspective, the university administration required quantitative data and an explanation of a variety of course, student, and faculty outcomes, often with time frames. The resources in Institutional Research and Assessment were shrinking, and more often data had to be collected at the departmental level. The department decided data and analyses available within the department were much preferable to obtaining data via the university administrative route. The selection of a person to analyze and interpret the data had particular implications since learning outcomes are sensitive in nature and are used within the department; another problem was that educational statistics and measurement design skills, distinctly different from mathematical expertise, were needed for the analysis and interpretation of data. Networking with the university assessment office on campus provided an acceptable option — an educational research graduate student who could work part-time with the department in the planning and implementation of the assessment program (Scruggs).

Method

Data Gathering and Organization: The departmental assessment effort roots itself in obtaining forms of data, organized by semester, with students coded for anonymity:

> High school data are obtained from admissions for entering freshmen and include SAT verbal and math scores, high school GPAs, high school attended, and initial choice of major.

Background survey data from the Cooperative Institutional Research Program (CIRP) are acquired during freshmen summer orientation.

Course Data including overall course enrollment, sectional enrollment, student identification numbers by section, instructors of each section, and times and locations of courses.

Achievement Data including grades from common final for about 8 core courses, ranging from college algebra to engineering calculus and differential equations. (Initially the purpose of these examinations was to provide a mechanism to evaluate Mathematica as a component of engineering calculus.) These examinations are multiple choice and include three (3) cognitive question types: skills, concepts, and applications. These examinations help determine the extent to which students have mastered mathematical skills and concepts; and secondly, allow comparisons between and among sections, in light of instructional modalities, methods and innovations. After common exams are administered and scored, each instructor receives a printout detailing the scores for their class, as well as the mean and standard deviation for all students taking the common examination. The assessment coordinator receives all scoring and test information electronically, including the individual item responses for all students. Test reliability, validity, and item analyses are performed for each course common exam. This data is then made available to mathematics faculty to aid in the interpretation of the current test results, as well as for the construction and refinement of future test questions.

Survey Data: Virginia Tech has administered the Cooperative Institutional Research Project (CIRP) Survey every year since 1966. Student data, specific to our institution, as well as for students across the United States, is available to the mathematics department for research, questions, and analyses. The mathematics department assessment program is actively involved in the process of identifying variables from the CIRP surveys which are associated with student success. Additionally, several in-house survey instruments have been designed to augment this general data and gauge specific instructional goals and objectives. The departmental surveys use Likert rating scales to accommodate student opinion. With this procedure, affective student variables can be merged with more quantitative data.

Methodology Overview: Data is stored in electronic files, with limited access because of student privacy concerns, on the mathematics department server. SPSS-Windows is used for statistical analyses. Specific data sets can be created and merged, using variables singularly or in combination, from academic, student, grade, and survey files. More information on this process is available from the author.

Findings and Use of Findings

The department's role on campus ranges from teaching students and providing information for individual instructors to furnishing information to external constituencies including other departments, colleges, university administration and state agencies. A comprehensive program of mathematical assessment must be responsive to this diverse spectrum of purposes and groups. Departmental assessment then refers to the far-reaching accounting of the students and departmental functioning within the department and throughout the university. Assessment has different purposes for different groups, and given this range of applications, the following discussion incorporates selected examples of data analyses, outcomes, and decisions.

Academic Measurement: While tests are important, their usefulness is contingent on the quality of the instrument, course goals, and the intended purposes. In the common final examinations, close attention is given to the construction and evaluation of the tests themselves. Common finals are constructed by faculty committees using questions submitted by individual faculty members who have taught the course. The tests are then evaluated for appropriateness of content, individual question format and style, and item difficulty. Post-test analyses are performed by the assessment coordinator that include test reliability coefficients, item analyses, overall and sectional means and standard deviations, and score distributions. With each administration, faculty and student feedback regarding the finals has become increasingly positive, indicating that the tests are more representative of course content, and the questions have greater clarity. During this iterative process, faculty knowledge and involvement in assessment has grown with increasing dialogue among faculty a welcome outcome.

Departmental assessment practices have provided a mechanism for monitoring and analyzing student outcomes as innovative and different teaching methods have been introduced and technology added to existing courses, such as engineering calculus and college algebra. Did these changes have a positive effect on student learning? What effects, if any, did the changes have on long-term learning and performance in other courses? These questions were posed from inside the department and from other departments and university administration. The departmental data base allows rapid access to grade and common final data for mathematics courses, and grade outcomes for engineering and science courses.

For the freshmen students enrolled in the fall semesters of 1993 and 1994, student academic background data in conjunction with course grades were used to examine longitudinal outcomes for traditional and "Mathematica" calculus students who subsequently enrolled in advanced mathematics and engineering courses, such as differential

Table I
Technology and Traditional Teaching
General Engineering Majors Engineering
Calculus Sequence, Fall 1994

	SATM	HSGPA	CALC I	CALC II	DIFF EQU	MULTI VAR	STAT	DYNAM
traditional n=324	629.9	3.48	2.77	2.19	2.58	2.55	2.28	1.83*
with technology n=165	624.0	3.50	2.72	2.43	2.48	2.43	2.38	2.06*

* indicates statistically significant difference between groups (t-test)

equations, statics, and dynamics. Mean comparison studies with t-tests were performed, comparing grade outcomes of the traditional and Mathematica students. (No statistically significant differences noted except for the dynamics course [Table 1].)

A designated section of differential equations was taught in the Fall of 1996 as a response to a request from civil engineering and was nicknamed the "Green" version. Targeted toward Civil Engineers, the course utilized environmental and pollution examples to support differential equation concepts and theory. Table 2 summarizes "Green" course outcomes as compared to sections taught in a traditional format with a variety of majors in each section.

Integrating teaching methods and theory application and use as they apply to specific major areas offers intriguing opportunities. During the fall of 1997, faculty in the college algebra sequence collaboratively with other departments and individual faculty outside of the mathematics department. So designed, the sets help students recognize that mathematics is a valuable aspect of all that they do, not just a core university requirement .

Student Placement: To keep up with a changing student population and changing expectations of the university, students, parents, state legislatures, and the media (as evidenced in the charge for academic and fiscal accountability), the departmental "menu" of courses and teaching methods and student support options have been expanded, as has the requirement for assessing and justifying the changes. Appropriate placement of students under these circumstances becomes both an educational and accountability issue.

Since 1988, Mathematics Readiness Scores have been calculated for entering freshmen. Institutional Research devised the initial formula using multiple regression analysis. The formula for the calculation has gone through several iterations, with scores currently calculated from student background variables available through university admissions: high school mathematics GPA; College Board Mathematics scores; and a variable which indicates whether

or not the student had taken calculus in high school. A decision score was determined above which students are placed in the engineering calculus course, and below which, in pre-calculus courses. Each semester, using the course grades, the scores are validated and the formula modified to maximize its predictive capability.

Special Calculus Sections: Even though student success in engineering calculus improved after the math readiness scores were utilized, academic achievement remained elusive for many capable students who had enrolled in the pre-calculus course. Grade averages were low in this course, and longitudinal studies indicated that many students who had earned a grade of C or above, failed to complete the second course in the engineering calculus sequence with a comparable grade. In the fall of 1996, based on a model developed by Uri Treisman at the University of California, Berkeley, a pre-Calculus alternative was piloted within the mathematics department. An augmented version of the engineering calculus sequence was begun — Emerging Scholars Program (ESP) calculus, now operating for the first and second semesters of calculus. The traditional three-hour lecture course was accompanied by two, two-hour required

Table 2
Green Differential Equation Approach, Fall 1996

	SATM	Mean Common Final Score	Mean Course Grade
Green section n=15	640	63	3.1
Composite section* n=22	643	47	2.3
All sections except for Green n=685	640	44	2.0

* 4% of the students not participating in the Green section were randomly selected and descriptive statistics calculated.

Table 3
Traditional and ESP Calculus Student Outcomes, Spring 1997

	%A	%B	%C	%C- or below	mean common final°	mean course grade
ESP calculus n=128	14.9	32.3	24.7	29.9	9.58*	2.32*
Traditional calculus n=155	10.3	25.0	23.1	41.0	8.58*	1.86*

°average number of correct items

* *t*-test indicates that the differences between the scores and grades for the ESP and traditional groups were statistically significant (*p*<.01).

problem-solving sessions, supervised by faculty and using undergraduate teaching assistants as tutors. Due to the academic success of the students, as well as faculty, tutor, and student enthusiasm for the approach, 6 sections of ESP calculus were incorporated into the spring course schedule. Students enrolled in the spring ESP sections had previously been enrolled in traditional calculus or pre-calculus in the fall. Average course grades for the traditional calculus students was 0.8 (out of a possible 4.0). Comparisons of the student outcomes for the traditional and ESP versions from the spring of 1997 are shown in Table 3. The fall of 1997 has 17 sections of ESP calculus on the schedule, with a number of sections of traditional engineering calculus. Previous assessment efforts, both quantitative and qualitative, supported the departmental decision to proceed toward the ESP approach and away from pre-calculus.

Developmental Courses: The college algebra/ trigonometry course enrolls approximately 1300–1400 students each fall semester. This non-major service course, serving primarily freshmen students, requires significant departmental resources. In the fall of 1995, a computer-assisted, self-paced approach was pilot tested, involving a cohort of 75 students from the 1300 total enrollment. At the beginning of the semester, all students were given a departmental survey that is designed to ascertain student perceptions of their learning skills and styles, motivation, and mathematical ability. At the conclusion of the semester, these non-cognitive items, determined from the factor analyses of survey data, were analyzed with student grades using regression analysis. The goal was to identify predictors of success in the computer-assisted version of the course. Significantly related to success were the self-reported attributes of being good to very good in math, organized and factual in learning new material.

Two very different means of placement have been described above. One utilized cognitive achievement data, while the second made use of non-cognitive student reported information. Both approaches have provided valuable information, for student placement and for course evaluation

and modification. Since the introduction of technology as the primary instructional modality in 1995, the college algebra course has undergone several iterations in response to quantitative and qualitative departmental data analyses. At the present time, this course maintains its technology-driven, self-paced instructional core. As a response to student survey responses which indicated a need for more personal and interactive experiences, a variety of instructional alternatives, such as CD lectures, have been incorporated into the course.

Technology: A variety of student outcomes and background variables were used as a means of assessing the incorporation of computer technology into the college algebra and engineering calculus courses. Assessment results and outcomes are generally positive with some concerns. One finding indicated that the use of technology allowed students to pace themselves, within a time frame beneficial to student schedules. Also the downstream results for engineering calculus indicated that students receiving technological instruction during the regular course time did as well, if not better, in the more advanced course work. Negative findings were related to computer and network functioning, certain aspects of the software, and the lack of congruity between lecture and computer assignments. Using the outcomes as a guide for modifying courses each semester, technology use within the department has increased. In fall of 1997, the mathematics department opened a Mathematics Emporium, with 200 computers and work stations, soon to be expanded to 500. Assessment has played and will continue to play a role in ideas, decisions, and educational innovation regarding technology.

Learning/Teaching: Our data base enables our department to effectively respond to issues raised from within the department and externally from other departments and university administration. For example, the common final examinations in many departmental service courses have given additional information regarding student, sectional, and course outcomes. Scores, in conjunction with course

Table 4
Selected Examples of Sectional Outcomes,
Engineering Calculus, Fall 1995

Section	mean SATM	mean Common Final*	mean course grade
1	620	6.6	2.1
2	648	6.6	1.7
3	618	5.2	2.7
4	632	6.3	2.1
5	640	6.7	2.4
overall	630	6.8	2.2

grades, have been used to examine the connection between grading practices and student learning. In the fall of 1995, there were 31 sections of engineering calculus, all relying on the same course goals and text book. After common finals were taken and grades assigned, Pearson Correlations were used to ascertain the association between sectional final scores and grades. For all sections taken together, the correlation was calculated to be a 0.45. Though statistically significant, the magnitude of the result was lower than expected, prompting further study, as sectional mean scores and grades were examined individually. The following table affords examples of the variety of sectional outcomes. Actual data is used, though the sections are identified only by number and in no particular order [Table 4].

One can note that Sections 2 and 3 are disturbing in the incongruence demonstrated between course grades and common final scores.

Making this data available anonymously to instructors offers them the opportunity to compare and analyze for themselves. The department head promoted the use of assessment data to generate an informed and potentially collaborative approach for the improvement of teaching.

Focus Groups: Seeking to evaluate the ESP calculus program, a focus group component was included to obtain students' views and feelings regarding the course format, philosophy, and expectations. Student responses were uniformly positive. About this Calculus approach, freshmen students suggested an unanticipated aspect of its value — the sense of community they experienced within the

mathematics department, and by extension, the university as a whole. Student comments show that they feel the value of esprit de corps in a school that uses their input. As one student remarked, "Learning math takes time and resources. ESP is what makes Tech a good school."

Success Factors

Assessment within the mathematics department is the reflection of a variety of factors, many planned, others serendipitous. But how can success be gauged? What is the evidence of the value added to students, the department, and the institution? Who has gained? Answers to these relate to the department, the individual faculty and the students. The ongoing assessment of the department allows the department to be public, share concerns and answer questions, and allow it to better identify, compete for, and manage available resources within the department, the university, and beyond. The faculty is more able to monitor their students' outcomes, as well as that of curriculum and instructional techniques. Ensuing program planning provides faculty the opportunity for increased ownership and distinctly defined roles in instructional development. Students have certainly received the benefits of assessment by feelings of enhanced involvement and contribution to their educational process. Though probably unaware of the scope and extent of quantitative information which impacts their educational experiences, students interact with assessment and the department through opinion surveys regarding their courses and occasionally through participation in focus groups. Through the realization that their opinions matter, there is the opportunity for a strengthened sense of affiliation with mathematics, the department, and the university.

A mathematics department faculty member recently asked the question, "Whatever happened to the ivory tower?" The answer of course is that it no longer exists, or that it has been remodeled. Departments are no longer concerned primarily with their discipline. In today's educational climate, valid thoughtful information must be readily available regarding student learning and success, program development and improvement. Stewardship of faculty and financial and space resources must be demonstrated to a variety of constituents beyond the department. As a matter of performance and outcomes, everyone gains from the assessment process on the departmental level.

Assessment in One Learning Theory Based Approach to Teaching: A Discussion

Ed Dubinsky

Georgia State University

In this discussion piece, the author explains an approach to teaching based on Learning Theory, particularly examining a Calculus course to ask how assessment can best feed back into the learning environment.

I am engaged in a number of curriculum development projects (see [2], [4], [6]) based on theoretical and empirical research in how mathematics can be learned. The research is done in connection with a loosely organized group of mathematicians and mathematics educators known as the Research in Undergraduate Mathematics Education Community, or RUMEC. (For more about RUMEC, visit our web site at http://rumec.cs.gsu.edu/.) The educational strategy which arises out of this research involves a number of innovations including: cooperative learning, students constructing mathematical concepts on the computer, de-emphasizing lectures in favor of problem solving and discussions designed to stimulate student constructions of mathematical concepts.

Implementing these innovations raises a number of assessment questions. How do we estimate what individual students have learned if most of their work is in a group? If students construct mathematical concepts on the computer, how can we tell if they have made similar constructions in their minds? If our theoretical perspective implies that a student may know something quite well but not necessarily display that knowledge in every instance, what is the meaning of answers to specific questions on a timed test?

I will describe how the curriculum development projects relate to these issues, beginning with a very brief sketch of the theoretical framework in which the research takes place and the overall pedagogical strategies it leads to. Then I will describe some ways in which research has influenced the assessment component of the curriculum development. Finally I will outline our approach to assessment.

A theoretical framework

Our theory begins with an hypothesis on the nature of mathematical knowledge and how it develops. An individual's mathematical knowledge is her or his tendency to respond to perceived mathematical problem situations by reflecting on them in a social context and constructing or reconstructing mathematical actions, processes and objects and organizing these in schemas to use in dealing with the situations. [1]

There are a number of important issues raised by this statement, many relating to assessment. For example, the fact that one only has a "tendency" rather than a certainty to respond in various ways brings into question the meaning of written answers in a timed exam. Another issue is that often the student perceives a very different problem from what the test-maker intended and it is unclear how we should evaluate a thoughtful solution to a different problem. The position that learning occurs in response to situations leaves very much open the sequence of topics which a student will learn. In fact, different students learn different pieces of the material at different times, so the timing of specific assessments becomes important. Finally, the position that learning takes place in a social context raises questions about how to assess individual knowledge.

The last part of our hypothesis relates directly to how the learning might actually take place. It is the role of our research to try to develop theoretical and operational understandings of the complex constructions we call actions, processes, objects and schemas (these technical terms are fully described in our publications) and then to relate those understandings to specific mathematical topics. (See [1] and some of our research reports which are beginning to appear in the literature, and visit our web site.)

Given our understandings of the mental constructions involved in learning mathematics, it is the role of pedagogy to develop strategies for getting students to make them and apply them to the problem situations. Following is a list of the major strategies used in courses that we develop. For more information see [1], [3], [7].

- Students construct mathematical concepts on the computer to foster direct mental constructions and provide an experiential base for reflection.
- Students work in cooperative groups that are not changed for the entire course.
- Lectures are de-emphasized in favor of small-group problem solving to help students reflect on their computer constructions and convert them to mental constructions of mathematical concepts.
- Students are repeatedly confronted with the entire panorama of the material of the course and have various experiences that help different students learn different portions of this material at different times. We refer to this arrangement as an holistic spray.

Some inputs to assessment from research

The position on assessment which follows from our theoretical framework is that assessment should ask two kinds of questions: Has the student made the mental constructions (specific actions, processes, objects and schemas) which the research calls for? and: Has the student learned the mathematics in the course? Positive answers to the first kind of question allow the assertion that the mathematics based on these mental constructions has been learned. This permits us to test, albeit indirectly, for knowledge that the second kind of question may not get to.

Unfortunately, it is not practical in a course setting to test students for mental constructions. In our research, we use interviews, teaching experiments and other methods, all of which require enormous amounts of time and energy, to get at such questions. So we must introduce another indirect component to our assessment. This involves two stages: design and implementation of a very specific pedagogical approach, referred to as the ACE teaching cycle, designed to get students to make certain mental constructions and use them to construct mathematical knowledge; and application, to a particular group of students, of certain assertions, based on research, about the effect of this pedagogical strategy on students' making mental constructions.

The ACE teaching cycle is a course structure in which there is a weekly repetition of a cycle of (A) activities in a computer lab, (C) classroom discussion based on those activities, and (E) exercises. The computer activities are intended to directly foster the specific mental constructions which, according to our research, can lead to understanding the mathematics we are concerned with; the classroom discussions are intended to get students to reflect on these constructions and use them to develop understandings of mathematical concepts; and the exercises, which are fairly traditional, are expected to help the students reinforce and extend their developing mathematical knowledge. (For more details, see [1].)

The second stage of this component is an application of our ongoing research. Our investigations use laborious methods combining both quantitative and qualitative data to determine what mental constructions students appear to be making, and which mental constructions appear to lead to development of mathematical understanding. One outcome of these studies is to permit us to assert, not with certainty, but with some support, that if the pedagogy operated as we intended, that is, the student participated in all of the course activities, cooperated in her or his group, completed the assignments, did reasonably well in exams, etc., then the mental constructions were made.

Because this last point is somewhat different from the kinds of assessments most of us have been used to, perhaps an example will help communicate what we have in mind. Consider the chain rule. We would like students to be able to use this to compute the derivative of a "function of a function" in standard examples, but we would also like the student to understand the rule well enough so that later it can be used to understand (and perhaps even derive, from the Fundamental Theorem of Calculus) Leibnitz' formula for the derivative of a function defined by an integral whose endpoints are functions.

Our research suggests that a key to understanding the chain rule might be an understanding that certain definitions of functions amount to describing them as the composition of two functions, which itself is understood as the sequential coordination of two processes. Our research also suggests that if students successfully perform certain computer tasks and participate in certain discussions, then they are likely to construct such an understanding of the chain rule and also will be reasonably competent in applying this rule in traditional examples.

In principle we could simply perform the same research on the students in our classes and get the assessment directly. But this would be vastly impractical since the research involves interviews and transcribing and analyses that could take years. Instead we ask if the students did perform the computer tasks, did participate in the discussions, and did cooperate in their groups (we determine this by keeping records of their written work, classroom participation, and meetings with groups). We also ask (by testing) if they can use the chain rule to compute various derivatives. If the answer to these questions is yes, then, given the research we have reason to hope that the students not only learned to use the chain rule, but also developed an understanding that could help them understand Leibnitz' formula in a subsequent course.

There is a second consequence of our theoretical position which moves us away from thinking of the course as a set of material which the students must learn so that assessment must measure how much of it they did learn. Rather we think of the students as beginning with a certain knowledge and the goal of the course is to increase that knowledge as much as possible. Thus, in making up an examination, for example, we don't think so much of questions that cover as large a portion of the material as possible, but we try to ask the hardest possible questions about the material we believe the students have learned. The expectation is that the students will do well on such tests and a part of our assessment of how well the course went in terms of what was intended (in the sense of the previous paragraphs) consists of assessing how hard the tests were and how much material they covered.

It could be argued that in this second consequence we are throwing out the requirement that, for example, everyone must learn a certain amount of material in order to get an A. We would respond that, in fact, such a requirement cannot be, and is not, implemented. It is simply impossible, given the realities in which we work, to take a course such as Calculus I, list a set of material and then determine with any degree of accuracy that a given student has learned this or that portion (i.e., numerical percentage) of it. We accept this reality, for example, when we give an exam limited to one, or even two hours and, of necessity, select only a portion of the material to test. We are making an assumption that students who score x on such a test understand x amount of the selected material and also x amount of the material not tested! We don't see this as a more compelling conclusion about how much of the material was learned than the conclusions we draw using our research.

We also accept the reality when we curve our results, basing our grades not on a given amount of material which we judge to warrant an A, but based on how well the brightest students in the class perform on the exam. Again, assumptions are being made that are not more certain than ones being made in our approach to assessment. As an aside, I would like to forestall an argument that curving grades is a practice not used very often today. I think it may be used more than we think, perhaps implicitly. For example, consider a large engineering oriented school with thousands of students each year taking calculus to satisfy engineering requirements. The grades in such a course generally fall along a certain bell shaped distribution. Imagine, for example, what would be the reaction if the student performance were significantly lower (three-quarters of the class failed) or higher (more than half the class got an A). Are we prepared to deny that there is (perhaps implicit) curving here? Do we think that this situation represents a reasonable standard of a given amount of material for an A? If so, what would a list of that material — as determined by what is on the tests — look like?

Finally, let me mention one other input, this time from general research in cooperative learning. The results regarding this pedagogical strategy are mixed. There are reports showing large gains as well as others that do not show much advantage from it, and there do not appear to be many results in which cooperative learning was harmful. Studies that have taken a closer look report that there are conditions under which cooperative learning is more likely to be beneficial. One of the most important conditions, according to Slavin [8] is that students are rewarded individually for the performance of their group. (There are some opposing views in the literature ([5]) but they are more about the general question of using rewards, such as tests, to motivate students.) As will be seen in the next section, we make heavy use of this principle.

An approach to assessment

In our courses, students are assigned to permanent groups (of 3 or 4) very early in the course and they do most of their work in these groups, including some of the tests. We use the following assessment items. Because the first of these, computer assignments, are designed to stimulate mental constructions and often ask students to do things that are new and different for them, the grading is relatively lenient and tries to measure mental effort as much as correctness. All of the other instruments are graded in standard ways. Each of the first two exams listed below is held throughout an entire day so that students do not have time limits. They are allowed to leave and return during the day, on the honor system that nothing related to the course will be done during that day except when they are in the exam room.

Weekly computer assignments. Students have lab time to work on these in their groups, but not enough for the whole assignment and they must spend large amounts of time on their own, either individually or in collaboration with their group. The assignment is submitted as a group.

1. Weekly exercises. These are almost entirely apart from the computer and are fairly traditional. They are done entirely on the students' own time, and again the submission is by group.
2. First exam. This is a group exam. It comes about 40% through the course and the students take it as a group, turning in only one exam for the entire group. Every student in a group receives the same grade.
3. Second exam. This comes half way between the first exam and the end of the course. It is taken individually, but each student receives two grades: her or his score on the exam, and the average of the scores of all of the members of the student's group.
4. Final Exam. This exam is given in the standard way during the standard time period. It is taken individually and students receive only their individual score.
5. Classroom participation. Much of the class time is taken up with small group problem solving and discussion of the problems and their solutions. Both individual and group participation are recorded.

For the final grade, all but the last item are given equal weight and the last is used to resolve borderline cases. Thus an individual student's grade is determined, essentially, by six scores, four of which are group scores and two are individual. This imbalance between group and individual rewards is moderated by one other consideration. If the student's individual scores differ sharply from her or his group scores, then as much as a single letter upgrade or downgrade in the direction of the individual scores will be given.

Uses of assessments

We can summarize the uses we make of the various assessment activities as follows: assigning grades, reconsidering specific parts of the course, and reconsidering our entire approach. We described in the previous section how assignment scores, exam scores and classroom participation of individuals and groups are combined in determining the grade of an individual student. As always, such measures leave some students on the borderline between two possible grades. In these cases we apply our feeling, based on our research, that when all of the components of our course work as we intended them to, then learning takes place. Thus, if the course seemed to go well overall, in its own terms, and students appeared to buy into our approach, then we will tend to choose the higher grade. Otherwise, we will not give students much "benefit of the doubt" in determining the final grade.

The combination of course information about each student and research on students in general provides a sort of triangulation that can be used in a formative way in some cases. If the research tells us that students who experience our approach are likely to learn a particular concept, but they do not do well on this point in the actual course, then we look hard at the specific implementation as it regards that concept. If, on the other hand, students perform well on a particular concept but research suggests that their understanding leaves much to be desired, then we worry that the performance might be due to memorization or other superficial strategies. Finally, if both student performance in courses and subsequent research suggests that they "are not getting it," then we think about local revisions to our approach.

This latter occurs from time to time. For example, in the C4L calculus reform project, we have a set of computer activities designed to help students develop an understanding of the limit concept by looking at the problem of adjusting the horizontal dimension of a graphics window so as to keep a particular curve within a specified vertical dimension. Students don't like these problems very much and don't do exceptionally well on related exam questions. Moreover, there is nothing in the research to suggest anything striking in their understanding of the limit concept. Therefore we have reconsidered and adjusted our approach to limits.

Finally, there is the possibility that, over a period of time, performance of students in courses and research could lead us to a more general feeling of malaise with respect to our overall methods. In this case, more systemic changes, including the possibility of rejecting the entire approach would be considered. So far, this has not happened.

Conclusion

There are two comments to make in evaluating our approach. One is that it is clear that this approach to assessment reflects the principles we espouse and what we think research tells us. In particular, we have addressed the questions raised at the beginning of this article. The second is that we cannot say with certainty how effective is our assessment. We do have research reports that encourage us but, in the end, teaching, like parenting, is an activity in which we can never really know how effective were our efforts. We can only try as hard as we can to determine and implement what seems to us to be the most effective approaches, and then hope for the best.

References

[1] Asiala, M., Brown, N., DeVries, D., Dubinsky, E., Mathews, D. and Thomas, K. "A Framework for Research and Development in Undergraduate Mathematics Education," *Research in Collegiate Mathematics Education II*, CBMS Issues in Mathematics Education, 6, 1996, pp. 1–32.

[2] Dubinsky, E. "A Learning Theory Approach to Calculus," in Karian, Z., ed. *Symbolic Computation in Undergraduate Mathematics Education*, MAA Notes Number 24, The Mathematical Association of America, Washington, DC, 1992, pp. 48–55.

[3] Dubinsky, E. "ISETL: A Programming Language for Learning Mathematics," *Comm. in Pure and Applied Mathematics*, 48, 1995, pp. 1–25.

[4] Fenton, W.E. and Dubinsky, E. *Introduction to Discrete Mathematics with ISETL*, Springer, 1996.

[5] Kohn, A. "Effects of rewards on prosocial behavior," *Cooperative Learning*, 10 (3), 1990, pp. 23–24.

[6] Leron, U. and Dubinsky, E. "An Abstract Algebra Story," *American Mathematical Monthly*, 102 (3), 1995, pp. 227–242.

[7] Reynolds, B.E., Hagelgans, N.L., Schwingendorf, K.E., Vidakovic, D., Dubinsky, E., Shahin, M., and Wimbish, G.J., Jr. *A Practical Guide to Cooperative Learning in Collegiate Mathematics*, MAA Notes Number 37, The Mathematical Association of America, Washington, DC, 1995.

[8] Slavin, R.E. "When does cooperative learning increase student achievement?" *Psychological Bulletin* 94, 1983, pp. 429–445.

An Evaluation of Calculus Reform: A Preliminary Report of a National Study

Susan L. Ganter
American Association for Higher Education

This is a preliminary report of a study at NSF dealing with what NSF has looked for, what it has found, and directions for future study. One such direction will be to shift from "teaching" to "student learning" and the learning environment.

Background and Purpose

In 1986, the National Science Board (NSB) published a report from their Task Committee on Undergraduate Science and Engineering Education, whose charge was to consider the role of the National Science Foundation (NSF) in undergraduate education. The Neal Report, as it is commonly known, outlined the problems in undergraduate mathematics, engineering, and science education that had developed in the decade prior to 1986. Specifically, the Task Committee discussed three major areas of the undergraduate environment that required the highest level of attention: (1) laboratory instruction; (2) faculty development; and, (3) courses and curricula. Their primary recommendation to the NSB was the development of a plan for "new and innovative program approaches that will elicit creative proposals from universities and colleges...as part of the National Science Foundation" ([5], p. v). As a response to this and other reports (e.g., [2], [10]), NSF published their first program announcement [6] for calculus in 1987, with the first awards implemented in 1988. This program has served as a driving force for the national effort in the mathematics community known as the calculus reform movement.

Institutions nationwide have implemented programs as part of the calculus reform movement, many of which represent fundamental changes in the content and presentation of the course. For example, more than half of the projects funded by NSF use computer laboratory experiences, discovery learning, or technical writing as a major component of the calculus course, ideas rarely used prior to 1986 [3]. The content of many reform courses focuses on applications of calculus and conceptual understanding as important complements to the computational skills that were the primary element of calculus in the past. It is believed by many that such change is necessary for students who will live and work in an increasingly technical and competitive society.

A critical part of this process of change is the evaluation of these programs and their impact on the learning environment. As early as 1991, NSF began receiving pressure from the academic community and Congress to place more emphasis on evaluating the impact of these developments on student learning and the environment in undergraduate institutions (see [7]). Although this pressure has resulted in a heightened awareness of the need for evaluation and financial support for a few such studies, the area of evaluation research in undergraduate reform is still in its infant stages, with much of the work done by graduate students in the form of unpublished doctoral dissertations and master's theses. Only through additional studies can the mathematics community continue to develop the calculus course in ways that are most conducive to the needs of their students, the profession, and society.

A number of reports that present programmatic information and indicators of success in the efforts to incorporate technology and sound pedagogical methods in calculus courses have indeed been written (e.g., [8], [9], [12]). Reform has received mixed reviews, with students seemingly faring better on some measures, while lagging behind students in traditional courses on others. However, these reports present only limited information on student learning in reform courses, primarily because the collection of reliable data is an enormous and complicated task and concrete guidelines on how to implement meaningful evaluations of reform efforts simply do not exist ([12]). The need for studies that determine the impact of these efforts, in combination with the increase in workload brought on by reform, is creating an environment of uncertainty. Funding agencies, institutions, and faculty require the results of such studies to make informed decisions about whether to support or withdraw from reform activities.

This study is being conducted as a part of a larger effort by NSF to evaluate the impact of reform in science, mathematics, engineering, and technology (SMET) education at the undergraduate level. This study has been designed to investigate what is currently known about the effect of calculus reform on (1) student learning, attitudes, and retention; (2) use of mechanisms that historically have been shown to improve the learning environment, i.e., faculty development activities, student-centered learning, and alternative methods of delivery and assessment of knowledge; and, (3) the general educational environment. Preliminary results from this project will be reported here, including information from NSF projects, insights from the mathematics community, and anticipated implications for future efforts in calculus.

Method

The calculus reform initiative that NSF encouraged through awards from 1988 to 1994 set the direction for much of the undergraduate reform that has followed. Therefore, it is especially important that a thorough study of this pioneering effort be conducted in order to make informed decisions not only about the future of calculus, but also regarding all reform efforts in undergraduate SMET education. To this end, the project was designed to synthesize what is currently known about the impact of calculus reform on the learning environment, including student learning, attitudes, and retention. The information to be presented is not intended to be a definitive end to the evaluation process, but rather a progress report of what is known to date. It is expected that this information will generate discussion within the academic community, resulting in additional evaluation studies.

For the purpose of this evaluation study, information was gathered using the following methods:

1. A search of literature was conducted, including journal articles, conference proceedings, dissertations, and other relevant publications, to identify evaluations of calculus reform that are available. The information was compiled and synthesized to determine what is currently known about calculus reform from these evaluations.

2. Folders containing all information that has been submitted to NSF for each NSF-funded calculus project were searched for any proposed evaluation of the project and corresponding findings, as well as dissemination information. This evaluation information has been summarized in a qualitative database and cross-analyzed between projects. A framework for the information to be documented in this database was developed in consultation with NSF program officers prior to the development of the database. Precise definitions to be used in determining the existence of various evaluation, dissemination, and reform activities for each project were also developed and used to guide the data entry.

3. A letter was developed and sent to approximately 600 individuals who have participated in the reform efforts. The letter requested assistance in the compilation of existing evaluation studies in calculus reform. The mailing list included not only principal investigators from NSF calculus projects, but also individuals from non-funded efforts and others who have been involved in the evaluation of calculus projects. The names of those who have not been affiliated with an NSF project were obtained through the literature search. The letter was sent via email to a select group of mathematicians and mathematics educators for feedback and comments prior to the mailing.

Preliminary Findings

1. Analysis of NSF Projects

NSF folders have been obtained for the 127 projects awarded to 110 institutions as part of the calculus initiative (1988–94). Each folder was reviewed and analyzed as discussed above, yielding the following information:

- NSF funding for individual projects ranged from $1,500 per year to $570,283 per year, with a mean annual award of $186,458; the duration of the awards was usually two years or less;

- computer use/laboratory experience, applications, and conceptual understanding are the objectives of reform deemed most important in the projects;

- pedagogical techniques most often cited as part of the projects are technical writing, discovery learning, use of multiple representations of one concept, and cooperative learning;

- the most popular methods for distributing results to the community were conference presentations, journal articles, organization of workshops, and invited presentations at college colloquia;

- evaluations conducted as part of the curriculum development projects mostly concluded that students in reform courses had better conceptual understanding, higher retention rates, higher confidence levels, and greater levels of continued involvement in mathematics than those in traditional courses; however, scores on common traditional exams yielded mixed results, making it unclear whether there was any significant loss of traditional skills in reform students.

2. *Information obtained from the Community*

Almost 200 evaluation documents have been obtained from the academic community, including published papers and curricular materials, dissertations, printed conference presentations and proceedings, letters describing results from projects, and internal reports submitted within various colleges and universities. All information is being summarized in qualitative data files that will be discussed in conjunction with the data on the NSF projects in the full report. A preliminary review of the information submitted has revealed the following trends:

- the reform effort has motivated many (often controversial) conversations among faculty about the way in which calculus is taught; these conversations are widespread and continuous and they have resulted in a renewed sense of importance about undergraduate mathematics education;

- in general, regardless of the reform method used, the attitudes of students and faculty seem to be negative in the first year of implementation, with steady improvement in subsequent years while making continuous revisions based on feedback;

- most faculty believe that the "old" way of teaching calculus was ineffective; however, there is much debate about whether reform is moving in the right direction;

- there does not seem to be any consistency in the characteristics of faculty who are for/against reform; however, the students who seem to respond most positively to reform are those with little or no experience in calculus prior to entering the course, with the students who excel in the traditional environment having the most negative reactions to reform methods;

- the requirements of standardized tests seem to have the greatest effect on the adoption of reform practices by secondary AP calculus teachers; i.e., the recently implemented requirement of graphing calculators on the AP calculus exam has ensured their use in virtually every secondary calculus classroom, even though many AP teachers are opposed to such a requirement;

- faculty universally agree upon the importance of evaluating the effect of the reform efforts on student learning, faculty and student attitudes, and curriculum development; they believe that this information is needed to justify their work within academic departments, to understand the impact of various methods, and to give them motivation to continue in the struggle.

Implications of Findings for the Reform Environment

The existence of common elements in many of the calculus reform projects with varying levels of success implies that the impact of reform is perhaps not so dependent upon what is implemented, but rather the educational environment that is created in which to implement it. This environment is determined in part by the level of departmental commitment and the amount of faculty involvement in the reform efforts. Specifically, departments in which the reform courses are developed and supported by only a small portion of the faculty inevitably confront difficulties, whether those difficulties be the inconsistency between their courses or simply the exhaustion of those involved as they work to keep the program going, often at the expense of a positive relationship with their colleagues.

However, even departments that are very committed to reform will confront problems. It is the anticipation of these problems as well as the construction of methods for handling them that seems most critical to the continuing success of a program. For example, the expectations of a reform calculus course are often ones with which students (and their instructors) have had very little experience. How can students be engaged in productive group learning situations? What are the most effective and appropriate uses of computers? These and the multitude of other questions that reform implies about the educational experience in calculus have made apparent the need for additional support from a variety of sources as both students and faculty adjust their styles in the classroom. These necessary resources include, for example, technical and educational support for faculty and special study sessions and computer assistance for students.

Directions for Further Study

A detailed report of the results from this evaluation project will be published and distributed as appropriate to the broader community. This report will provide information that can be used not only by NSF, but also by the mathematics community at large to inform educational improvements at the undergraduate level. In addition, the report will recommend a plan for the continuing evaluation of calculus reform. The need for studies that employ more in-depth, rigorous data collection, including studies of the long-term impact of the reform efforts, is clearly an area for further research.

Specifically, projects that contribute to the sparse literature addressing the effects of undergraduate reform on student learning, attitudes, and retention are much needed. Although there has been considerable work in the area of student learning in calculus (e.g., [1], [4], and [11]), most of these investigations have looked at how students learn. It is imperative to understand not only how students learn, but also the actual impact of different environments created by the calculus reform movement on their ability to learn. This information can be used to inform allocation of resources and to help the faculty involved in reform to continue developing their ideas in ways that will be most productive for them and their students.

In addition, this study has made evident the need to investigate further the impact of reform on faculty, departments, and institutions, including methods for developing the teaching styles of current and future instructors in ways that are conducive to reform. Such work and the implementation of the resulting findings could help faculty to better understand their students and the pedagogical methods that best suit their needs at various times in their undergraduate experience. It will also help departments and institutions to provide the most effective educational experience for their students.

This research was supported by a grant from the American Educational Research Association which receives funds for its "AERA Grants Program" from the National Science Foundation and the National Center for Education Statistics (U.S. Department of Education) under NSF Grant #RED-9452861. Opinions reflect those of the author and do not necessarily reflect those of the granting agencies. The author would like to thank John Luczak, Patty Monzon, Natalie Poon, and Joan Ruskus of SRI International for their assistance in compiling the information for this report.

Please send all correspondence regarding this article to:
    Dr. Susan L. Ganter
    Director, Program for the Promotion of Institutional
      Change
    American Association for Higher Education
    One Dupont Circle, Suite 360
    Washington, DC 20036
    sganter@aahe.org
    202-293-6440, ext. 32

References

[1] Davidson, N.A. *Cooperative Learning in Mathematics: A Handbook for Teachers*, Addison-Wesley Publishing Company, Inc., Menlo Park, CA, 1990.

[2] Douglas, R. *Toward a Lean and Lively Calculus*, Report of the Tulane Conference on Calculus, Mathematical Association of America, Washington DC, 1987.

[3] Ganter, S.L. "Ten Years of Calculus Reform and its Impact on Student Learning and Attitudes," *Association for Women in Science Magazine*, 26(6), Association for Women in Science, Washington, DC, 1997.

[4] Heid, M.K. "Resequencing Skills and Concepts in Applied Calculus using the Computer as a Tool," *Journal for Research in Mathematics Education*, 19(1), National Council of Teachers of Mathematics, Reston, VA, 1988, pp. 3–25.

[5] National Science Board. *Undergraduate Science, Mathematics and Engineering Education: Role for the National Science Foundation and Recommendations for Action by Other Sectors to Strengthen Collegiate Education and Pursue Excellence in the Next Generation of U.S. Leadership in Science and Technology,* Report of the Task Committee on Undergraduate Science and Engineering Education, Neal, H., Chair, Washington DC, 1986.

[6] National Science Foundation. *Undergraduate Curriculum Development in Mathematics: Calculus.* Program announcement, Division of Mathematical Sciences, Washington, DC, 1987.

[7] National Science Foundation. *Undergraduate Curriculum Development: Calculus,* Report of the Committee of Visitors, Treisman, P. Chair, Washington, DC, 1991.

[8] Roberts, A.W., ed. *Calculus: The Dynamics of Change*, Mathematical Association of America, Washington, DC, 1996.

[9] Solow, A., ed. *Preparing for a New Calculus*, Conference Proceedings, Mathematical Association of America, Washington, DC, 1994.

[10] Steen, L.A., ed. *Calculus for a New Century*, Mathematical Association of America, Washington, DC, 1987.

[11] Tall, D. "Inconsistencies in the Learning of Calculus and Analysis," *Focus on Learning Problems in Mathematics*, 12 (3 and 4), Center for Teaching/Learning of Mathematics, Framingham, MA, 1990, pp. 49–63.

[12] Tucker, A.C. and Leitzel, J.R.C., eds. *Assessing Calculus Reform Efforts: A Report to the Community*, Mathematical Association of America, Washington, DC, 1995.

Increasing the Dialogue About Calculus with a Questionnaire

Darien Lauten, Karen Graham, and Joan Ferrini-Mundy
Rivier College in Nashua and University of New Hampshire at Durham

A practical survey is provided here which might be found useful for departments looking to initiate discussion about goals and expectations in a Calculus course.

Background and Purpose

Many mathematics departments across the country are considering change in the teaching of calculus. In our work on the Calculus Consortium Based at Harvard Evaluation and Documentation Project (CCH EDP) [2], we developed a set of questionnaires. Items from these questionnaires can provide mathematics departments with a vehicle for beginning discussions about possible change in the teaching of calculus or for assessing changes already made. Throughout the project we were constantly reminded of the wide variation in participants' perspectives and local situations. The implementation of reform-based calculus should be different at each institution because implementation patterns reflect local student needs, faculty attitudes and beliefs, and characteristics unique to institutions.

Method

The CCH EDP began in the fall of 1994. In this project, we sought (a) to investigate faculty perceptions of student learning and faculty attitudes and beliefs towards calculus reform and (b) to examine and describe the evolution of efforts to reform the teaching of calculus in the context of CCH Curriculum Project materials ([3]). As part of the CCH EDP, we surveyed 406 faculty members, who were using or had used CCH Curriculum Project materials, at 112 institutions of higher education. In the surveys, we asked about types of courses, types of students using the materials, faculty interpretation and use of the materials, pedagogical approaches, the extent of technology use, institution and department reaction to the use of the materials, factors influencing the initiation of reform efforts, and faculty attitudes and beliefs.

Findings

The questionnaire (see Appendix 1) contained in this article represents a subset of the original CCH EDP survey items. The items were chosen because we found, based on responses to CCH EDP survey, that they revealed significant information about how faculty interpret and implement reform-based calculus. Many of the questionnaire items were based on goals for reform in calculus instruction established at the Tulane Conference in 1986 [1].

Use of Findings

Our experience indicates that casual and formal discussions within a department are critical in assessing a calculus program. There are several ways you might use the questionnaire. You could use the questions as the basis for calculus-instructor or full-department discussions without the faculty completing the survey beforehand. You may decide to modify the items to develop a questionnaire more suited to your own situation or to include fewer or more comment-type questions. On the other hand, you might ask faculty members to anonymously complete and return the questionnaire prior to a meeting. You could collate and analyze the responses and distribute an analysis of the quantitative responses and a listing of the comments.

The responses to each set of items could serve as the foundation for more formal calculus-instructor or department discussions. To get a discussion rolling, you might ask the following useful discussion questions:

- How would you characterize your department's approach to calculus instruction? What are the more important features, the less important?
- Is using technology the basis for your efforts? Why or why not?
- Do your efforts encourage student engagement with calculus? How do you know?
- What role do the following pedagogical methods play in your approach: cooperative groups, projects, student presentations, using concrete materials to teach calculus? What solid evidence do you have that these techniques work or that the lecture method works?
- What does "rigor" mean to you, and what importance do you give "rigor" in the teaching of calculus?

Success Factors

Continued use of this questionnaire will refine the issues that are important to you as a faculty and the process could initiate bringing about changes in the questions you ask. The questionnaire provides a format for meetings in which the faculty, and possible student representation, can come together periodically to discuss the results. In this way, your department can create a beginning for reforming the learning environment. Above all, your faculty and students will become more sensitized to the goals of the department and the issues surrounding the teaching of calculus.

References

[1] Douglas, R., ed. *Toward a Lean and Lively Calculus: Report of the conference/workshop to develop curriculum and teaching methods for calculus at the college level* (MAA Notes Series No. 6), Mathematical Association of America, Washington, DC, 1986.

[2] Ferrini-Mundy, J., *CCH Evaluation and Documentation Project*, University of New Hampshire, Durham, NH, 1994.

[3] Hughes-Hallett, D., Gleason, A.M., Flath, D.E., Gordon, S.P., Lomen, D.O., Lovelock, D., McCallum, W.G., Osgood, B.G., Pasquale, A., Tecosky-Feldman, J., Thrash, J.B., Thrash, K.R., & Tucker, T.W. *Calculus*, John Wiley & Sons, Inc., New York, 1992.

Appendix

CALCULUS QUESTIONNAIRE
Please answer all questions and comment as often as you wish.

1. CONTENT:

1A. Please indicate the level of emphasis that you feel should be placed on the following topics in a first-year calculus course.

	Amount of emphasis little or none ——— heavy
a. Preliminaries (functions, absolute value, etc.)	1 2 3 4 5
b. Limits (lengthy treatment, rate "heavy")	1 2 3 4 5
c. Derivative as a rate, slope of tangent line, etc.	1 2 3 4 5
d. Using definition to find derivative	1 2 3 4 5
e. Techniques of differentiation	1 2 3 4 5
f. Applications of the derivative (max/min, related rates, etc.)	1 2 3 4 5
g. Fundamental Theorem of Calculus (lengthy treatment, rate "heavy")	1 2 3 4 5
h. The definite integral as area (lengthy treatment, rate, "heavy")	1 2 3 4 5
i. Techniques of integration	1 2 3 4 5
j. Applications of integration (arc length, volumes of solids, surface area, etc.)	1 2 3 4 5
k. Applications of exponential/logarithmic functions (growth, decay, etc.)	1 2 3 4 5
l. Solving differential equations	1 2 3 4 5
m. Applications of differential equations	1 2 3 4 5
n. Series (lengthy treatment, rate "heavy")	1 2 3 4 5
o. Series — techniques to determine convergence	1 2 3 4 5
p. Taylor series (lengthy treatment, rate "heavy")	1 2 3 4 5
q. Applications of Taylor Series	1 2 3 4 5
r. Parametrizations, Vectors	1 2 3 4 5

1B. Please add any other topics that you feel should be included, and any other remarks concerning *content*. *Technology* will be addressed separately.

2. THEORETICAL METHODOLOGY

2A. Please indicate the level of emphasis that you feel should be placed on the following methods in a first-year calculus course. It might be helpful to think about how much emphasis you place on these items in your assessment of students. *Technology* will be addressed separately.

	Amount of emphasis				
	little or none				heavy
a. Formal definitions	1	2	3	4	5
b. Statements of theorems, counterexamples, etc.	1	2	3	4	5
c. Proofs of significant theorems	1	2	3	4	5
d. Historical themes in mathematics	1	2	3	4	5
e. Writing assignments	1	2	3	4	5
f. Student practice of routine procedures	1	2	3	4	5
g. Applications of real world problems	1	2	3	4	5
h. The analysis and solution of non-routine problems	1	2	3	4	5

2B. Please state your own definition of the phrase "mathematical rigor."

2C. Based on your definition, please describe what level is appropriate in first-year calculus and how you would assess (grade, for example) at that level.

3. TECHNOLOGY FOR LEARNING:

3A. Please state your feelings about the role of technology in the classroom, in promoting or hindering learning in a first year calculus course.

3B. To what extent do you feel technology should be integrated throughout the course versus for special projects, if at all.

3C. Please select the response that best represents your views about the ideal use of calculators or computers in the classroom.

	Amount of emphasis				
	little or none				heavy
a. Calculators for numerical purposes	1	2	3	4	5
b. Calculators for graphing purposes	1	2	3	4	5
c. Calculators for symbolic manipulation	1	2	3	4	5
d. Computer courseware (Maple, Mathematica, etc.)	1	2	3	4	5
e. Modifying existing programs/Programming	1	2	3	4	5
f. Spreadsheets or tables	1	2	3	4	5

4. CLASSROOM TEACHING APPROACHES

4A. Please respond: In an ideal calculus course, how frequently would your students use the following instructional systems?

	not frequent				very frequent
a. Use lecture notes as basis for learning	1	2	3	4	5
b. Participate in a specially designed calculus laboratory	1	2	3	4	5
c. Use concrete materials/equipment to explore calculus ideas	1	2	3	4	5
d. Work in small groups on mathematics problems	1	2	3	4	5
e. Work in small groups on projects that take several class meetings to complete	1	2	3	4	5
f. Practice calculus procedures in the classroom	1	2	3	4	5
g. Make conjectures, explore more than one possible method to solve a calculus problem	1	2	3	4	5

4B. Please add any additional comments. Perhaps you would like to address the phrase "ideal classroom." What types of support might your department in an ideal world provide you with to help you accomplish your teaching goals?

5. STUDENT ASSESSMENT/ EVALUATION

5A. In your calculus course, what importance to course grade do you assign to each of the following items? (1) Please rate on the scale. (2) Please circle the methods of assessment that you would like to discuss.

	unimportant				very important
a. Quizzes, tests, or examinations that measure individual mastery of content material	1	2	3	4	5
b. A final examination that measures individual mastery of content material	1	2	3	4	5
c. Individual tests of mastery of content material	1	2	3	4	5
d. Small group tests of mastery of content material	1	2	3	4	5
e. Lab reports — individual grades	1	2	3	4	5
f. Lab reports — group grades	1	2	3	4	5
g. Quizzes, tests, or examinations of material learned in labs	1	2	3	4	5
h. Homework exercises — individual grades	1	2	3	4	5
i. Homework exercises — group grades	1	2	3	4	5
j. Projects — individual grades	1	2	3	4	5
k. Projects — group grades	1	2	3	4	5
l. Journals	1	2	3	4	5
m. Class participation	1	2	3	4	5
n. Portfolios	1	2	3	4	5
o. Other: Please describe:	1	2	3	4	5

6. PERSPECTIVES ON CALCULUS REFORM

6A. How aware are you of "calculus reform" issues and efforts? Please explain.

6B. What do you find encouraging about the directions of calculus reform?

6C. What are your concerns about the directions of calculus reform?

7. PERSPECTIVES ON THE CURRENT TEXT:
Please write your comments about the text in use in your department? Do you want to change the current text? Please explain.

8. PERSPECTIVES ON THE CURRENT STUDENTS:
It is important to share perspectives about the students we teach. Please give your impressions of the latest Calculus class you have taught, and share any comments from the students that you would like to pass on to members of the department.

Workshop Calculus:
Assessing Student Attitudes and Learning Gains

Nancy Baxter Hastings
Dickinson College

This assessment program stresses breadth — pre- and post-testing, journals, comparative test questions, student interviews and questionnaires, and more.

Background and Purpose

Dickinson College is a four-year, residential, liberal arts institution serving approximately 1,800 students. During the past decade, the introductory science and mathematics courses at the college have been redesigned to emphasize questioning and exploration rather than passive learning and memorization. The Workshop Calculus project[1], now in its sixth year of development, is part of this college-wide effort.

Workshop Calculus is a two-semester sequence that integrates a review of pre-calculus concepts with the study of fundamental ideas encountered in Calculus I: functions, limits, derivatives, integrals and an introduction to integration techniques. The course provides students who have had three to four years of high school mathematics, but who are not prepared to enter Calculus I, with an alternate entry point into the study of higher-level mathematics. It seeks to help these students, who might otherwise fall through the cracks, develop the confidence, understanding and skills necessary to use calculus in the natural and social sciences and to continue their study of mathematics. After completing both semesters of Workshop Calculus, workshop students join their peers who have completed a one-semester Calculus I course, in Calculus II.

All entering Dickinson students who plan to take calculus are required to take the MAA Calculus Readiness Exam. Students who score below 50% on this exam are placed in Workshop Calculus, while the others enter Calculus I. The two strands serve the same clientele: students who plan to major in mathematics, physics, economics or other calculus-based disciplines. While both courses meet 5 hours a week, Calculus I has 3 hours of lecture and 2 hours of laboratory sessions, while there is no distinction between classroom and laboratory activities in Workshop Calculus, which is primarily hands-on.

A forerunner to Workshop Calculus consisting of a review of pre-calculus followed by a slower-paced version of calculus was developed in response to the fact that 30–40% percent of the students enrolled in Calculus I were having difficulty with the material, even though they appeared to have had adequate high school preparation. Students' reactions to the course, which was lecture-based with an emphasis on problem solving, were not positive. They did not enjoy the course, their pre-calculus skills did not improve and only a few ventured on to Calculus II. In addition, colleagues in client departments, especially economics and physics, continued to grumble about the difficulty students had using calculus in their courses. Workshop Calculus was

---

[1] The Workshop Calculus project has received support from the Knight Foundation, the US Department of Education Fund for Improvement of Postsecondary Education (FIPSE) and the National Science Foundation (NSF).

the department's response.

With the Workshop approach, students learn by doing and by reflecting on what they have done. Lectures are replaced by interactive teaching, where instructors try not to discuss an idea until after students have had an opportunity to think about it. In a typical Workshop class, the instructor introduces students to new ideas in a brief, intuitive way, without giving any formal definitions or specific examples. Students then work collaboratively on tasks in their Workshop Calculus Student Activity Guide [1]. The tasks in this manual are learner-centered, including computer tasks, written exercises and classroom activities designed to help students think like mathematicians — to make observations and connections, ask questions, explore, guess, learn from their errors, share ideas, and read, write and talk mathematics. As they work on assigned tasks, the instructor mingles with them, guiding their discussions, posing additional questions, and responding to their queries. If a student is having difficulty, the instructor asks the student to explain what he or she is trying to do and then responds using the student's approach, trying not to fall into let-me-show-you-how-to-do-this mode. The instructor lets — even encourages — students to struggle, giving only enough guidance to help them overcome their immediate difficulty. After completing the assigned activities, students participate in class discussions, where they reflect on their own experiences. At this point, the instructor can summarize what has been happening, present other theoretical material or give a mini-lecture.

Method

Assessment activities are a fundamental part of the Workshop Calculus project. With the help of external collaborators[2], we have analyzed student attitudes and learning gains, observed gender differences, collected retention data and examined performance in subsequent classes. This information was used to help make clearer the tasks expected of students laid out in our Study Guide. More significantly, it has provided the program with documented credibility, which has helped the Workshop program gain the support of colleagues, administrators, outside funding agencies, and even the students themselves. The following describes some of the tools we have used. For a more in-depth description of these assessment tools and a summary of some of the results, see [2].

a. *Collecting Baseline Data.* During the year prior to introducing Workshop Calculus, baseline information was collected concerning students' understanding of basic calculus concepts. Workshop Calculus students were asked similar questions after the new course was implemented. For example, students in the course prior to the Workshop version were asked:

- What is a derivative?
- If the derivative of a function is positive in a given interval, then the function is increasing. Explain why this is true.

In response to the first question, 25% of the students stated that the derivative represented the slope of the tangent line, and half of these students used this fact to give a reasonable explanation to the second question. The remaining students answered the first question by giving an example and either left the second question blank or wrote a statement that had little relationship to the question; they could manipulate symbols, but they didn't understand the concepts.

This data showed the need to emphasize conceptual understanding of fundamental concepts and to have the students write about these ideas. Consequently, Workshop students learn what an idea is — for example, what a derivative is and when it is useful — before they learn how to do it — in this case, the rules of differentiation — and they routinely write about their observations.

b. *Administering Pre- and Post-tests.* Workshop Calculus students answer questions prior to undertaking particular activities and then are re-asked the questions later. For example, on the first day of class, students are asked to write a short paragraph describing what a function is, without giving an example. Although they all claim to have studied functions in high school, many write gibberish, some leave the question blank, and only a few of students each year describe a function as a process. After completing the activities in the first unit of their activity guide (where they do tasks designed to help them understand what a functions is, without being given a formal definition of "function"), nearly 80% give correct, insightful descriptions of the concept of function.

c. *Analyzing Journal Entries.* At the end of each of the ten units in the Workshop Calculus activity guide, students are asked to write a journal entry, addressing the following questions:

- Reflect on what you have learned in this unit. Describe in your own words the concepts that you studied and what you learned about them. How do they fit together? What concepts were easy? Hard? What were the main, important ideas? Give some examples of these ideas.

---

[2] Ed Dubinsky, from Georgia State University, and Jack Bookman, from Duke University, helped assess student learning gains and attitudes in Workshop Calculus. They were funded by FIPSE.

- Reflect on the learning environment for the course. Describe the aspects of this unit and the learning environment that helped you understand the concepts you studied.
- If you were to revise this unit, describe the changes you would make. What activities did you like? Dislike?

The students' responses provide the instructor with valuable insight into their level of understanding; their candid replies also provide important feedback about what works and doesn't work and about changes that might need to be made.

d. Asking Comparative Test Questions. Student performance in Workshop Calculus has been compared to students at other institutions. For instance, on the final exam for Workshop Calculus, in the spring of 1994, students were asked four questions from an examination developed by Howard Penn at the US Naval Academy to assess the effectiveness of using the Harvard materials at the USNA versus using a traditional lecture-based approach [3]. (See the Appendix for these questions.) We are pleased to report that the Workshop Calculus students did about as well as the USNA students who were using the Harvard materials, even though the Academy is certainly more selective than Dickinson College.

e. Conducting Critical Interviews. A representative group of Workshop students have been interviewed in structured ways, to help determine what they are thinking as they work on a given problem and to determine their level of understanding of a particular concept. For this, a questionnaire is administered to students, and we categorize responses and approaches used, even taping one-on-one "critical interviews" with a representative group of students, transcribing the interview sessions, and analyzing the results. After using the first version of the Workshop Calculus materials in the spring of 1992, this approach was used, for instance, to analyze students' understanding of "definite integral." After analyzing students' responses on the questionnaire and during the interview sessions, we realized that some students could only think about a definite integral in terms of finding the area under a curve and had difficulty generalizing.[3] Based on this observation, the tasks pertaining to the development of definite integral in the Student Activity Guide were revised.

f. Scrutinizing the End-of-Semester Attitudinal Questionnaire. At the end of each semester, students are asked to rate the effectiveness of various activities, such as completing tasks in their activity guide, participating in discussions with peers, using a computer algebra system. They are also asked to rate their gains in understanding of fundamental ideas — such as their understanding of a function, or the relationship between derivatives and antiderivatives — to compare how they felt about mathematics before the course with how they feel after completing the course, and to describe the most and least useful activities in helping them learn mathematics.

Student responses are analyzed for gender differences. For instance, both male and female students who took Workshop Calculus in 1993-1994, claimed, on the average, that they felt better about using computers after completing the course than before. Men showed a greater increase in confidence, however, and even after the course, women were not as comfortable using computers as the men were initially.

g. Gathering Follow-up Data. Information about student attitudes is also collected from students who took either the Workshop Calculus sequence or the regular Calculus I course, one or two years after they complete the course (irrespective of whether they have gone ahead with mathematics or not).[4] Their responses are used to determine the impact of the course on their attitudes towards mathematics and their feelings about the usefulness and applicability of calculus in any follow-up courses. For instance, students are asked whether they strongly agree, somewhat agree, somewhat disagree, or strongly disagree to 46 statements, including:

- I am more confident with mathematics now, than I was before my calculus course.
- I feel that I can apply what I learned in calculus to real world problems.
- On the whole, I'd say that my calculus class was pretty interesting.

In addition, we gather data about Workshop Calculus students who continue their study of mathematics and/or take a course outside the Mathematics Department that has calculus as a pre-requisite, asking questions such as:

- How many Workshop students continued their study of mathematics by enrolling in Calculus II?
- How do the Workshop students perform in Calculus II in comparison with those who entered via the regular Calculus I route?
- How many Workshop students became mathematics majors?
- How do Workshop students do in courses outside the Department that have calculus as a pre-requisite?

Findings and Use of Findings

With the exception of the critical interviews, the assessment tools used for Workshop Calculus are easy to administer and the data can be easily analyzed. Moreover, these tools can be used in a variety of courses at a variety of institutions. In general, our assessment process requires developing a

---

[3] Ed Dubinsky helped design the questionnaire and analyze the results.
[4] Items on the follow-up questionnaire were developed by Jack Bookman, Project Calc, Duke University.

clear statement of intended outcomes, designing and utilizing supportive assessment tools, and analyzing the data and summarizing the results in ways that can be easily and readily understood (for instance, bar charts or graphs are helpful).

Success Factors

Something should be said here for the advantages of using a variety of assessment measures. We feel that our program benefits from the breadth of our methods for collecting information about performance, teaching, learning, and the learning environment. In particular our focus is on understanding how students are thinking, learning, and processing the courses. Often assessment measures raise more questions than they answer, and when students are asked questions, they may be reading and responding on different wavelengths from those we are broadcasting on. We have found that an advantage to using a "prismatic" assessment lens is that we obtain a variety of ways of exploring issues with students, and therefore we become closer to understanding students and more inspired to make the changes that will be beneficial to all of us. Good assessment offers students feedback on what they themselves report. We feel that our methods are appreciated by students, and that the department is perceived as caring as reforms are instituted. Good assessment also promotes ongoing discussion. And our measures have certainly helped to stimulate ongoing faculty dialogue, while unifying the department on the need for further assessment. It has been six years since the Workshop Calculus project began and we are still learning!

References

[1] Hastings, N.B. and Laws, P. Workshop Calculus: Guided Exploration with Review, Springer-Verlag, New York, vol. 1, 1996; vol. 2, 1998.

[2] Hastings, N.B. "The Workshop Mathematics Program: Abandoning Lectures," in D'Avanzo, C. and McNeal, A., *Student-Active Science: Models of Innovation in College Science Teaching*, Saunders Publishing Co., Philadelphia, 1997.

[3] Penn, H., "Comparisons of Test Scores in Calculus I at the Naval Academy," in *Focus on Calculus*, A Newsletter for the Calculus Consortium Based at Harvard University, 6, Spring 1994, John Wiley & Sons, Inc., p. 6.

Appendix

Q1 showed five graphs and asked which graph had $f'(x) < 0$ and $f''(x) < 0$.

Q2 showed the graph of a function h and asked at which of five labeled points on the graph of h is $h'(x)$ the greatest.

Q3 showed the graph of a function which was essentially comprised of two line segments—from $P(-1,0)$ to $Q(0,-1)$ and from $Q(0,-1)$ to $R(3,2)$—and asked to approximate the integral of the function from -1 to 3.

Q4 showed the graph of the derivative of a function, where f' was strictly positive, and asked at which of five labeled points on the graph of f' is $f(x)$ the greatest.

Does Calculus Reform Work?

Joel Silverberg
Roger Williams University

This large study examines data over a long period of time regarding a calculus "reform" course. Grades and attitude are studied, with advice for novices at assessment.

Background and Purpose

For many years now the mathematical community has struggled to reform and rejuvenate the teaching of calculus. My experience with one of these approaches is in the context of a small comprehensive college or university of 2400 students. Our university attracts students of modest academic preparation and achievement. Most freshmen come from between the 30th and 60th percentiles of their high school class. The central half of our students have mathematics SAT scores between 380 and 530 (prior to the re-norming of the exams). We offer support courses in elementary algebra, intermediate algebra, trigonometry, and precalculus for any student whose placement examinations show they lack the requisite skills for the first mathematics course required by their major.

In view of the characteristics of our student body, the nature of the improvements seen in calculus performance are extremely important. One or two sections of the six normally offered for first semester calculus were taught using a reform calculus approach and text developed by Dubinsky and Schwingendorf at Purdue University. The remaining sections were taught using a traditional approach and text supplemented by a weekly computer lab.

The goals we established for this course were to improve the level of understanding of fundamental concepts in calculus such as function, limit, continuity, and rate of change, and to improve student abilities in applying these concepts to the analysis of problems in the natural sciences and engineering. We hoped to reduce the number of students who withdrew from the course, who failed the course, or who completed the course without mastering the material at a level necessary to succeed in further mathematics courses or to successfully apply what they had learned in courses in science and engineering. Performance in these courses has frequently been sharply bimodal, and it was hoped that we could find an approach which would narrow the gap between our better students and our weaker students. Our past experience had been that between 20 and 45% of beginning calculus students failed to perform at a level of C– or better. Approximately half of that number withdrew from the course before completion, the remainder receiving grades of D or F. Furthermore, students passing with a C– or D in Calculus I seldom achieved a passing grade in Calculus II.

In this reform approach, the way in which material is presented is determined by extensive research into how mathematics is learned. The very order in which the topics appear is a reflection not of the finished mathematical structure, but of the steps students go through in developing their understanding from where they start to where we wish them to end. Mathematics is treated as a human and social activity, and mathematical notation is treated as a vehicle which people use for expressing and discussing mathematical ideas. A cooperative group learning environment is created and nurtured throughout the course. Students work in teams of 3–4 students in and out of class and stay with the same group for the entire semester. Group pride and peer pressure seem to play a significant motivational role in encouraging students to engage with the material in a nontrivial way.

The computer is used not for practice or for drill, but rather for creating mathematical objects and processes. The objects and processes which the students create are carefully chosen to guide the students through a series of experiences and investigations which will enable them to create their own mental models of the various phenomena under investigation. Basic concepts are developed in depth and at leisure, looking at them from many viewpoints and in many representations. ISETL and Maple are used to describe actions in mathematical language and to generate algorithmic processes which gradually come to be viewed as objects in their own right upon which actions can be performed. Objects under study are dissected to reveal underlying processes and combined with other processes to form new objects to study. The time devoted to what the student views as "real calculus" — the mechanics of finding limits, determining continuity, taking derivatives, solving applications problems, etc. is perhaps only one-half that in a traditional course. We hoped that the learning of these skills would go far more easily, accurately, and rapidly, because the underlying ideas and concepts were better understood, and in fact this did seem to be the case.

Method

Two assessment models were used, each contributing a different viewpoint and each useful in a different way. One was a quantitative assessment of student performance and understanding based upon performance on the final examination for the course. A parallel qualitative assessment was obtained through careful observation and recording of student activities, discussions, insights, misconceptions, thought patterns, and problem solving strategies. Some data came from student interviews during and after the classes. Other data came from the frequent instructor/student dialogues generated during classroom time, laboratory and office hours.

The quantitative assessment provided a somewhat objective measure, unclouded by the instructor's hopes and aspirations, of the general success or failure of the effort. Although rich in numerical and statistical detail, and valuable in outlining the effectiveness (or lack thereof) for various sub-populations of students, it offered few clues, however, as to directions for change, modification, and improvement.

The goal of the qualitative assessment was to gain a more complete idea of what the student is actually thinking when he or she struggles with learning the material in the course. Without a clear understanding of the student's mental processes, the instructor will instinctively fill in gaps, resolve ambiguities, and make inferences from context that may yield a completely inappropriate impression of what the student understands. If we assume that the student reads the same text, sees the same graph, ponders the same expression as we do, we will be unable to effectively communicate with that student, and are likely to fail to lead them to a higher level of understanding. The qualitative assessment, although somewhat subjective, proved a rich and fertile source for

ideas for improvement, which were eventually reflected through both assessment vehicles.

The two methods, used in tandem, were quite effective, over time, in fine tuning the course delivery to provide an enhanced learning environment for the vast majority of our students. By clarifying what it is that was essential for a student to succeed in the course, and by constantly assessing the degree to which the essential understandings were being developed (or not) and at what rate, and in response to which activities, the delivery of the course was gradually adjusted.

Findings

The instructors of the traditional sections prepared a final examination to be taken by all calculus sections. The exam was designed to cover the skills emphasized in the traditional sections rather that the types of problems emphasized in the reform sections. In an attempt to minimize any variance in the ways individual instructors graded their examinations, each faculty member involved in teaching calculus graded certain questions for all students in all sections.

At the start of this experiment, the performance of students in the reform calculus sections, as measured by this exam, lagged significantly behind students in traditionally taught sections. During the third semester of the experiment the students in the reform sections were scoring higher than their cohorts in the traditional sections. Furthermore, the segment of our population who most needed support, those in the lower half or lower two-thirds of their calculus section (as measured by performance on the final examination) proved to be those most helped by these changes in pedagogy. Although every class section by definition will have a lower half, it is quite a different story if the average grade of those in the lower half is a B than it is if it is a D. By the third semester of the experiment, the bottom quartile score of reform sections was higher than the median score of the traditional sections (see the graphs on the next page).

Student interviews, discussions, and dialog quickly revealed that what the student sees when looking at a graph is not what the teacher sees. What students hear is not what the instructor thinks they hear. Almost nothing can be taken for granted. Students must be taught to read and interpret the text, a graph, an expression, a function definition, a function application. They must be taught to be sensitive to context, to the order of operations, to implicit parentheses, to ambiguities in mathematical notation, and to differences between mathematical vocabulary and English vocabulary when the same words are used in both. Interviews revealed that the frequent use of pronouns often masks an ignorance of, or even an indifference to, the nouns to which they refer. The weaker student has learned from his past experience, that an instructor will figure out what "it" refers to and assume he means the same thing.

Final Exam Grades/ Traditional Sections

Percentile Range:	0–24	25–49	50–74	75–100
semester 1	C	C	B	A
semester 2	F	C	C+	A–
semester 3	D	C	B	B

Final Exam Grades /Reform Sections

Percentile Range:	0–24	25–49	50–74	75–100
semester 1	F	D	C	B–
semester 2	F	C–	C	B
semester 3	C	B	B	B+

Summary of Final Examination Scores (out of 160 points)

	Term 1		Term 2		Term 3	
	Reform	Trad	Reform	Trad	Reform	Trad
Maximum:	146	159	103	114	154	158
3rd Quartile	84	115	83.5	91	123	118.5
Median	86	91.5	66.5	75	105	93
1st Quartile	43.5	71	44.5	54.5	100.5	74
Minimum	16	43	31	2	55	9

Use of Findings

When first implemented, students resisted the changes, and their final exam grades lagged behind those of the other sections. Some of the students found the method confusing. On the other hand, others found this the first satisfying experience with mathematics since fourth grade, and reported great increases in confidence. Withdrawals in these courses dropped to zero. Analyzing the program, we found that among the students with a negative reaction, the responsibility and independence required of them was a factor. As a result, in future offerings I hired lab assistants; morale improved and student reactions became highly positive; scores meanwhile rose modestly. The challenge of the following semester was to convert student enthusiasm within the classroom setting to longer periods of concentration required outside the classroom. Here I myself became more active in guiding students. Student explorations were replaced by teacher-guided explorations, but simultaneously, I pushed students harder to develop a mastery of skills and techniques which the problems explored. Each group was required to demonstrate their problem solving skills in front of the other groups. This led to a drop in morale to more normal levels but also to sharp increases in student performance as demonstrated throughout the semester and also on the common final examination.

The dramatic changes in student performance were not observed the first or even the second time the course was offered. It required an ongoing process of evaluation, modification, revision, before achieving a form which appears to work for our students with their particular backgrounds. A break-in period of 1-1/2 to two years may be necessary for experimental approaches to be adapted to local conditions before the full effectiveness of the changes made can be determined and measured. The "curricular material" was essentially the same through all deliveries of the course. What was refined and addressed over time was the pedagogy and classroom approach, as well as student attitudes and morale issues.

Success Factors

The approach to teaching and assessment outlined in this paper requires that the professor take the time to figure out what is going on inside his or her student's minds. The professor should make no assumptions and let the students explain in their own words what it is that they see, read, hear, and do. The professor must then take the time to design activities that will help students replace their naive models with more appropriate ones. The syllabus and use of class time must be adapted to allow this to happen. Some students may be resistant to thinking about the consequences of their mathematical action and will be resistant to reasoning about what the mathematical symbols communicate. Some faculty may have difficulty understanding what you are doing, and prefer that you "train" or "drill" them to perform the required manipulations. But if approached with sensitivity, perseverance, and helpfulness most students can be encouraged to expand their horizons and indulge in some critical thinking. Some advice:

- Be patient: changes take time, changes in long established habits take a long time.
- Advise your students to be patient: learning is challenging, success in this course will require a lot of time on the part of the teacher and on the part of the student.
- Persevere: it make take you several tries to get it right.
- Maintain your sense of humor: some people (and they can be students, faculty, or administrators) don't like change.

- You will not please everybody, but you may be a welcome breath of fresh air and a new chance for success to many.
- Make no assumptions: be open to many revelations about the student's world view and his or her view of the material.
- Be creative in what you do to help develop the student's view of the material.
- Enjoy what you are doing and share that pleasure with your students.

Assessing the Effectiveness of Innovative Educational Reform Efforts

Keith E. Schwingendorf

Purdue University North Central

This study explains the creation of a calculus reform program, its objectives, and philosophy and provides an in-depth comparison of reform-trained students with traditional students.

Background and Purpose

The "Calculus, Concepts, Computers and Cooperative Learning," or C4L Calculus Reform Program is part of the National Calculus Reform Movement. The initial design of the C4L program began in 1987 under the leadership of Ed Dubinsky and Keith Schwingendorf on the West Lafayette campus of Purdue University, a large Midwestern Land Grant University. The C4L program has received NSF funding for its design and development during eight of the past nine years, most recently from Fall 1994 through Summer 1997 (grant #DUE-9450750) to continue and expand the program's assessment and evaluation efforts on two fronts: (1) qualitative research into how students learn calculus (mathematics) concepts, and (2) quantitative research and the development of a (national) analytical model for assessment and evaluation of the effectiveness of innovative educational reform efforts as compared to other pedagogical treatments. The C4L program is co-directed by Dubinsky (Georgia State University), Schwingendorf (Purdue University North Central), and David Mathews (Southwestern Michigan College, Dowagiac, MI).

This paper will describe the twofold assessment and evaluation process* developed as part of the C4L Calculus Reform Program. The major focus of this paper will be the analytical model designed to address differences within the study population of students and deal effectively with the limitation of self-selection — a key problem encountered by virtually all control versus alternative treatment studies in education and similar studies related to (among others) health and industrial issues. The analytical method allows researchers to make more meaningful comparisons and conclusions regarding the effectiveness of innovative curriculum reform treatments and other pedagogical treatments. (A detailed description of the C4L Study together with references on qualitative research studies regarding the C4L Program can be found in [3].) The qualitative research phase and its impact on curriculum development will be briefly addressed. (A detailed description of the research framework used in the qualitative phase of assessment of the C4L Calculus Reform Program can be found in [1].)

The C4L Program is based on a constructivist theoretical perspective of how mathematics is learned (see [1]). According to the emerging learning theory on which the C4L Program courses are based, students need to construct their own understanding of each mathematical concept. The framework and step-by-step procedure of the qualitative research phase is described in detail in [1].

*The C4L Study and subsequent research paper could not have been completed without the expert advice and generous time provided by our statistical consultants George and Linda McCabe, and Jonathan Kuhn (Purdue University North Central). The author wishes to thank Professor Kuhn for his insightful comments and suggestions for the completion of this paper.

C4L Calculus courses differ radically from traditionally taught courses and from courses in most other Calculus reform programs in many fundamental ways. Traditional courses, delivered primarily via the lecture/recitation system, in general, attempt to "transfer" knowledge, emphasize rote skill, drill and memorization of paper-and-pencil skills. In contrast, the primary emphasis of the C4L program is to minimize lecturing, explaining, or otherwise attempting to "transfer" mathematical knowledge, but rather to create situations which foster student to make the necessary mental constructions to learn mathematics concepts. The emphasis of the C4L program is to help students gain a deeper understanding of concepts than can be obtained via traditional means, together with the acquisition of necessary basic skills through the Activity-Class-Exercise, or ACE, learning cycle developed for the C4L Program, [1].

Each unit of the ACE learning cycle, which generally lasts about a week, begins with students performing computer investigations in a laboratory setting in an effort to help students construct their own meaning of mathematical concepts and reflect on their experiences with their peers in a cooperative learning environment. Lab periods are followed by class meetings in which a modified Socratic approach is used in conjunction with cooperative problem solving in small groups to help the students to build upon their mathematical experiences from the computer laboratory. Finally, relatively traditional exercises are assigned to reinforce the knowledge students are expected to have constructed during the first two phases of the learning cycle.

Method

(1) Qualitative Research Phase of the C4L Program. A critical aspect of the qualitative aspect of the C4L program is a "genetic decomposition" of each basic mathematical concept into developmental steps following a Piagetian theory of knowledge. Genetic decompositions, initially hypothesized by the researchers based on the underlying learning theory, are modified based on in-depth student interviews together with observations of students as they attempt to learn mathematical concepts. Qualitative interviews have been designed and constructed by the program's research team resulting to date in over 125 completed interviews on the concepts of limit, derivative, integral, and sequences and series. A complete description of the C4L Program's qualitative research procedure is carefully and completely described in [1].

(2) Quantitative Research Phase of the C4L Program. A longitudinal study (which will be referred to as the "C4L Study") of the C4L Calculus Reform Program was designed in an effort to make meaningful comparisons of C4L (reform) calculus students with traditionally taught students (TRAD) via lecture and recitation classes.

The students in the C4L Study were those enrolled on the West Lafayette campus of Purdue University from Fall 1988 to Spring 1991, in either the C4L (reformed) Program three semester calculus sequence, or in the traditionally taught lecture/recitation three semester calculus sequence.

The student population consisted primarily of engineering, mathematics and science students. The average Math SAT score for first semester calculus students in both three semester calculus sequences was about 600. Comparisons of the 205 students who completed the C4L calculus sequence were made with the control group of 4431 students from the traditionally taught courses. Only data on 4636 students enrolled for the first time in: (i) first semester calculus during the Fall semesters of 1988, 1989 and 1990; (ii) second semester calculus during the Spring semesters of 1989, 1990, and 1991; and (iii) third semester calculus during the Fall semesters of 1989, 1990, and 1991 were included in the longitudinal study. This was done in order to make meaningful comparisons with C4L and TRAD students having as similar as possible backgrounds and experiences. The caveat of self-selection is a limitation of the C4L Study design. However, there are no practical alternatives to the C4L Study design that would provide a sufficient amount of data on which to draw meaningful conclusions. In our situation, random assignment of students to the C4L and TRAD courses would have been preferable to self-selection, since by doing so any possible confounding factors which may have influenced outcomes of the C4L Study design would have been eliminated. For example, the possibility that academically better prepared students would enroll in the C4L courses would be offset by random assignment, thus preventing bias in favor of the C4L program. However, a study involving random assignment of students is not pragmatic, since students dropping out of either program would adversely effect the study results. Moreover, the C4L and TRAD programs are so different that students would quickly become aware of these differences, resulting in possible resentment, which would again detract from the C4L Study results.

The C4L Study, under the supervision of Professor George P. McCabe (Professor of Statistics and Head of Statistical Consulting at Purdue University), featured response and explanatory variables; the first measuring outcomes, and the latter, attempting to explain these outcomes. For example, in a comparison of two Hospitals A and B, we might find a better success rate for all surgeries in Hospital A, whereas, on a closer analysis, Hospital B has a better rate for persons entering surgery in poor condition. Without explanatory variables, any statistical study becomes dubious. Our basic idea was to compare C4L and TRAD students using explanatory variables as controls, a method similar to that found in epidemiological studies. Confounding factors in the C4L Study, indeed in any such observational study of this kind, are often accounted for by a "matching" procedure.

However, a "traditional matching procedure" would have been problematic for the C4L Study, since the number of students enrolled in the C4L program was so small as compared to the number of TRAD students. In other words, a traditional matching procedure would have used only a very small portion of the available study data. So, in order to use all the available data, the C4L Study used linear models which were designed to determine whether or not each of the explanatory variables had a statistically significant effect on each of the response variables, and the linear models themselves performed a matching of C4L and TRAD students with as "identical" as is possible characteristics for comparison. The explanatory variables used in the linear models represented various confounding and interaction variables in addition to the one explanatory variable of interest, namely, whether or not the C4L or TRAD teaching method was used. A complete discussion of the analytical method can be found in [3].

The set of response variables used in the C4L Study linear models were as follows:

- The number of calculus courses taken from first, second and third semester calculus
- The number of mathematics courses taken beyond calculus
- The average grade in calculus courses
- The average grade in mathematics courses beyond calculus
- The last available (overall) grade point average

The set of explanatory variables used in the C4L Study linear models were as follows:

- Predicted grade point average (PGPA) — a statistic computed by the registrar's office for entering freshman at Purdue to predict student success (at the C grade level or higher). Multiple linear regression analysis was performed to determine which combination of available predictor variables were most strongly associated with a students' first semester grade index. The best predictors were found to be SAT-Verbal, SAT-Math, average high school grades, and the high school percentile rank cubed (cubed rank was used since it spreads the distribution and reduces the skew). Prediction equations involving these four variables were developed for each of the various schools within Purdue, and in some cases, specific programs within schools. The equations provided the weights needed to combine a given student's scores on the four variables to arrive at a student's PGPA.
- An indicator variable for a missing PGPA was included as a categorical variable.
- The number of semesters since the first calculus course was included as a categorical variable.
- Major was included as a categorical variable with values corresponding to engineering, math, science and other.
- The interaction between semester and major was included as a quantitative variable to account for the fact that the

effect of the number of semesters since the first calculus course was taken on the response variables can depend on major and conversely.
- Gender was included as a categorical variable.
- An indicator variable to distinguish C4L students from TRAD students was used as a categorical variable.

Each response variable was modeled as a linear function of the explanatory variables. Statistical tests were performed for the general linear model to determine whether or not the explanatory variables had statistically significant effects on each response variable. The focus of the C4L Study was to draw meaningful conclusions and comparisons by answering questions like the following:

- Are C4L students equally likely to take one or more courses beyond calculus?
- Do C4L students take more calculus courses than TRAD students?
- Do C4L students take more math courses after calculus than do TRAD students?
- How do Calculus course grades of C4L students compare to those of TRAD students?
- How do grades in courses beyond Calculus compare for C4L and TRAD students?
- How does the last grade point average compare for C4L and TRAD students?

Findings

The following summarizes the comparisons and conclusions made in the C4L Study:

- C4L students are equally likely as TRAD students to take at least one more math course beyond calculus.
- C4L students take more calculus courses than TRAD students.
- C4L students take more math courses beyond calculus than do TRAD students.
- C4L students' average grades in courses beyond calculus versus those of TRAD students have no statistically significant differences.
- C4L students' grades in calculus are better than those of TRAD students.
- C4L students' last (overall) grade point average versus TRAD students have no statistically significant differences.

These results indicate that not only did C4L students do just as well as their TRAD counterparts in mathematics courses beyond calculus, but a larger number of C4L students went on to do just as well as traditional students in higher mathematics courses.

A recently published study [2] suggests that C4L calculus students appear to be spending more time studying calculus than do their traditionally taught students counterparts. But, as the results of the C4L longitudinal study suggest, C4L students may indeed reap the rewards for studying more than traditional students in that they often receive higher grades in calculus. Moreover, C4L students appear not to be adversely affected in their other courses by the increased study time spent on calculus, since no statistically significant differences in C4L students' grade point averages as compared to traditionally taught students were found in the C4L Study.

Use of Findings

No significant changes in the C4L Program three semester calculus sequence were made based on the findings of the C4L Study. However, the findings of the study do provide potential implementers of the C4L program and the program directors with useful information regarding the effects of the use of such a radical approach to calculus reform. Concerns regarding the increased study time required of students and the possible detrimental effects on their overall performance in other classes seem to have been be effectively addressed. Most faculty might agree that students across the nation do not spend enough time studying calculus which may be one of the critical factors which contributes to attrition and poor performance in calculus.

The results of the qualitative research phase of the C4L assessment/evaluation process have been used to make revisions in curriculum design, the text and other C4L Program course materials. A critical aspect of the C4L program is a decomposition of each mathematical concept into developmental steps (which is often referred to as a "genetic decomposition" of a concept) following a Piagetian theory of knowledge based on observations of and in-depth student interviews as they attempt to learn a concept [1]. The results of the qualitative research phase of the C4L assessment program is used to modify and adjust the C4L pedagogical treatment. In particular, the development steps proposed by researchers may, or may not, be modified. For example, the C4L text treatment and genetic decomposition of the concept of the limit of a function at a point was modified and an alternative treatment was proposed based on analyses of 25 qualitative student interviews on the concept of limit. When possible an attempt is made to make meaningful comparisons of the analyses of interviews of both C4L and TRAD students. This was possible with a set of 40 interviews on derivatives and an in-depth analyses of two interview questions on the students' understanding of the derivative as slope. However, whether such comparisons can be made or not, sometimes the research results confirm that the C4L treatment of a particular concept appears to be doing what is expected regarding the outcomes of student understandings. In this case no change in the C4L treatment is made, as was the case with results of analyses of the interview questions on derivative as slope.

Success Factors

To plan and design a longitudinal study like the C4L Study and then carry out the analysis of the results requires numerous hours of planning, brainstorming and consultation with statistical consultants. Such a major study cannot be taken lightly and does require expert advice and counsel.

Regarding qualitative interviews, we will only say here that to design and construct each interview, the questions and the guidelines for what interviewers should probe for during an interview requires detailed and careful planning. Pilot interviews must be carried out and analyzed prior to the completion of the entire set of interviews. A particular theoretical perspective on how students learn mathematical concepts, such as that used in the C4L program, provides a solid foundation on which the interview process can be based. Students are paid $10-$15 for each qualitative interview, each of which lasted between one and two hours in duration. We note that this need not always be done, as we have conducted interviews on courses other than calculus where student volunteers for interviews were obtained. Transcribing each interview usually requires at least six hours. In the C4L Program, in addition to the researchers doing transcriptions, undergraduate and graduate students are often paid to do transcriptions. The analysis phase of a set of interviews requires many hours of dedicated researchers, not to mention the writing of research papers. The process becomes more time consuming if video taping is involved. However, the whole interview process from design to the writing of a research paper can be a very rewarding experience which contributes to pedagogical design (not to mention the contribution to professional development, and possible tenure and promotion). Once again, we caution that a qualitative research program should not be taken lightly. Such a program requires that the researchers become knowledgeable of qualitative research procedures through the necessary training in order to do a competent job.

References

[1] Asiala, M., Brown, N., DeVries, D., Dubinsky, E., Mathews, D. and Thomas, K. "A Framework for Research and Development in Undergraduate Mathematics Education," *Research in Collegiate Mathematics Education II*, CBMS Issues in Mathematics Education, 6, 1996, pp. 1–32.

[2] Mathews, D.M. "Time to Study: The C4L Experience," *UME Trends*, 7 (4), 1995.

[3] Schwingendorf, K.E., McCabe, G.P., and Kuhn, J. "A Longitudinal Study of the Purdue C4L Calculus Reform Program: Comparisons of C4L and Traditional Students," *Research in Collegiate Mathematics Education*, CBMS Issues in Mathematics Education, to appear.

The Evaluation of Project Calc at Duke University, 1989-1994

Jack Bookman
Duke University

Charles P. Friedman
University of Pittsburgh

This in-depth study analyzes a Calculus reform program. It looks not only at analytical gains in student understanding but affective gains as well.

Background and Purpose

As part of the National Science Foundation's Calculus Initiative, Lawrence Moore and David Smith developed a new calculus course at Duke University. The course, called Project CALC, differs from the traditional Calculus course in several fundamental ways. The traditional course emphasizes acquisition of computational skills whereas the key features of Project CALC are real-world problems, activities and explorations, writing, teamwork, and use of technology. The explicitly stated goals for Project CALC are that students should

- be able to use mathematics to structure their understanding of and to investigate questions in the world around them;
- be able to use calculus to formulate problems, to solve problems, and to communicate the solution of problems to others;
- be able to use technology as an integral part of this process of formulation, solution, and communication;
- learn to work cooperatively [11, 12].

Project CALC classes meet for three 50-minute periods with an additional two-hour lab each week. Classroom projects are worked in teams. For approximately five of the labs or classroom projects each semester, teams of two students submit a written report. In the traditional Calculus class taught at Duke, classes (maximum size of 35) meet three times per week for 50 minutes. The lectures usually closely follow a textbook like Thomas and Finney's *Calculus and Analytic Geometry*. Though there is, naturally, some overlap, the two courses do not have the same mathematical content.

Method

Paralleling Moore and Smith's development of the course, the authors of this paper designed and implemented an evaluation of Project CALC. The evaluation had two phases. During the first years of the project, the emphasis was on formative evaluation, during which the evaluator

- taught both Project CALC (PC) and traditional students (TR) calculus;
- observed, once a week, a TR class and a PC class;
- conducted a dinner discussion, on alternate weeks with a regular group of students from the TR class that was being observed and with a regular group from the PC class being observed;
- read students' comments about both the TR and PC calculus courses;
- held extensive conversations with the faculty who taught PC.

During the later years, the emphasis changed to comparing traditionally taught students with experimentally taught students on a set of outcomes. The outcome based phase had three main components

- a problem solving test given to both PC and TR while they were enrolled in Calculus II (See [4].)
- a "retention" study of sophomores and juniors, both PC and TR
- a follow-up study, conducted during the fourth and fifth years of the project, which focused on the question, "Do PC students do better in and/or take more courses that require calculus?"

Findings

Formative Evaluation. During the 90–91 and 91–92 academic years, PC was taught to about one-third of the students enrolled in beginning calculus. The primary focus of the evaluation during the second year of the project's development was on formative evaluation and on observing and describing qualitatively the differences between PC and TR. By the end of the first semester certain strengths and problems became apparent [1,5,6]. Students were upset and unhappy about Lab Calculus I. They complained about the unfamiliar course content, pedagogy and style. They felt that the course required an unreasonable amount of time outside of class and they had difficulty learning how to read mathematics. The student response to traditional Calculus I was better but not glowing. The TR students felt that the course was too hard and went too fast. They also felt that the course was not useful and presented "too much theory."

The attitudes of the students in Project CALC improved remarkably from the fall to the spring. In the spring semester, though they still complained that the material was too vague and complicated, they also stated that they understood the material rather than just having memorized it and that it was interesting to see the connections to the real world. In the second semester, the responses of the TR students were much harsher. Their most positive comments were that the course "forces students to learn lots of material" and the "basics of calculus were taught." One difference was remarkably clear from the classroom observations: The level of attention and concentration of the PC students was much higher. In every TR class observed, at least some of the students fell asleep and most of them started packing their books before the lecture was finished. These behaviors were not observed in PC classes. In fact, often the class ran five minutes overtime without the students even noticing. In summary, PC students were engaged and relatively interested in their work. The TR students were largely passive in class and alienated. On the other hand, TR students knew what to expect and what was expected of them, whereas the PC students probably expended a lot of energy figuring out what the course was about.

Problem Solving Test. In year two of the study, a five-question test of problem solving was administered to Calculus II students (both PC and TR). An effort was made to make the testing and grading conditions for the students as uniform as possible. The test consisted of five items novel to both groups of students and that reflected topics that were taught in both courses. The items were selected with contexts from various fields: biology, chemistry, economics as well as mathematics. On both versions of the test, PC students outperformed the TR students, with highly statistically significant differences, indicating that Project CALC made some progress towards meeting its goals. A more detailed discussion of both the method and the findings can be found in [4].

Retention Study. Also during the third year of the project, in the spring of 1992, the five-part test developed during the first year was administered to a group of sophomores and juniors half of whom had a year of PC in their freshman year and half of whom had a year of TR in their freshman year. Approximately one third of randomly selected students agreed to participate in the study. The five sections of the test addressed writing, basic skills, problem-solving, concepts, and attitudes. The test items are briefly described below.

Attitudes. Each student was asked to indicate the extent to which he or she agreed or disagreed with 40 items such as "I've applied what I've learned in calculus to my work in non-mathematics courses" and "When I was in calculus, I could do some of the problems, but I had no clue as to why I was doing them."

Skills. Ten items such as: (1) Compute $\int x^2 \sqrt{x^3 + 1}\, dx$ and (2) Find the maximum and minimum values of $y = \sin x \cos x$ on $[0, \pi]$

Writing. A short essay addressing the question, "Illustrate the concept of derivative by describing the difference between the average velocity and instantaneous velocity."

Problem solving. Two non-routine word problems.

Conceptual understanding. Ten items such as, "The graphs of three functions appear in the figure below [omitted here]. Identify which is f, which is f', and which is f''."

The results showed a highly statistically significant difference in attitude favoring PC students. This was somewhat surprising in light of the fact that the same students who had often complained, sometimes bitterly, about PC exhibited, a year and two years later, much more positive attitudes about calculus and its usefulness than their traditionally taught peers. A more complete discussion of attitudes is given in [5]. Not surprisingly, the regular students outperformed PC students on the skills calculations, though not statistically significantly. This finding that TR students were better at these kind of skills was verified in much of the interview data collected during the five year study. PC students out-performed TR students on the writing test but not statistically significantly. The results of the problem solving were somewhat consistent with the problem solving test described above (see [3]). The scores on the conceptual understanding test were poor for both groups. It is highly possible that fatigue could have been a factor in their poor performance, since, during the previous two hours, the subjects had taken four other subtests. In retrospect, the subjects were probably asked to do too much in one sitting. The most useful information was gotten from the attitude and skills subtests. The results of the attitude test were powerful and dramatic. The results of the skills test, though not quite statistically significant were indicative of a weakness in the experimental course already perceived by students and faculty.

Follow-up Study. The evaluation in the fourth and fifth years of the study focused on the question, do experimental students do better in and take more subsequent courses that require calculus? The grades of two groups of students were examined. in a set of math related courses. The detailed results of this study are available elsewhere ([3,6]). To better understand these results seven pairs of these students — matched by major and SAT scores, with one subject from the experimental course and one from the traditional course — were interviewed.

It was found that on average, the TR students did better by 0.2 out of a 4-point GPA. This is a small but statistically significant difference. It seems that though there were few differences between the grades of the two groups in the most mathematical courses, the significant differences overall may be explained partially by differences in the performance in other classes. In particular, there were significant differences in the grades of TR and PC students in Introductory Biology with TR students outperforming others. On average, PC students took about one more math, physics, engineering, and computer science course than the TR students and PC students took significantly more math courses. It is not clear why we got these results. One possible explanation is that during the first two weeks of classes, when students were allowed to add and drop courses, the most grade oriented and risk aversive students, who are often pre-medical students, may have switched from the PC to TR course. Though the students who were the subjects of this study were randomly assigned to PC and TR classes, it was not possible to force them to remain in their assigned sections.

In the interview data, most of the students felt that they were adequately prepared for future courses. In general, the PC students were less confident in their pencil and paper computational skills, but more confident that they understood how calculus and mathematics were used to solve real world problems [5].

Project CALC has made some modest headway in keeping more students in the mathematical "pipeline," that is increasing continuation rates in mathematics and science courses. As mentioned above, PC students took more of the most mathematically related courses. In addition, retention rates from Calculus I to Calculus II have improved slightly.

Use of Findings

The evidence, from both the qualitative and quantitative studies, indicates several strengths and weaknesses of the project. The observations and interviews, as well as the problem solving tests and attitude instrument, indicate that the students in PC are learning more about how math is used than their counterparts in the TR course. On the other hand, as reported in interviews, students and faculty in the early version of the PC course did not feel that enough emphasis was placed on computational skill, whereas this has been the main emphasis of the TR course. This view is supported by the results of the skills test. Although, there is a danger in overemphasizing computational skills at the expense of other things, PC students should leave the course with better abilities in this area. The course has been revised to address these problems and in more recent versions of the course, there is considerably more practice with basic skills such as differentiation, integration and solving differential equations. (See [2].)

Students in PC became better problem solvers in the sense that they were better able to formulate mathematical interpretations of verbal problems and solve and interpret the results of some verbal problems that required calculus in their solution. Although PC appears to violate students' deeply held beliefs about what mathematics is and asks them to give up or adapt their coping strategies for dealing with mathematics courses, their attitudes gradually change. When surveyed one and two years after the course, PC students felt, significantly more so than the TR students, that they better understood how math was used and that they had been required to understand math rather than memorize formulas. Observations of students in class showed that PC students are much more actively engaged than were TR students. The evidence gathered indicates some improvements in continuation rates from Calculus I to Calculus II and from there into more advanced mathematics classes.

Success Factors

The heart of program evaluation is assessment of student learning, but they are not the same thing. The purpose of program evaluation is to judge the worth of a program and, in education, that requires that assessment of student learning take place. But program evaluation may include other things as well, for example, problems associated with implementation, costs of the program, communication with interested parties. Assessment of student learning is a means to achieve program evaluation, but assessment of student learning is also a means of understanding how students learn and a means of evaluating the performance of individual students [8].

Educational research in a controlled laboratory setting is critical for the development of our understanding of how students learn which in turn is critical for learning how to assess student learning. Program evaluation must rely on this research base and may use some of the same methods but it must take place in a messier environment. In this evaluation for example, while it was possible to randomly assign students to classes, it was not possible to keep them all there. The calculus program at a university is a much more complex setting than would be ideal for conducting basic educational research. On the other hand, program evaluation can contribute to an understanding of what actually happens in real, complex learning situations.

This study used both qualitative and quantitative methods. For example, the interview data collected during the formative evaluation pointed out concerns about student's level of computational skills. The quantitative data collected the next year corroborated this. In addition, the interviews conducted with students two years after they took the course, helped provide insight into the quantitative attitude instrument given in the previous year. Traditionally, these two methodologies were seen as diametrically opposed but this attitude has changed as more evaluators are combining these approaches [7].

There is not always a clear distinction between formative and summative evaluation though Robert Stake summed up the difference nicely when he stated that: "When the cook tastes the soup, that's formative; when the guests taste the soup, that's summative." [9] The summative aspects of this evaluation were used to inform changes in the course and became, therefore, a formative evaluation as well [2].

Inside evaluators (evaluators who work in the program being evaluated) are subject to bias and influence. On the other hand, outside evaluators are less familiar with the complexities of the situation and can rarely spend as much time on an evaluation as an insider can. To overcome these problems, this evaluation used both an inside and outside evaluator. The insider conducted most of the evaluation; the outsider served as both a consultant and auditor, whose job, like that of certified public accountants, was to periodically inspect the "books" to certify that the evaluation was being carefully, fairly and accurately executed.

Finally we offer some advice for those readers who are planning on conducting program evaluations at their institutions. It is both necessary and difficult to communicate the value and difficulties of program evaluation. Beware of administrators who say: "We're going to apply for a continuation of the grant, so we need to do an evaluation." Be prepared for results that are unexpected or that you don't like. Always look for sources of bias and alternative explanations. Develop uniform, replicable grading schemes and where possible use multiple graders. Be clear and honest about limitations of the study, without being paralyzed by the limitations. Get an outsider to review your plans and results.

References

[1] Bookman, J. Report for the Math Department Project CALC Advisory Committee — A Description of Project CALC 1990-91, unpublished manuscript, 1991.

[2] Bookman, J. and Blake, L.D. Seven Years of Project CALC at Duke University — Approaching a Steady State?, PRIMUS, September 1996, pp. 221–234.

[3] Bookman, J. and Friedman, C.P. Final report: Evaluation of Project CALC 1989–1993, unpublished manuscript, 1994.

[4] Bookman, J. and Friedman, C.P. "A comparison of the problem solving performance of students in lab based and traditional calculus" in Dubinsky, E., Schoenfeld, A.H., Kaput, J., eds., *Research in Collegiate Mathematics Education I*, American Mathematical Society, Providence, RI, 1994, pp. 101–116.

[5] Bookman, J. and Friedman, C.P. Student Attitudes and Calculus Reform, submitted for publication to *School Science and Mathematics*.

[6] Bookman, J. and Friedman, C.P. The Evaluation of Project CALC at Duke University — Summary of Results, unpublished manuscript, 1997.

[7] Herman, J.L., Morris, L.L., Fitz-Gibbon, C.T. *Evaluators Handbook*, Sage Publications, Newbury Park, CA. 1987.

[8] Schoenfeld, A.H. et al. *Student Assessment in Calculus*. MAA Notes Number 43, Mathematical Association of America, Washington, DC, 1997.

[9] Stevens, F. et al. *User-Friendly Handbook for Project Evaluation*, National Science Foundation, Washington D.C., 1996.

[10] Smith, D.A. and Moore, L.C. "Project CALC," in Tucker, T.W., ed., *Priming the Calculus Pump: Innovations and Resources*, MAA Notes Number 17, Mathematical Association of America, Washington, DC, 1990, pp. 51–74.

[11] Smith, D.A. and Moore, L.C. (1991). "Project CALC: An integrated laboratory course," in Leinbach, L.C., et al, eds., *The Laboratory Approach to Teaching Calculus*, MAA Notes Number 20, Mathematical Association of America, Washington, DC, 1991, pp. 81–92.

[12] Smith, D.A. and Moore, L.C.(1992). "Project CALC: Calculus as a laboratory course," in *Computer Assisted Learning*, Lecture Notes in Computer Science 602, pp. 16–20.

Part IV
Assessing Teaching

Introduction

Bill Marion
Valparaiso University

For many of us the primary means for assessing how well we teach have been student evaluation instruments and possibly, on occasion, peer visits of our classes. Yes, we faculty still make the final judgments about teaching effectiveness since in most cases the student evaluations and peer visit observations are summarized and the results are put into some kind of context. But in the end, all too often judgments about the quality of our teaching come down to the observations students make about what goes on in our classrooms. We feel uneasy about the over-reliance on student evaluations in measuring teaching effectiveness. We even wonder whether something as qualitative as teaching can be measured, yet we know that somehow our teaching must be assessed.

But, what is it that we really want to access? Reform of the mathematics curriculum, changes in our style of teaching — no longer does the lecture method reign supreme — and variations in our own methods for assessing student learning have led us to address this question more urgently than we have done in the past. Some tentative answers are now beginning to emerge and along with them a variety of methods with which to assess teaching effectiveness are being explored. What follows in this section are descriptions of some of those experiments.

In the first article, Alan Knoerr, Michael McDonald and Rae McCormick describe a departmental-wide effort to assess teaching, both while a course is in progress as well as an on-going practice. This effort grew out of curriculum reform, specifically the reform of the calculus sequence at the college.

Next, Pat Collier tells us of an approach his department is taking to broaden the definition of what it means to be an effective teacher and then, how to assess such effectiveness. This effort grew out of dissatisfaction with the use of student evaluations as the primary measure of the quality of one's teaching.

Hamkins discusses using video and peer feedback to improve teaching. While his experience with these techniques involved graduate students, the methods can be modified for use with faculty members, both junior and senior.

Lastly, we have two articles which give us some guidance as to how we might use peer visitation more effectively in the assessment of teaching. Deborah Bergstrand describes her department's class visitation program which is designed to be of benefit to both junior and senior faculty. Pao-sheng Hsu talks about her experience with a peer review team, consisting of faculty from within and outside the mathematics department

Departmental Assistance in Formative Assessment of Teaching

Alan P. Knoerr   Michael A. McDonald   Rae McCormick
Occidental College

At a small liberal arts college in the West a department-wide program has been developed to help faculty assess and improve their teaching while courses are in progress. Descriptions of what led to this effort, the steps already taken, the resources involved and plans for the future are presented.

College teaching is usually formally assessed at the end of a course by students and occasionally through observation of classes by another (more senior) faculty member. These assessments, along with other documentation of teaching, are evaluated as part of annual or multiyear reviews. This approach to assessment and evaluation does not, however, provide a teacher with timely feedback on his or her performance and effectiveness.

This article discusses what a department can do to help its members improve their teaching while courses are in progress. We draw on over six years of experience with extensive curricular and pedagogical reform in mathematics at Occidental College, a small, residential liberal arts college with strong sciences and a diverse student body. Our specific concern is assessment — ways in which teachers can get information about their teaching and their students' learning.

From Curriculum Reform to Formative Assessment

Our department's interest in formative assessment is a natural outgrowth of our work in curriculum reform. As the calculus reform movement of the past decade began to develop, we examined our own program and decided to make some changes. We wanted to accomplish the following:

- improve persistence rates through the first and second year courses;

- integrate technology into the teaching and learning of mathematics;
- strengthen ties to other fields which use mathematics;
- engage all students in mathematics as a living science rather than as an archive of procedures

Within two years, all our calculus courses were using reform materials and had weekly computer lab sections. Dissatisfaction with published reform texts led us to create a great deal of our own supplementary material. New courses were introduced at the second-year level and others were modified. Subsequent conversion from a term to a semester academic calendar gave us an opportunity to reformulate the major and to make improvements in our upper-division courses as well.

Although we began with a focus on curricular reform, we soon began to also change how we taught mathematics. From the outset we formed teaching teams for our calculus courses. This was initially done to facilitate development of our own materials and to enable us to staff computer lab sections. Extensive conversations about pedagogy as well as content were a happy, if unexpected, result.

Modifying what we were teaching was not enough to meet the goals of our reform. By also changing how we teach, we have made substantial and encouraging progress towards these goals. Many of us have made some effort to become acquainted with the literature on pedagogy and educational psychology, both in general and as it applies to mathematics. Our growing interest in assessment parallels increased attention to student outcomes in educational theory and administrative policy.

The Role of the Department

It is certainly possible for an individual teacher to learn and adopt new methods of teaching on their own. However, it is unusual for many to make significant changes in their professional practice without some institutional support. The department can provide leadership and support for improving teaching. The expectations of one's colleagues and the resources the department can offer are especially important.

Expectations

Our department believes it is possible and important to improve teaching, and shares a strong expectation that its members will do so. This expectation is communicated in daily conversations, through policy decisions such as adopting calculus reform and team teaching, by directing resources to improving teaching, and through the exemplary practice of senior members of the department. It has also been an important criterion in hiring new members. The climate created by these expectations is especially welcomed by junior faculty who care deeply about teaching and still have much to learn. This climate enables them to improve their teaching while also advancing in other areas of professional development.

Resources

There are many ways a department can help its members obtain resources needed to improve their teaching. Many of these involve investments of time rather than money. Here are some of the things our department does internally:

- senior faculty mentor junior faculty, particularly in team-teaching reformed calculus;
- the department maintains a common computer network drive for sharing teaching materials we create;
- we keep faculty informed of conferences, workshops, and funding opportunities related to improving teaching;
- we encourage publishing scholarship related to teaching mathematics;
- senior and junior faculty collaborate to secure course development grants;
- and we hold departmental retreats for sharing teaching innovations, planning curricula, and discussing policy related to teaching. These retreats have been inexpensive, one day affairs held either on campus or at the home of a faculty member.

Interaction between the department and the rest of the college has been important in developing and maintaining college-wide resources for improving pedagogy. The Dean of Faculty provides support for faculty attending conferences and workshops related to teaching and provides matching funds for grants related to teaching. We also benefit from the resources of Occidental's Center for Teaching and Learning. This center offers faculty opportunities to learn more about college teaching, including:

- monthly interdisciplinary lunch discussions;
- individual and small group diagnosis and consultation;
- videotaping and debriefing services;
- a library of resources on teaching and learning;
- publicity about conferences and workshops on college teaching;
- and Teaching Institutes for faculty at Occidental and other area colleges. Themes are Cooperative Learning, Women's Ways of Knowing, and Teaching With Technology.

Several Mathematics faculty were also instrumental in developing LACTE, the Los Angeles Collaborative for Teacher Excellence. This National Science Foundation supported collaborative of ten area colleges and community colleges sponsors many activities aimed at improving the training and continuing education of elementary and secondary teachers. These include faculty development workshops and course development release time for college professors who are educating future teachers.

A Departmental Retreat and Assessment Workshop

Several of us in the Mathematics Department began learning more about formative assessment through meetings, workshops, and reading. Our program of curricular and pedagogical reform had reached a point where we felt assessment could be very helpful. Together with the Center for Teaching and Learning, we planned a one day departmental retreat and workshop at the start of this academic year, focussed on assessment.

Prior to the workshop, the Center for Teaching and Learning provided each member of the department with a personal Teaching Goals Inventory [3] and readings on classroom assessment. In the morning, the department discussed a number of issues related to teaching and curriculum development. Having devoted a lot of energy in previous years to lower-division courses, we focussed this time on students making the transition to upper-division courses. A shared concern was students' abilities to engage more abstract mathematics.

In the afternoon, the director of the Center for Teaching and Learning led a workshop on formative assessment. We began by discussing individual and departmental goals, as revealed by the Teaching Goals Inventory. This was an interesting follow-up to the morning's discussion and modeled the value of assessment. A number of points were raised by the Teaching Goals Inventory which had not emerged in the earlier conversation. In the second part of the workshop, we learned about specific assessment techniques and set specific individual goals for ourselves on using assessment.

Some techniques concerned assessing teaching per se. The course portfolio was described as a way to assess and document one's work for subsequent evaluation and to support scholarship on teaching. Several of us decided to develop a portfolio for one or more courses we were teaching this year. (See [4] in this volume for further information on course portfolios.)

We also learned about several techniques for classroom assessment of student learning, drawn from the excellent collection by Angelo and Cross [1]. An "expert-groups strategy" was used during the workshop to teach these techniques. Individuals set goals to use some of these techniques, to learn more about other classroom assessment techniques, and to improve assessment in cooperative learning. Other goals included using assessment to help answer certain questions: what difficulties do students have in writing about mathematics, what difficulties do first-year students have adjusting to academic demands in college, what factors affect student confidence, and how can we help students develop a more mature approach to mathematics?

Findings and Use of Findings

We frequently discuss our teaching with each other. A midsemester follow-up meeting with the director of the Center for Teaching and Learning gave the department a specific opportunity to discuss its experience with assessment. A number of us have gone beyond the goals we set for ourselves at the start of the year.

Some of us did develop course portfolios. These portfolios document materials prepared for the course, such as class notes and exercises, projects, and computer labs. They may also include copies of representative student work. A journal, reflecting on teaching and student learning as the course progresses, is another important component. Other efforts in assessing teaching include the following:

- Several of us are asking students to complete midcourse evaluations and many of us use supplementary evaluations at the end of a course.
- Members of some teaching teams are helping each other make specific improvements in their teaching by observing and providing specific feedback.

Student learning is the most important outcome of teaching, and more of us have focussed on this. Reflection on the results of this sort of assessment helps us improve teaching. Here are some of the things we are doing.

- The "muddiest point" and variations on this technique are widely used.
- Weekly self-evaluated take-home quizzes are given with solutions on the back. Students are required to report how they did by e-mail, and to ask questions about difficulties. Participation is recorded but grades are not.

- Concept maps are being used for both pre-unit assessment and for review.
- Several of us have students keep journals about their experiences in a course.
- A number of us have created e-mail servers for particular courses. This provides a forum outside of class hours for discussing course topics. It also can be used to create a class journal in which the instructor requests responses to particular prompts.
- A "professional evaluation" model of assessment is being used in a second-year multivariable calculus course. Its multiple components are designed to foster mathematical maturity while ensuring that fundamental skills are mastered [see 4].
- Students in some courses are being asked to write on the question, "What is abstraction and what does it have to do with mathematics?" We hope to gain some insight into how student perspectives on abstraction change as they gain experience with mathematics.

While departmental activities have helped us each learn more about assessment, the process of evaluating the information we acquire through assessment has been largely informal and private. With the assistance of the Center for Teaching and Learning we are hoping to learn how to make better use of this information. Both teaching portfolios and colleague partnerships seem to be especially promising and to suit the culture of our department.

Volunteer faculty pairs would agree to help each other work towards specific goals concerning teaching and student learning. Among other ways, this assistance could be provided by observing teaching, reviewing course materials and student work, and meeting regularly to discuss teaching and learning. These partnerships would be an extension of the work of our teaching teams.

Success Factors

Our experience with formative assessment is one facet of a serious and long-term departmental commitment to improving the teaching and learning of mathematics. We believe that the depth of this commitment has been essential to the progress we have made. Critically reviewing the undergraduate mathematics curriculum to decide what to teach, and training ourselves to teach mathematics in ways which best promote student learning, are challenging and ongoing processes. The cooperation of department members helps us take risks and learn from each others' experience and strengths.

In reflecting on how this departmental commitment has been achieved and maintained, we find the elements present which have been noted in studies of successful institutional change in secondary education [2]. We began with a history of commitment to high quality traditional instruction at an institution which allowed individuals and departments to take some risks. The decision to become involved in calculus reform was supported by a majority of senior faculty in the department and enthusiastically adopted by the junior faculty and new hires. Teams for teaching and curriculum development soon led to a culture in which mathematics, pedagogy, and the interaction of these two are our constant focus. The college and the National Science Foundation provided adequate financial support for equipment, release time, conferences and workshops. Through the Center for Teaching and Learning, the college also provided a critical source of expertise. Finally, we have been given time to learn and create new ways to work as teachers. For us, a cooperative departmental approach to learning about assessment has been natural, efficient, and has further strengthened our program.

References

[1] Angelo, T.A. and Cross, K.P. *Classroom Assessment Techniques: A Handbook for College Teachers*, 2nd edition, Jossey-Bass Publishers, San Fransisco, 1993.

[2] Fullan, M.G. *The New Meaning of Educational Change*, 2nd edition, Teachers College Press, New York, 1991.

[3] Grasha, A.F. *Teaching with Style: A practical guide to enhancing learning by understanding teaching and learning styles*, Alliance Publishers, Pittsburgh, 1996.

[4] Knoerr, Alan P. and Michael A. McDonald. "Student Assessment Through Portfolios," this volume, p. 123.

Assessing the Teaching of College Mathematics Faculty

C. Patrick Collier
University of Wisconsin Oshkosh

At a regional, comprehensive university in the Midwest the mathematics faculty have been grappling with what it means to be an effective teacher and how to evaluate such effectiveness. Their conclusion is that student evaluations should not be the primary means of evaluating teaching. Hence, the department is in the process of articulating a statement of expectations for teaching from which appropriate assessment instruments will be developed.

Background and Purpose

The University of Wisconsin Oshkosh is one of thirteen four-year campuses in the University of Wisconsin System. UW Oshkosh typically has an undergraduate student body of 9,000–10,000 plus another 1,000–1,300 graduate students. It offers masters degrees in selected areas, including an MS Mathematics Education degree administered by the Mathematics Department.

The mathematics department is primarily a service department, providing mathematics courses for general education, for business administration and for prospective and inservice teachers. Over 80% of the credits generated by the department are earned by students who are not majors or minors in mathematics. The largest number of credits are generated in intermediate and college algebra. About one-third of all students enrolled in service courses earn less than a C grade or withdraw, with permission, after the free drop period.

The system for evaluating teaching effectiveness, particularly evaluation for the purpose of determining merit pay increments, has not had wide support for many years. Minor changes have been proposed with the intention of making the policy more widely acceptable. More recently changes that are more substantive have been considered. These will be described after providing some background on personnel decisions, on the role of teaching effectiveness in those decisions, and on the role of student input into those decisions.

Faculty are involved in four kinds of personnel decisions: renewal, promotion, merit review and post tenure review. Each of these decisions involves an evaluation of teaching, professional growth and service. Our experience has probably been similar to that of many departments of our size. There has been dissatisfaction for many years with the manner in which teaching has been evaluated but there has not been any consensus on how to change it. We have generally required probationary faculty to demonstrate they were effective teachers without giving them a working definition of effective teaching. Instead we have generally relied on how their student "evaluations" compared with those of other teachers.

The UW System Board of Regents has, since 1975, mandated the use of "student evaluations" to assess teaching effectiveness. The mathematics department responded to the Regent mandate by constructing a nine-item form. Students were given nine prompts (e.g. "prepared for class," "relates well to students," "grading") and were asked to rate each instructor on a five point scale ranging from "very poor" to "outstanding." The ninth item was "considering all aspects of teaching that you feel are important, assign an overall rating to your instructor." Almost all of the items were designed to elicit subjective responses. Department policy called for deriving a single number for each class by taking a weighted average in which the ninth "overall item" was weighted at one-half and the remaining items as a composite group weighted half. The numbers assigned to an instructor by course were then averaged over courses to determine a single numerical rating for each instructor. For some number of years the single numerical score on the student evaluation survey represented the preponderance of evidence of teaching effectiveness used in retention decisions and merit allocation. Peer evaluation was a factor in some instances, but those instances were most often situations in which the peer evaluation was to "balance" or mitigate the effects of the student evaluation. Over a period of time, the department identified several faults in this process. Some of those were faults with the survey instrument; some were faults in interpreting and using the results. A partial listing follows.

1. Most of the items in the survey were written from the perspective that an effective teacher is essentially a presenter and an authority figure. This is not the image of an effective teacher that is described in the current standards documents.

2. The student opinions were not entirely consistent with other measures of teaching effectiveness. For example, some faculty found they could improve their student evaluation ratings by doing things they did not consider to be effective practice or by ceasing to do things that they believed were good instructional practices.

3. Faculty objected to calling the data collected from students "evaluations" but did not object to it being called "opinions," since the data were to be input to an evaluation performed by faculty members.

4. The Regents had mandated the use of student evaluation data but had not mandated that it represent the preponderance of evidence of effective teaching.

5. We had been led to evaluate teaching effectiveness by comparing single numerical scores. Eventually each score was translated into a decile. So, half of the students could rate you as "average" and the other half as "good" for a 3.5 on the five point scale. But that may have placed you in the 20th percentile in the departmental ranking. What started out as a fairly good evaluation (between average and good) was ultimately translated into a rating which suggested poor showing (since you were in the bottom fifth of all teachers).

Probationary faculty are subject to renewal every two years. The decision to renew is made by a committee of tenured faculty who rate each candidate on a five point scale with "meets expectations" in the center. The decision to promote in rank is made by a committee of faculty in the upper ranks who rate each candidate on a five point scale with "meets expectations" in the center. Merit reviews have been conducted every two years and have been independent of other reviews. Each faculty member has been rated on a nine point scale by separate committees which evaluate respectively, teaching, professional growth and service. The post tenure review, conducted by a committee of tenured faculty, is based on merit reviews over four years and results in a designation of "meets expectations" or "needs a plan of improvement."

In renewal and promotion decisions, the appropriate personnel committee evaluates teaching after examining evidence submitted by the candidate. Candidates are required to submit student evaluations from each course taught, including free response comments of students. There will also be about two reports of class visitations each year from peers. Candidates also write a summary of their teaching accomplishments and may supplement it with copies of syllabi and sample exams.

In merit review, a rating for teaching has been determined by taking 0.4 times a weighted average of student evaluations and 0.6 times a rating determined by a committee. The committee has not had access to student evaluations, but bases its ratings on two-page statements from the faculty. The faculty are directed to include in their statement: curriculum activity (course and instructional materials development), classroom activities different from traditional lecture, assessment (exams, quizzes, assignments), grading standards, accommodations made (new preparations, more than two preparations, night classes, adapting to schedule changes). The committee rates each faculty member on a scale of 1–9.

Method

The department has not been satisfied with the merit policy and has taken some steps to reform it. The reforms deal with the following items.

1. Rather than have the committee rate each individual on a scale of 1–9, the department would develop a statement of what is expected and the committee would rate faculty on a five point scale with "meets expectations" in the middle. The expectation would be stated in observable terms. This would make the form of the decision of the teaching evaluation committee in the merit process consistent with the form of the decision for renewal and promotion. It would also make the decision process less arbitrary.

2. The old policy did not promote improvement in teaching. Faculty were given a rating (which they usually perceived as being too low relative to colleagues) and that rating did not provide a direction for improvement. But with the "expectations" approach we can ask for evidence that the instructor has thought about teaching, has done some honest reflection on teaching, and has a realistic plan for improvement. For a department it may be more important over time that each member has some identifiable program for improving teaching than it is to compare "performance" and determine whose is in the top quartile and whose is in the bottom quartile.

3. The old policy did not promote improvement in programs offered by the department. For example, we are required to have an assessment plan. That plan is supposed to ensure that data are collected and used to make decisions to improve programs. But none of our evaluation of teaching is connected to promoting the assessment plan. A revised

merit policy should recognize individual efforts that help the department meet its collective responsibilities.

In summary, our experience with "student evaluation" has led to general acceptance of the following principles.

1. The opinions of students should be collected systematically as input to an evaluation of teaching to be made by faculty. The input should be properly referred to as "student opinion" or as "student satisfaction" and not be referred to as "student evaluation."

2. The survey instrument should, as much as possible, ask students to record observations of what they perceive in the classroom; it should not ask for direct evaluative judgment on those observations.

3. Faculty should have "ownership" of the items in the survey instrument, by selecting them from a larger bank of items or by constructing them to match with teaching standards they have adopted.

4. Student opinions most often reflect classroom performance. Classroom performance is one aspect of teaching effectiveness. Other very important aspects are curriculum planning, materials development and assessment. Generally, students are not reliable sources for input into evaluating those components of effective teaching.

Having articulated these principles, we began to develop a new survey instrument. The general strategy was to create a consensus on a statement of teaching standards and construct a new survey instrument which would ask students whether they observed behavior that reflected those standards. After about a year's work had produced a rough draft of a statement of standards, the task was temporarily abandoned because the University developed a new instrument. The new SOS, as it was called, contained 30 items. The department policy committee recommended we choose a subset of the new SOS. It asked department members to examine each of the SOS items and to choose a subset of ten or more items on which they would prefer to be judged. After several rounds of surveying the faculty, there was general consensus on ten items. These ten items became the New Mathematics Department SOS. [See Appendix.]

The new survey is probably not what would have resulted if we had developed teaching standards and then developed a survey to match them. However, the new survey had several properties that we sought. It does have a separate scale for each item and most items ask students to make an observation, rather than an evaluation. For example, consider this item: "During most class periods, this teacher raises thought-provoking ideas and asks challenging questions _____." The choices for the blank are: "several times, more than a few times, a few times, one or more times, zero times." Of course, we have an implied value system that "several times" is preferable to "zero times." But students may not share that value system. Indeed, some may respond (accurately for their class) "more than a few times" and believe that is not the mark of an effective teacher. As another example consider the prompt: "The teacher is _____ attentive and considerate when listening to students' questions and ideas." The choices for the blank are "never or almost never, often not, usually, almost always, always." The implied value is that teachers are to encourage dialogue and student participation.

Findings

The new survey instrument, consisting of the selected subset of the university instrument items, took about a year to get into place. Meanwhile, work on developing a set of department standards for teaching has not progressed. Rather, we will be redirecting that effort to developing a statement of expectations for teaching. The expectations are different from standards in the sense that standards relate exclusively to classroom management and performance, while the expectations will include aspects of teaching (e.g., planning, assessing, developing curriculum, experimenting with new strategies) that occur outside the classroom, aspects which students are not generally able to observe. The expectations will be outcomes that will be judged by peers, not by students. Some sample proposed statements of expectations include the following:

(1) Every faculty member should be involved, perhaps with a group of others in the department, in a program of improvement. That program can and, perhaps, should include making visits to classes of colleagues and having colleagues visit their classes.

(2) Every faculty member should be involved in the assessment of students, courses and programs and in the formulation of reasonable strategies for improving those students, courses and programs.

(3) Every faculty member should be involved in curriculum building by designing new courses, revising old courses, and constructing meaningful activities for students.

(4) Every faculty member should collect student opinions of his/her teaching and should consider those opinions as part of a program for improving his/her teaching.

After work on the expectations has been completed we expect to resume the quest to construct a statement of teaching standards that can be translated into prompts for a department SOS instrument. That is likely to take at least two more years.

Use of Findings

The merit evaluation for the last two decades had been independent of all other evaluations. The department decided that the merit policy should not continue as an independent process but take as input the results of review for renewal of probationary faculty and post tenure review of tenured faculty. Expectations for probationary faculty had already been articulated in the Renewal/Tenure Policy. It remains to complete a list of expectations for tenured faculty in the Post Tenure Review Policy. We have broadened the definition of effective teaching to include performance on tasks that usually take place outside of the classroom. We have placed more emphasis on encouraging improvement than on making comparisons between one faculty member and another.

We have taken steps to reduce the amount of weight given to student opinions. In fact student opinions will not factor directly into merit points in the future. Rather, student opinion is one of several factors that are considered by the Renewal Committee or the Post Tenure Review Committee as they make an evaluation. At the same time we have tried to gather student opinions on items that we believe they can observe and report with some objectivity. We have also taken steps to describe aspects of teaching that students do not usually observe. As a result we hope to have a policy that focuses more on professional development than on forcing faculty to compete in a zero-sum game.

Success Factors

As a result of this experience we recommend the following to all departments confronted with the very difficult task of evaluating teaching.

1. Focus as much as possible on improving teaching and as little as possible on rewards and punishments. Faculty can be motivated to participate in a program of professional development much more readily than they can be persuaded to participate in a system designed to reward or punish.

2. Focus as much as possible on developing and building consensus for your own statement of standards of teaching as a foundation for evaluating teaching. Evaluation should be based on an explicit standard, a standard with which all faculty can identify.

3. Focus on a definition of teaching that extends to tasks that are performed outside the classroom. Teaching effectiveness is much more than a "performance" in the classroom.

4. Find a realistic role for student input. Determine the role that you want student data to play in the evaluation process.

5. Do not expect anything more than temporary, partial solutions but continue to seek more permanent and complete solutions.

APPENDIX
Student Opinion of Instruction Items

1. This teacher makes _____ use of class time.

    5 - very good; 4 - good; 3 - satisfactory; 2 - poor; 1 - very poor

2. The syllabus (course outline) did _____ job in explaining course requirements.

    5 - a very good; 4 - a good; 3 - an adequate; 2 - not do an adequate; 1 - (there is no syllabus/course outline)

3. This teacher raises thought-provoking ideas and asks challenging questions _____ during most class periods.

    5 - several times; 4 - more than a few times; 3 - a few times; 2 - one or two times; 1 - zero times

4. This teacher provides _____ feedback on my progress in the course.

    5 - a great deal of; 4 - more than an average amount of; 3 - an average amount of; 2 - little; 1 - very little or no

5. The assignments (papers, performances, projects, exams, etc.) in this class are _____ used as learning tools.

    5 - always; 4 - almost always; 3 - sometimes; 2 - seldom; 1 - never

6. Tests _____ assess knowledge of facts and understanding of concepts instead of memorization of trivial details.

    5 - always; 4 - almost always; 3 - usually; 2 - often do not; 1 - almost never

7. The quality of teaching in this course is giving me _____ opportunity to gain factual knowledge and important principles.

    5 - a very good; 4 - a good; 3 - an average; 2 - a poor; 1 - a very poor

8. This teacher is _____ attentive and considerate when listening to students' questions and ideas.

    5 - always; 4 - almost always; 3 - usually; 2 - often not; 1 - never or almost never

9. The work assigned contributes _____ to my understanding of the subject.

    5 - a very significant amount; 4 - a significant amount; 3 - a good amount; 2 - little; 1 - very little

10. The difficulty level of this course is _____ challenging to me.

    5 - constantly; 4 - often; 3 - sometimes; 2 - almost never; 1 - never

Using Video and Peer Feedback to Improve Teaching

Joel David Hamkins
CUNY-College of Staten Island

This article discusses a program at Berkeley of using videotaping of actual classes, and peer feedback, to improve teaching. While the program was aimed at graduate students, it can be adapted to use with faculty members.

Background and Purpose

Diverse and informed criticism can improve teaching, especially when coupled with video feedback. In the Graduate Student Instructor (GSI) training program which I directed in the UC Berkeley Mathematics Department, such criticism helped all the GSIs improve their teaching. At UC Berkeley, the introductory mathematics courses are typically divided into large lectures which meet with the professor, and then are further subdivided into recitation sections of about 25 students each of which meets twice weekly with a GSI. All first time GSIs are required, concurrently with their first teaching appointment, to enroll in the training program (as a course), and in the Fall of 1994 we had about 35 such GSIs. Though the course was intended specifically to improve the teaching of novice graduate student instructors, teachers and professors at all levels could benefit from the critical feedback at the heart of our program.

Method

After an initial university-wide training workshop, the course met for the first half of the semester every Monday afternoon for about 2-1/2 hours. Each meeting consisted largely of an extended discussion of various topics on teaching effectiveness, such as the importance of knowing whether the students understand your explanations, the best ways to encourage student involvement in the classroom and how to manage groups, as well as more mathematical matters, such

as how best to explain continuity, the chain rule, or inverse functions. Many of the topics we discussed are found in the text [1], which I highly recommend. Outside the weekly discussion session, the GSIs were required to visit each other's sections and evaluate the teaching they observed, meeting individually with the observed GSI to discuss directly their criticisms. Also, the GSIs were required to experiment in their own sections with some of the dozens of the more unusual or innovative teaching techniques we had discussed (e.g. using name-tags on the first few days of class, group exams, games such as Math Jeopardy, or having students contribute exam questions), and report back to the rest of us on the success or failure of the technique, or how the technique might be improved. These analyses were posted publicly in the mathematics department for the benefit of all GSIs and professors. Finally, twice during the semester, GSIs, while teaching their classes, were videotaped by the Teaching Consultant (a talented, experienced GSI). This video was the basis of an intensive evaluation and consultation session with the Teaching Consultant. First the novice GSI and the Consultant watched the video alone, looking for strengths and weaknesses while recalling the actual class performance. Then, while watching the video together, the Teaching Consultant led the GSI through a sort of guided self-analysis of his of her performance in the classroom. In sum, the training program relied essentially on three methods: the program's classroom discussions, peer observation and criticism, and a video consultation session with the Teaching Consultant.

Findings

The goal of the program was to improve the educational experience of undergraduates at UC Berkeley by improving GSI teaching. We therefore discussed and analyzed a wide variety of teaching techniques and topics, many of which grew out of situations that the GSIs had encountered in their own classrooms. Let me list here a few of the most important topics that arose:

- How to generate enthusiasm in the students by using interesting or unusual examples.
- How to get more useful feedback from students by phrasing questions in a positive manner (e.g. "Raise your hand if you understand this part" rather than "Does anyone still not get this?").
- How to use a chalkboard effectively.
- How to engage the students with leading questions to actively involve them in the classroom ("Who can suggest what we should do at this step?").
- How to encourage collaboration among students.
- How to give useful comments on homework.
- How to deal with disruptive students.
- How to simultaneously challenge the students to think and also obtain critical feedback about what the students know by asking the right kinds of questions.
- How best to explain specific mathematical topics such as partial differentiation or the ratio test, etc.
- How to manage various unusual teaching techniques such as group quizzes, special 'all-theory' office hours or student presentations.
- How to deal with cheating on exams.
- How to run a review session.
- How to teach simultaneously to students of varying ability.

All the GSIs benefitted from these suggestions and the chance to discuss their ideas about teaching in an open supportive forum.

Use of Findings

Most of the GSIs were able to immediately improve their teaching using the ideas and techniques they had encountered in the program's class meetings. Indeed, after experimenting with some of the more unusual techniques, many of the GSIs reported them to be so successful that they were now using them regularly. By attending each other's sections and giving critical feedback to each other, the GSIs became aware of certain shortcomings, such as their need to lead the discussion more, their need to write more on the chalkboard, or their need to engage more of the students. After the videotaped observations, the experienced Teaching Consultant was able to point out specific strengths and weaknesses in each GSI's teaching. One GSI, for example, was having trouble getting students to respond; the Teaching Consultant pointed out

simply that he was rushing, and not allowing them enough time to answer. Afterwards, this GSI simply waited a bit longer after a question and found that his students were perfectly willing to contribute. Other GSIs were helped by the Consultant's suggestions concerning boardwork or how to recognize cues that the students have not followed an explanation.

Success Factors

Overall, the GSIs and I were very pleased with the structure and effectiveness of the program. The use of an experienced Teaching Consultant, I believe, was an essential aspect of it. I have, however, several suggestions on how to implement an even better program.

- *Let each GSI observe and evaluate many other GSIs.* This was one of the most effective and easy-to-organize means of giving a lot of diverse criticism to each GSI, and, simultaneously, expose each GSI to various teaching styles, which they viewed with a critical eye. Thus, it benefitted both the evaluator and the GSI who was evaluated.
- *Use videos only to augment actual observations, and have the Teaching Consultant make the video while observing the class.* Because a videotape only imperfectly records what went on in the classroom — sounds are distorted, student reactions are lost — one should never evaluate a teacher using a videotape of a class one didn't attend. Rather, use videos to augment actual observations. We found it very convenient for the Teaching Consultant to simply arrive a few minutes early to set up equipment, and then occasionally check the camera during the class while taking notes and recording points to be discussed later in the consultation session. In past years, each GSI was videotaped by a technician who knew nothing about mathematics, but who only delivered the tape to the mathematics department to be viewed by the GSI and perhaps the program's class, none of whom actually observed the class in person. Worse, in some years, the tapes were made by film students, who attempted to make the tapes more interesting with close-up shots of students and whatnot, when it would have been more desirable to record the GSIs' board-work. These suggestions have the following corollary.
- *Do not show the videos as a part of the discussion session.* In previous years, a large part of program consisted of viewing videos. But this is a waste of time. It is far better to simply send the GSIs into each other's classrooms and then discuss the issues that arise in the GSI discussion section.
- *Avoid mini-teaching sessions or other forms of simulated teaching.* Other programs rely on this technique, but I find it to have little value. In past years, for example, the

GSIs would prepare calculus lessons to be presented to the other GSIs in the course, who were encouraged to pretend not to understand the material and ask questions as though they were in an actual calculus class. But this is both absurd and boring for everyone. Important parts of teaching well — such as generating enthusiasm in the students, getting feedback from the students by asking questions, sometimes easy, sometimes difficult, and paying attention to nonverbal cues — simply cannot be simulated; there is no substitute for actual students.

- *Insist on a rigorous schedule for the consultation sessions.* We had problems scheduling all the sessions and would have benefitted from a rigorous initial policy.

- *Videotape the GSIs twice.* Once is not enough. The first time one sees oneself on videotape is a bit disconcerting. Time is spent merely getting over how one's hair looks on camera, and so forth; it is difficult to pay attention to the teaching. With a second tape, it is much easier to focus on the teaching. Also, taping twice may show improvements in the instructor's teaching effectiveness.

- *Encourage experienced GSIs and professors to participate in the video consultation program.* Experienced faculty can benefit from the critical feedback which lies at the heart of our program. I therefore suggest that all teaching faculty be encouraged to participate in the peer observation and videotape consultation sessions; they could simply be included into the schedule. A long-term continuing program would perhaps focus on the novice GSIs every fall semester, when most are starting their first teaching assignment, and in the spring semester, when there are fewer novice GSIs, focus on more experienced teachers and professors.

Reference

[1] Davis, B.G. *Tools for Teaching*, Jossey-Bass, 1993.

Exchanging Class Visits: Improving Teaching for Both Junior and Senior Faculty

Deborah Bergstrand
Williams College

A peer visitation program for both junior and senior faculty at a small, liberal arts college in the East has been put into place to help improve the quality of teaching. Every junior faculty member is paired with a senior colleague to exchange class visits. This program is designed to foster discussion of teaching, and the sharing of ideas and to provide constructive criticism about the teaching effectiveness of each member of the pair.

Background and Purpose

Williams College is a small liberal arts college in rural Massachusetts with an enrollment of 2000 students. The Mathematics Department has 13 faculty members covering about 9.5 FTE's. Williams students are bright and very well prepared: the median Math SAT score is 700. The academic calendar is 4-1-4: two twelve-week semesters are separated by a four-week January term called Winter Study. The standard teaching load is five courses per year; every other year one of the five is a Winter Study course.

We put a lot of effort into our teaching, as well as into evaluating teaching and supporting the improvement of teaching. Many of the more formal aspects of teacher evaluation exist in large part to inform decisions on reappointment, promotion, and tenure. Some of our evaluative procedures also serve as good mechanisms to foster discussion among colleagues about teaching. In particular, we find exchanging class visits between junior and senior faculty a highly collegial way to learn about and from each other.

To put the evaluative aspects into context, I will briefly describe the general procedures Williams uses to evaluate teaching. As at many other institutions, Williams has a structured, college-wide protocol for evaluating teaching. The Student Course Survey (SCS) is required in all courses. Students give numerical ratings anonymously on various aspects of the course and instructor. All data are gathered and analyzed by the Vice Provost, who produces detailed comparisons of individual results with departmental, divisional, peer group, and college-wide results. Teachers receive the analysis of their own results, with results for nontenured faculty also going to the department chair and the Committee on Appointments and Promotion (CAP). The SCS also includes a page for descriptive comments from each student; these are passed directly to the teacher after grades are submitted.

Following specific guidelines, departments themselves must also gather student opinion on all nontenured faculty members annually, using either interviews, letters, or departmental questionnaires. Though not required, the college encourages class visits, in accordance with specified guidelines. All information on a nontenured faculty member's teaching gathered through the above means is discussed with the faculty member, and summarized in an annual report to the CAP.

Method

In the Mathematics Department, senior faculty have been observing classes of junior faculty for many years. The goals were to evaluate nontenured faculty and to offer comments and constructive criticism. Even if being observed was somewhat awkward, the system worked well and was seen by all as an important and useful complement to student evaluations of teaching. During the 1980's, as anxieties about tenure decisions rose along with the complexity of procedures for evaluating teaching and scholarship, junior faculty in many departments at Williams felt more and more "like bugs under a microscope." In an effort to alleviate some of the anxiety, and in turn expand the benefits of class visits, our department decided to have pairs of junior and senior faculty exchange class visits each semester. Thus junior faculty are now also observers as well as the observed. They appreciate the opportunity to see their senior colleagues in the classroom, and we all benefit from increased exposure to different teaching styles.

Every semester, each junior faculty member is paired with a senior member to exchange class visits. The department chair arranges pairs so that over the course of a few years, every senior colleague observes each junior colleague. The chair also insures that someone observes every course taught by a junior colleague at least once. (So if the junior faculty member teaches Calculus III several times, at least one of those offerings will be observed.) Special requests are also considered, such as a senior faculty member's curiosity about a particular course offered by a junior colleague.

Because visits to classes are part of the department's evaluation of junior faculty, they follow a formal structure under college guidelines. Early in the semester the two faculty members meet and go over the course syllabus. The junior colleague may suggest particular classes to visit or might leave the choice open. Two or sometimes three consecutive classes are visited. In the discussion following a visit, it is important to put observations into context. What looks to the observer like an inadequate answer to a question might make sense once the visitor learns that the same question had been addressed at length the previous day. What looks like an awkward exchange between a student and the teacher might be the result of some previous incident in which the student was disrespectful. The two come to an understanding of the visitor's evaluation of the class, which the senior colleague then conveys to the junior colleague in writing, with a copy to the department chair. This letter becomes part of the junior faculty member's file. The write-up includes a general description of the class, specific comments both positive and negative, and the suggestions or ideas discussed after the visit.

Visits by junior faculty to senior faculty classes are less structured. As with senior visits to junior classes, they are intended to foster discussion of teaching, sharing ideas, and constructive criticism. Unlike senior visits to junior classes, they are not evaluative. Because these visits are not a formal part of departmental evaluation procedures, sometimes they don't even take place. (We encourage but do not require junior faculty to visit classes.) Those visits that do occur are followed by an informal discussion of the visitor's impressions, comments and criticisms. We recognize that not all junior faculty are going to be comfortable criticizing their senior colleagues, no matter how valid those criticisms may be.

Findings

Some outcomes of these class visits are predictable. The visitor might have suggestions about organizing the lecture, improving blackboard technique, or responding to and encouraging student questions. Sometimes more subtle observations can be made. A visitor in the back of the room can watch student reactions to various aspects of the class in a way the teacher running the class may not be able to do. On the positive side, I have seen classes where the students were so engaged and excited that while the teacher was facing the blackboard some students would look at each other and smile, clearly enjoying both the material and the teacher's style. On the negative side, I have seen classes where some students were engaged but many others were not, and some were even doing other work. In each example it was helpful to share my observations with my colleague.

Another subtle effect of a class visit might be hearing a colleague's impression of how one comes across to the class. In a basically positive report of a colleague's visit to one of my classes (both of us tenured), he described my style as somewhat businesslike. Though not intended as a criticism, I was quite surprised. I had thought of my teaching style as quite warm, friendly, and encouraging. Not that "businesslike" was bad, it just wasn't how I thought I came across.

Use of Findings

Every faculty member reacts in their own way to comments about their teaching. Following the "businesslike" comment on my own teaching style referred to above, I tried to pay more attention to the tone I set in the classroom. For example, I used to take a fairly strict approach to collecting quizzes precisely at the end of the designated time period, in an effort both to be fair and to keep control of class time. Thinking about the atmosphere such a policy created, however, I realized that it was more strict and formal than I really wanted or needed to be. As a result, I'm now more relaxed about quiz time and about some other things as well. The students still respect me, I still have control of my class, but I feel we're all more relaxed and hence able to learn more.

One junior colleague's classroom style was quite formal, even though outside of class he was very friendly and engaging. While visiting classes of two senior colleagues he admired and respected, he noticed they were closer and friendlier with their students in class than he was. He has since incorporated some of that spirit in his own classes. For example, he now arrives in class a few minutes early so he can chat with his students, not only about mathematics, but about other things going on in their lives.

Not all faculty take the advice given. After being told his pace was slow, one junior colleague decided not to speed up the class. He decided his own pace was the appropriate one. Thus even in a department with lots of visiting, discussing, and evaluating, faculty members retain their teaching autonomy. Structuring class visits as a two-way street really helps. One junior colleague commented that while he knows he's being evaluated, he also knows that comments he makes to senior colleagues about their teaching will be taken seriously.

Success Factors

Classroom visits do have their limits. One sees only a few classes, not the entire course. In smaller classes, the very presence of a visitor can affect student behavior and class dynamics. It is our practice to have the teacher being observed decide whether to introduce visitors, explain their presence, invite them to participate, etc. In some cases the visitor remains unobtrusive and unacknowledged. The latter approach has the advantage of perhaps producing a "purer" observation. The former has the advantage of informing students about and including them more directly into the process of teacher evaluation and improvement.

Over the last few years, the tenure balance in the department has shifted. We now have only two nontenured members and eleven tenured. Being familiar with one-to-one functions, we recognize that a true "exchange" of class visits between junior and senior faculty will either exclude many senior faculty each year or will impose an unreasonable burden on the junior faculty. We now also encourage class visits between senior faculty. Such visits are both relaxed and stimulating. We have all taken ideas from our colleagues to use in our own classes; we have all benefitted from even the simplest of observations from another teacher about our own teaching. The result, we hope, is a set of junior and senior colleagues, all aware of each other's teaching efforts and challenges, and all ready to support and learn from each other's creative energy.

Peer Review of Teaching

Pao-sheng Hsu
University of Maine

At a land-grant university in the Northeast one mathematics faculty member has begun experimenting with a peer visitation program in which a team of faculty visits her class at least twice a semester. What's unique about this process is that the team, usually three in number, consists of faculty both within and without the department and all visit the same class at the same time. This technique provides the instructor with a diversity of views on her teaching effectiveness.

Background and Purpose

The University of Maine, the land-grant university and the sea-grant college of the State of Maine, enrolls approximately 8000 undergraduate and 2000 graduate students. In 1996-97 there were 24 full-time faculty and several part-time faculty in the Department of Mathematics and Statistics. The department had 42 majors and awarded seven degrees that year. Also, there were seven Master's degree students. The department uses "classroom observations" by department members on a one-time basis as part of an evaluation process for tenure considerations; it has no regular "peer review" practice.

Nationally, in the evaluation of teaching process, the pros and cons of having other faculty review a classroom have been widely debated. Proponents argue that peer reviews can provide the teacher with insights into the classroom learning environment unattainable in other ways, and that these reviews also strengthen the faculty's voice in personnel decisions. Opponents maintain that political and personal factors sometimes enter the evaluative process and the opportunity for misuse and abuse is real. While the debate continues, it is not uncommon that by default, the burden of evidence for arriving at a judgment of a faculty member's teaching effectiveness may fall entirely on student evaluations, since these are often mandated by university administrations. In an effort to generate discussion and broaden the perspective on evaluation of teaching, I initiated the experiment of peer review of my classes. I have experimented with inviting faculty members who have experience in ethnographic research[1] and are from outside of my own discipline, as well as colleagues from my own department.

Method

Observers in the classroom can document what actually goes on there. Having several observers in the class at the same time allows multiple perspectives on the class. Further, observers from other disciplines bring different perspectives and different expectations of teaching strategies; these enrich the ensuing discussions. A team of three observers is selected, either by the faculty member or (if part of an official departmental process) the department: one from within the department, the other two from outside the department. All observers should have an interest or previous experience in evaluating teaching. Observers visit the class at least twice during the term. Each time, all three observers are present, and the faculty member is informed in advance of the visit.

---

[1] In their book *Ethnography and Qualitative Design in Educational Research* [1], Goetz and LeCompte described this kind of research as the "holistic depiction of uncontrived group interaction over a period of time, faithfully representing participant views and meanings" (p.51)

Before the first visit, the faculty member meets with the observers to inform them of what has occurred so far in the course and provide them with written materials such as syllabi, assignments, and samples of student writing.

During the visit, some observers may arrive early to chat with some students or observe student interactions. Observers may sit in different parts of the room, so that some can watch students' level of engagement from the back, while others can hear the muttered comments of shy students in the front. Observers may also stay after class to talk with some students. The faculty member should meet with the observers immediately after the visit to give context to the observations. The instructor can provide background history about the class which explained some aspects of classroom dynamics and the direction of the discussion or what was not done in the class. Without that knowledge, an observer might have come to a different conclusion of what had taken place in the class due to the instructor's actions or inactions. The results of these conversations were sometimes incorporated into the observation reports.

After the visit, each observer writes a report of what was observed (either in a format that the observer chooses, to give the widest possible range of records, especially if the review is done at the faculty member's request; or on a departmentally constructed format if uniformity is required for an official process). The observers may also meet with the faculty member to discuss their observations, either as a group or individually.

Between visits, the faculty member provides the observers with written information such as the course tests, test results and correspondences with some students.

For example, during the Fall of 1994, I invited a trio of observers to a Calculus II class of about 30–35 students. The observers consisted of (1) a colleague from the Speech/Communication Department who had participated in system-wide Women-in-the-Curriculum activities, (2) a sociologist who had attended Writing-Across-the-Curriculum workshops and (3) a mathematician who oversaw graduate teaching assistants. They visited twice during the term and wrote a total of four reports. In the Fall of 1996, the observers were (1) an English Department colleague who had coordinated the campus Writing-Across-the-Curriculum Program and who has the responsibility of evaluating part-time/fixed length (i.e., nontenure track) faculty in her department, (2) a sociologist who had won a university teaching award, and (3) a mathematician with whom I had discussed evaluation of teaching. This group observed my Precalculus class (30–35 students) three times and wrote five sets of observation notes.

Findings

In the classes observed, I was using a very collaborative lecturing style, inviting students to participate in what was being discussed and to interact with me as well as with each other. Sometimes the discussions were led in a certain direction because of comments or questions by one or more students; sometimes I cut short a discussion because I wanted students to gather all information they had after class and think more before we discussed a problem further. Most observers found that learning environment positive. They also had suggestions for improvement. For example, one observer suggested that I write on transparencies so that I could face the students more; one observer suggested that I leave more time for discussions of new topics.

Students did not seem to behave much differently with observers in the room. However, the presence of observers did have quite a psychological impact on me. For example, it is difficult to avoid eye contacts with the observers.

Use of Findings

Observers' suggestions can lead to changes in teaching style or strategies, both immediately and after some reflection on how to incorporate the suggestions in a manner consistent with the instructor's personality. Sometimes suggestions do not work with a particular class. However, when this happens, the instructor may think of other ways to remedy an identified problem.

I have found both the encouragement and critical suggestions by the observers useful in helping me work out my practices. Thus, I felt encouraged to continue working with students in helping them think through the mathematics that they are learning and doing while I became more careful in weighing how much time I could spend in drawing students into a discussion versus the time I would need for new topics. However, I found the use of transparencies very restricting and asked the students for their opinion; they suggested that I use the board for writing.

Together with some student work and students' evaluations, the peer review reports can provide a fuller view of an instructor's work in the classroom for a teaching portfolio — a piece of much needed documentation of a faculty member's teaching. Above all, the peer review reports help inform *the instructor* from outside perspectives on changes that can be made. It also gives the instructor more colleagues with whom to discuss teaching.

Success Factors

This kind of peer review is very labor intensive and time-consuming: scheduling multiple visits for three people while avoiding test dates for the class is a logistic challenge and the writing of reports takes tremendous time and energy. In an institution where peer review is not an established practice, it will take persuasion to convince colleagues to participate.

Ideally, the purpose of observing a class is to assess how students are learning; nevertheless, a great deal of learning should be taking place outside the classroom, but most observers will be assessing teaching performance or "teaching effectiveness" on what is observed in the classroom. As views of what is "effective" may vary widely, the faculty member being reviewed should exercise judgment about the sensitivity of these colleagues to his or her teaching goals and make these goals explicit to the observers in advance. Providing the observers with any material such as syllabus, assignments, discussion topics given to students and student writings will afford the observers a context in which to view the sessions.

To avoid the misuse and abuse that opponents of peer reviews worry about, there needs to be some consensus in a department so that the practice is seen especially by the students as a departmental effort to help students learn. Faculty development workshops may provide training for faculty members in observing and commenting sensitively on teaching.

An alternative form of peer evaluation is having a faculty member visit the class, with the instructor absent, and hold focus group discussions with the class. For further information on this, see Patricia Shure's article in this volume, p. 187.

Reference

[1] Goetz and LeCompte, *Ethnography and Qualitative Design in Educational Research*, 1984.

Appendix
Reprint of "Assessment of Student Learning for Improving the Undergraduate Major in Mathematics"

Prepared by The Mathematical Association of America, Subcommittee on Assessment, Committee on the Undergraduate Program in Mathematics

Approved by CUPM at the San Francisco meeting, January 4, 1995

Preface

Recently there has been a series of reports and recommendations about all aspects of the undergraduate mathematics program. In response, both curriculum and instruction are changing amidst increasing dialogue among faculty about what those changes should be. Many of the changes suggested are abrupt breaks with traditional practice; others are variations of what has gone on for many decades. Mathematics faculty need to determine the effectiveness of any change and institutionalize those that show the most promise for improving the quality of the program available to mathematics majors. In deciding which changes hold the greatest promise, student learning assessment provides invaluable information. That assessment can also help departments formulate responses for program review or other assessments mandated by external groups.

The Committee on the Undergraduate Program in Mathematics established the Subcommittee on Assessment in 1990. This document, approved by CUPM in January 1995, arises from requests from departments across the country struggling to find answers to the important new questions in undergraduate mathematics education. This report to the community is suggestive rather than prescriptive. It provides samples of various principles, goals, areas of assessment, and measurement methods and techniques. These samples are intended to seed thoughtful discussions and should not be considered as recommended for adoption in a particular program, certainly not in totality and not exclusively.

Departments anticipating program review or preparing to launch the assessment cycle described in this report should pay careful attention to the MAA Guidelines for Programs and Departments in Undergraduate Mathematical Sciences [1]. In particular, Section B.2 of that report and step 1 of the assessment cycle described in this document emphasize the need for departments to have

a. A clearly defined statement of program mission; and
b. A delineation of the educational goals of the program.

The Committee on the Undergraduate Program in Mathematics urges departments to consider carefully the issues raised in this report. After all, our programs should have clear guidelines about what we expect students to learn and have a mechanism for us to know if in fact that learning is taking place.

James R. C. Leitzel, Chair, Committee on the Undergraduate Program in Mathematics, 1995

Membership of the Subcommittee on Assessment, 1995
Larry A. Cammack, Central Missouri State University, Warrensburg, MO
James Donaldson, Howard University, Washington, DC
Barbara T. Faires, Westminster College, New Wilmington, PA
Henry Frandsen, University of Tennessee, Knoxville, TN
Robert T. Fray, Furman University, Greenville, SC
Rose C. Hamm, College of Charleston, Charleston, SC
Gloria C. Hewitt, University of Montana, Missoula, MT
Bernard L. Madison (Chair), University of Arkansas, Fayetteville, AR

William A. Marion, Jr., Valparaiso University, Valparaiso, IN

Michael Murphy, Southern College of Technology, Marietta, GA

Charles F. Peltier, St. Marys College, South Bend, IN

James W. Stepp, University of Houston, Houston, TX

Richard D. West, United States Military Academy, West Point, NY

I. Introduction

The most important indicators of effectiveness of mathematics degree programs are what students learn and how well they are able to use that learning. To gauge these indicators, assessment — the process of gathering and interpreting information about student learning — must be implemented. This report seeks to engage faculty directly in the use of assessment of student learning, with the goal of improving undergraduate mathematics programs.

Assessment determines whether what students have learned in a degree program is in accord with program objectives. Mathematics departments must design and implement a cycle of assessment activity that answers the following three questions:

- What should our students learn?
- How well are they learning?
- What should we change so that future students will learn more and understand it better?

Each step of an ongoing assessment cycle broadens the knowledge of the department in judging the effectiveness of its programs and in preparing mathematics majors. This knowledge can also be used for other purposes. For example, information gleaned from an assessment cycle can be used to respond to demands for greater accountability from state governments, accrediting agencies, and university administrations. It can also be the basis for creating a shared vision of educational goals in mathematics, thereby helping to justify requests for funds and other resources.

This report provides samples of various principles, goals, areas of assessment, and measurement methods and techniques. Many of the items in these lists are extracted from actual assessment documents at various institutions or from reports of professional organizations. These samples are intended to stimulate thoughtful discussion and should not be considered as recommended for adoption in a particular program, certainly not in totality and not exclusively. Local considerations should guide selection from these samples as well as from others not listed.

II. Guiding Principles

An essential prerequisite to constructing an assessment cycle is agreement on a set of basic principles that will guide the process, both operationally and ethically. These principles should anticipate possible problems as well as ensure sound and effective educational practices. Principles and standards from several sources (see references 2,3,4,5,and 6) were considered in the preparation of this document, yielding the following for consideration:

a. Objectives should be realistically matched to institutional goals as well as to student backgrounds, abilities, aspirations, and professional needs.

b. The major focus of assessment (by mathematics departments) should be the mathematics curriculum.

c. Assessment should be an integral part of the academic program and of program review.

d. Assessment should be used to improve teaching and learning for all students, not to filter students out of educational opportunities.

e. Students and faculty should be involved in and informed about the assessment process, from the planning stages throughout implementation.

f. Data should be collected for specific purposes determined in advance, and the results should be reported promptly.

III. The Assessment Cycle

Once the guiding principles are formulated and understood, an assessment cycle can be developed:

1. Articulate the learning goals of the mathematics curriculum and a set of objectives that should lead to the accomplishment of those goals.

2. Design strategies (e.g., curriculum and instructional methods) that will accomplish the objectives, taking into account student learning experiences and diverse learning styles, as well as research results on how students learn.

3. Determine the areas of student activities and accomplishments in which quality will be judged. Select assessment methods designed to measure student progress toward completion of objectives and goals.

4. Gather assessment data; summarize and interpret the results.

5. Use the results of the assessment to improve the mathematics major.

Steps 1 and 2 answer the first question in the introduction — what should the students learn? Steps 3 and 4, which answer the second question about how well they are learning, constitute the assessment. Step 5 answers the third question on what improvements are possible.

Step 1. Set the Learning Goals and Objectives

There are four factors to consider in setting the learning goals of the mathematics major: institutional mission, background of students and faculty, facilities, and degree program goals. Once these are well understood, then the goals and objectives of the major can be established. These goals and objectives of the major must be aligned with the institutional mission and general education goals and take into account the information obtained about students, faculty, and facilities.

Institutional Mission and Goals. The starting point for establishing goals and objectives is the mission statement of the institution. Appropriate learning requirements from a mission statement should be incorporated in the departments goals. For example, if graduates are expected to write with precision, clarity, and organization within their major, this objective will need to be incorporated in the majors goals. Or, if students are expected to gain skills appropriate for jobs, then that must be a goal of the academic program for mathematics majors.

Information on Faculty, Students, and Facilities. Each institution is unique, so each mathematics department should reflect those special features of the institutional environment. Consequently, the nature of the faculty, students, courses, and facilities should be studied in order to understand special opportunities or constraints on the goals of the mathematics major. Questions to be considered include the following:

- What are the expectations and special needs of our students?
- Why and how do our students learn?
- Why and how do the faculty teach?
- What are the special talents of the faculty?
- What facilities and materials are available?
- Are mathematics majors representative of the general student population, and if not, why not?

Goals and Objectives of Mathematics Degree Program. A degree program in mathematics includes general education courses as well as courses in mathematics. General education goals should be articulated and well-understood before the goals and objectives of the mathematics curriculum are formulated. Of course, the general education goals and the mathematics learning goals must be complementary and consistent [6, pages 183-223]. Some examples of general education goals that will affect the goals of the degree program and what learning is assessed include the following:

Graduates are expected to speak and write with precision, clarity, and organization; to acquire basic scientific and technological literacy; and to be able to apply their knowledge.

Degree programs should prepare students for immediate employment, graduate schools, professional schools, or meaningful and enjoyable lives.

Degree programs should be designed for all students with an interest in the major subject and encourage women and minorities, support the study of science, build student self-esteem, ensure a common core of learning, and encourage life-long learning.

Deciding what students should know and be able to do as mathematics majors ideally is approached by setting the learning goals and then designing a curriculum that will achieve those goals. However, since most curricula are already structured and in place, assessment provides an opportunity to review curricula, discern the goals intended, and rethink them. Curricula and goals should be constructed or reviewed in light of recommendations on the mathematics major as contained in the 1991 CUPM report on the Undergraduate Major in the Mathematical Sciences [6, pages 225–247].

Goal setting should move from general to specific, from program goals to course goals to assessment goals. Goals for student learning can be statements of knowledge students should gain, skills they should possess, attitudes they should develop, or requirements of careers for which they are preparing. The logical starting place for discerning goals for an existing curriculum is to examine course syllabi, final examinations, and other student work.

Some samples of learning goals are:

Mathematical Reasoning. Students should be able to perform complex tasks; explore subtlety; discern patterns, coherence, and significance; undertake intellectually demanding mathematical reasoning; and reason rigorously in mathematical arguments.

Personal Potential. Students should be able to undertake independent work, develop new ideas, and discover new mathematics. Students should possess an advanced level of critical sophistication; knowledge and skills needed for further study; personal motivation and enthusiasm for studying and applying mathematics; and attitudes of mind and analytical skills required for efficient use, appreciation, and understanding of mathematics.

Nature of Mathematics. Students should possess an understanding of the breadth of the mathematical sciences and their deep interconnecting principles; substantial knowledge of a discipline that makes significant use of mathematics; understanding of interplay among applications, problem-solving, and theory; understanding and appreciation of connections between different areas of mathematics and with other disciplines; awareness of the abstract nature of theoretical mathematics and the ability to write proofs; awareness of historical and contemporary contexts in which mathematics is practiced; understanding of the fundamental dichotomy of mathematics as an object of study and a tool for application; and critical perspectives on inherent limitations of the discipline.

Mathematical Modeling. Students should be able to apply mathematics to a broad spectrum of complex problems and issues; formulate and solve problems; undertake some real-world mathematical modeling project; solve multi-step problems; recognize and express mathematical ideas imbedded in other contexts; use the computer for simulation and visualization of mathematical ideas and processes; and use the process by which mathematical and scientific facts and principles are applied to serve society.

Communication and Resourcefulness. Students should be able to read, write, listen, and speak mathematically; read and understand technically-based materials; contribute effectively to group efforts; communicate mathematics clearly in ways appropriate to career goals; conduct research and make oral and written presentations on various topics; locate, analyze, synthesize, and evaluate information; create and document algorithms; think creatively at a level commensurate with career goals; and make effective use of the library. Students should possess skill in expository mathematical writing, have a disposition for questioning, and be aware of the ethical issues in mathematics.

Content Specific Goals. Students should understand theory and applications of calculus and the basic techniques of discrete mathematics and abstract algebra. Students should be able to write computer programs in a high level language using appropriate data structures (or to use appropriate software) to solve mathematical problems.

Topic or thematic threads through the curriculum are valuable in articulating measurable objectives for achieving goals. Threads also give the curriculum direction and unity, with courses having common purposes and reinforcing one another. Each course or activity can be assessed in relation to the progress achieved along the threads. Possible threads or themes are numerous and varied, even for the mathematics major. Examples include mathematical reasoning, communication, scientific computing, mathematical modeling, and the nature of mathematics. The example of a learning goal and instructional strategy in the next section gives an idea of how the thread of mathematical reasoning could wind through the undergraduate curriculum.

Step 2. Design Strategies to Accomplish Objectives

Whether constructing a curriculum for predetermined learning goals or discerning goals from an existing curriculum, strategies for accomplishing each learning goal should be designed and identified in the curricular and co-curricular activities. Strategies should respect diverse learning styles while maintaining uniform expectations for all students.

Strategies should allow for measuring progress over time. For each goal, questions such as the following should be considered.

- Which parts of courses are specifically aimed at helping the student reach the goal?
- What student assignments help reach the goal?
- What should students do outside their courses to enable them to reach the goal?
- What should the faculty do to help the students reach the goal?
- What additional facilities are needed?
- What does learning research tell us?

The following example of a goal and strategy can be made more specific by referencing specific courses and activities in a degree program.

Learning goal. Students who have completed a mathematics major should be able to read and understand mathematical statements, make and test conjectures, and be able to construct and write proofs for mathematical assertions using a variety of methods, including direct and indirect deductive proofs, construction of counterexamples, and proofs by mathematical induction. Students should also be able to read arguments as complex as those found in the standard mathematical literature and judge their validity.

Strategy. Students in first year mathematics courses will encounter statements identified as theorems which have logical justifications provided by the instructors. Students will verify the need for some of the hypotheses by finding counterexamples for the alternative statements. Students will use the mathematical vocabulary found in their courses in writing about the mathematics they are learning. In the second and third years, students will learn the fundamental logic needed for deductive reasoning and will construct proofs of some elementary theorems using quantifiers, indirect and direct proofs, or mathematical induction as part of the standard homework and examination work in courses. Students will construct proofs for elementary statements, present them in both written and oral form, and have them critiqued by a mathematician. During the third and fourth years, students will formulate conjectures of their own, state them in clear mathematical form, find methods which will prove or disprove the conjectures, and present those arguments in both written and oral form to audiences of their peers and teachers. Students will make rational critiques of the mathematical work of others, including teachers and peers. Students will read some mathematical literature and be able to rewrite, expand upon, and explain the proofs.

Step 3. Determine Areas and Methods of Assessment

Learning goals and strategies should determine the areas of student accomplishments and departmental effectiveness that will be documented in the assessment cycle. These areas should be as broad as can be managed, and may include curriculum (core and major), instructional process, co-curricular activities, retention within major or within institution, and success after graduation. Other areas such as advising and campus environment may be areas in which data on student learning can be gathered.

Responsibility for each chosen area of assessment should be clearly assigned. For example, the mathematics faculty should have responsibility for assessing learning in the mathematics major, and the college may have responsibility for assessment in the core curriculum.

Assessment methods should reflect the type of learning to be measured. For example, the Graduate Record Examination (GRE) may be appropriate for measuring preparation for graduate school. On the other hand, an attitude survey is an appropriate tool for measuring an aptitude for life-long learning. An objective paper-and-pencil examination may be selected for gauging specific content knowledge.

Eight types of assessment methods are listed below, with indications of how they can be used. Departments will typically use a combination of methods, selected in view of local program needs.

1. *Tests.* Tests can be objective or subjective, multiple-choice or free-response. They can be written or oral. They can be national and standardized, such as the GRE and Educational Testing Service Major Field Achievement Test, or they can be locally generated. Tests are most effective in measuring specific knowledge and its basic meaning and use.

2. *Surveys.* These can be written or they can be compiled through interviews. Groups that can be surveyed are students, faculty, employers, and alumni. Students can be surveyed in courses (about the courses), as they graduate (about the major), or as they change majors (about their reasons for changing).

3. *Evaluation reports.* These are reports in which an individual or group is evaluated through a checklist of skills and abilities. These can be completed by faculty members, peers, or employers of recent graduates. In some cases, self-evaluations may be used, but these tend to be of less value than more objective evaluations. Grades in courses are, of course, fundamental evaluation reports.

4. *Portfolios.* Portfolios are collections of student work, usually compiled for individual students under faculty supervision following a standard departmental protocol. The contents may be sorted into categories, e.g., freshman or sophomore, and by type, such as homework, formal written papers, or examinations. The work collected in a student's portfolio should reflect the student's progress through the major. Examples of work for portfolios include homework, examination papers, writing samples, independent project reports, and background information on the student. In order to determine what should go in a portfolio, one should review what aspects of the curriculum were intended to contribute to the objectives and what work shows progress along the threads of the curriculum. Students may be given the option of choosing what samples of particular types of work are included in the portfolio.

5. *Essays.* Essays can reveal writing skills in mathematics as well as knowledge of the subject matter. For example, a student might write an essay on problem-solving techniques. Essays should contribute to learning. For example, students might be required to read four selected articles on mathematics and, following the models of faculty-written summaries of two of them, write summaries of the other two. Essays can be a part of courses and should be candidates for inclusion in portfolios.

6. *Summary courses.* Such courses are designed to cover and connect ideas from across the mathematics major. These may be specifically designed as summary courses and as such are usually called capstone courses, or they may be less specific, such as senior seminars or research seminars. Assessment of students performances in these courses provides good summary information about learning in the major.

7. *Oral presentations.* Oral presentations demonstrate speaking ability, confidence, and knowledge of subject matter. Students might be asked to prepare an oral presentation on a mathematics article. If these presentations are made in a summary course setting, then the discussion by the other students can serve both learning and assessment.

8. *Dialogue with students.* Student attitudes, expectations, and opinions can be sampled in a variety of ways and can be valuable in assessing learning. Some of the ways are student evaluations of courses, interviews by faculty members or administrators, advising interactions, seminars, student journals, and informal interactions. Also, in-depth interviews of individual students who have participated in academic projects as part of a group can provide insights into learning from the activities.

Student cooperation and involvement are essential to most assessment methods. When selecting methods appropriate to measuring student learning, faculty should exercise care so that all students are provided varied opportunities to show what they know and are able to do. The methods used should allow for alternative ways of presentation and response so that the diverse needs of all students are taken into account, while ensuring that uniform standards are supported. Students need to be aware of the goals and methods of the departmental assessment plan, the goals and objectives of the mathematics major and of each course in which they enroll, and the reason for each assessment measurement. In particular, if a portfolio of student work is collected, students should know what is going to go into those portfolios and why. Ideally, students should be able to articulate their progress toward meeting goals — in each course and in an exit essay at the end of the major.

Since some assessment measures may not affect the progress of individual students, motivation may be a problem. Some non-evaluative rewards may be necessary.

Step 4. Gather Assessment Data

After the assessment areas and methods are determined, the assessment is carried out and data documenting student learning are gathered. These data should provide answers to the second question in the introduction — how well are the students learning?

Careful record keeping is absolutely essential and should be well-planned, attempting to anticipate the future needs of assessment. Additional record storage space may be needed as well as use of a dedicated computer database. The data need to be evaluated relative to the learning goals and objectives. Evaluation of diverse data such as that in a student portfolio may not be easy and will require some inventiveness. Standards and criteria for evaluating data should be set and modified as better information becomes available, including longitudinal data gathered through tracking of majors through the degree program and after graduation. Furthermore, tracking records can provide a base for longitudinal comparison of information gathered in each pass through the assessment cycle.

Consistency in interpreting data, especially over periods of time, may be facilitated by assigning responsibility to a core group of departmental faculty members.

Ways to evaluate data include comparisons with goals and objectives and with preset benchmarks; comparisons over time; comparisons to national or regional norms; comparisons to faculty, student, and employer expectations; comparisons to data at similar institutions; and comparisons to data from other majors within the same institution.

If possible, students should be tracked from the time they apply for admission to long after graduation. Their interests at the time of application, their high school records, their personal expectations of the college years, their curricular and extracurricular records while in college, their advanced degrees, their employment, and their attitudes toward the institution and major should all be recorded. Only with such tracking can the long-term effectiveness of degree programs be documented. Comparisons with national data can be made with information from such sources as Cooperative Institutional Research Program's freshman survey data [7] and American College Testing's College Outcomes Measures project [8].

Step 5. *Use the Assessment Results to Improve the Mathematics Major*

The payoff of the assessment cycle comes when documentation of student learning and how it was achieved point the way for improvements for future students. Assessment should help guide education, so this final step in the cycle is to use the results of assessment to improve the next cycle. This is answering the third assessment question — what should be changed to improve learning? However, this important step should not be viewed solely as a periodic event. Ways to improve learning may become apparent at any point in the assessment cycle, and improvements should be implemented whenever the need is identified.

The central issue at this point is to determine valid inferences about student performances based on evidence gathered by the assessment. The evidence should show not only what the students have learned but what processes contributed to the learning. The faculty should become better informed because the data should reveal student learning in a multidimensional fashion.

When determining how to use the results of the assessment, faculty should consider a series of questions about the first four steps—setting goals and objectives, identifying learning and instructional strategies, selecting assessment methods, and documenting the results. The most critical questions are those about the learning strategies:

- Are the current strategies effective?
- What should be added to or subtracted from the strategies?
- What changes in curriculum and instruction are needed?

Secondly, questions should be raised about the assessment methods:

- Are the assessment methods effectively measuring the important learning of all students?
- Are more or different methods needed?

Finally, before beginning the assessment cycle again, the assessment process itself should be reviewed:

- Are the goals and objectives realistic, focused, and well-formulated?
- Are the results documented so that the valid inferences are clear?
- What changes in record-keeping will enhance the longitudinal aspects of the data?

IV. Conclusion

During an effective assessment cycle, students become more actively engaged in learning, faculty engage in serious dialogue about student learning, interaction between students and faculty increases and becomes more open, and faculty build a stronger sense of responsibility for student learning. All members of the academic community become more conscious of and involved in the way the institution works and meets its mission.

References

1. *Guidelines for Programs and Departments in Undergraduate Mathematical Sciences*, Mathematical Association of America, Washington, DC, 1993.
2. *Measuring What Counts*, Mathematical Sciences Education Board, National Research Council, National Academy Press, Washington, DC, 1993.
3. *Assessment Standards for School Mathematics*, National Council of Teachers of Mathematics, Reston, VA, circulating draft, 1993.
4. *Principles of Good Practice for Assessing Student Learning*, American Association for Higher Education, Washington, DC, 1992.
5. "Mandated Assessment of Educational Outcomes," *Academe*, November-December, 1990, pp. 34-40.
6. Steen, L.A., ed. *Heeding the Call for Change,* The Mathematical Association of America, Washington, DC, 1992.
7. Astin, A.W., Green, K.C., and Korn, W.S. *The American Freshman: Twenty Year Trends*, Cooperative Institutional Research Program, American Council on Education, University of California, Los Angeles, 1987. (Also annual reports on the national norms of the college freshman class.)
8. *College Level Assessment and Survey Services*, The American College Testing Program, Iowa City, 1990.

Suggestions for Further Reading

The following are several books on assessment which are not primarily concerned with assessment in mathematics. However, they are worth reading, for they contain a good number of assessment methods not discussed here, many of which can be adapted to the mathematics classroom.

Alverno College Faculty, *Student Assessment-as-Learning at Alverno College*, 3rd ed., Alverno College, Milwaukee, WI, 1994

Angelo, T.A. and Cross, K.P. *Classroom Assessment Techniques: A Handbook for College Teachers*, 2nd ed. Jossey-Bass, San Francisco, 1993.

Astin, A.W. *Assessment for Excellence: the Philosophy and Practice of Assessment and Evaluation in Higher Education*. Oryx Press, Phoenix, AZ, 1993.

Banta, T.W. *Implementing Outcomes Assessment: Promise and Perils*, Jossey-Bass, San Francisco, 1988.

Banta, T.W. and associates. *Making a Difference: Outcomes of a Decade of Assessment in Higher Education*, Jossey-Bass, San Francisco, 1993.

Banta, T.W., Lund, J.P., Black, K.E.,and Oblander, F.W. eds. *Assessment in Practice: Putting Principles to Work on College Campuses*, Jossey-Bass Publishers, San Francisco, 1996.

Braskamp, L.A., Brandenburg, D.C., and Ory, J.C. *Evaluting Teaching Effectiveness: a practical guide*, Beverley Hills Pub., Sage, 1984.

Braskamp, L.A. and Ory, J.C. *Assessing Faculty Work : Enhancing Individual and Institutional Performance*, Jossey-Bass, San Francisco, 1994.

Kulm, G. *Mathematics Assessment: What Works in the Classroom*, Jossey-Bass, San Francisco, 1994.

Light, R., *Harvard Assessment Seminars* (first and second), Harvard University, Cambridge, MA, 1990, 1992.

Wiggins, G. *Assessing Student Performance: Exploring the Purpose and Limits of Testing*, Jossey-Bass, San Francisco, 1993.

See also two journals, *AAHE Assessment,* published by the American Association for Higher Education, and *Assessment Update,* edited by Trudy Banta, published by Jossey-Bass.

Bibliography

Academic Excellence Workshops. *A Handbook for Academic Excellence Workshops,* Minority Engineering Program and Science Educational Enhancement Services, Pomona, CA, 1992.

Adams, C., Bergstrand, D., and Morgan, F. "The Williams SMALL Undergraduate Research Project," *UME Trends*, 1991.

Angel, A.R., and Porter, S.R. *Survey of Mathematics with Applications* (Fifth Edition), Addison-Wesley Publishing Company, 1997.

Angelo, T.A. and Cross, K.P. *Classroom Assessment Techniques: A Handbook for College Teachers,* 2nd ed. Jossey-Bass, San Francisco, 1993.

Asiala, M., Brown, N., DeVries, D., Dubinsky, E., Mathews, D. and Thomas, K. "A Framework for Research and Development in Undergraduate Mathematics Education," *Research in Collegiate Mathematics Education II*, CBMS Issues in Mathematics Education, 6, 1996, pp. 1–32.

Barnes, M. "Gender and Mathematics: Shifting the Focus," *FOCUS on Learning Problems in Mathematics*, 18 (numbers 1, 2, and 3), 1996, pp. 88–96.

Barnett, J. "Assessing Student Understanding Through Writing." *PRIMUS* 6 (1), 1996. pp. 77–86.

Battaglini, D.J. and Schenkat, R.J. *Fostering Cognitive Development in College Students: The Perry and Toulmin Models*, ERIC Document Reproduction Service No. ED 284 272, 1987

Bauman, S.F., & Martin, W.O. "Assessing the Quantitative Skills of College Juniors," *The College Mathematics Journal*, 26 (3), 1995, pp. 214–220.

Bonsangue, M.*The effects of calculus workshop groups on minority achievement and persistence in mathematics, science, and engineering*, unpublished doctoral dissertation, Claremont, CA, 1992.

Bonsangue, M. "An efficacy study of the calculus workshop model," *CBMS Issues in Collegiate Mathematics Education*, 4, American Mathematical Society, Providence, RI, 1994, pp. 117–137.

Bonsangue, M., and Drew, D. "Mathematics: Opening the gates—Increasing minority students' success in calculus," in Gainen, J. and Willemsen, E., eds., *Fostering Student Success in Quantitative Gateway Courses*, Jossey-Bass, New Directions for Teaching and Learning, Number 61, San Francisco, 1995, pp. 23–33.

Bookman, J. and Blake, L.D. Seven Years of Project CALC at Duke University — Approaching a Steady State?, PRIMUS, September 1996, pp. 221–234.

Bookman, J. and Friedman, C.P. Final report: Evaluation of Project CALC 1989–1993, unpublished manuscript, 1994.

Bookman, J. and Friedman, C.P. "A comparison of the problem solving performance of students in lab based and traditional calculus" in Dubinsky, E., Schoenfeld, A.H., Kaput, J., eds., *Research in Collegiate Mathematics Education I*, American Mathematical Society, Providence, RI, 1994, pp. 101–116.

Bookman, J. and Friedman, C.P. Student Attitudes and Calculus Reform, submitted for publication to *School Science and Mathematics*.

Buerk, D. "Getting Beneath the Mask, Moving out of Silence," in White, A., ed., *Essays in Humanistic Mathematics,* MAA Notes Number 32, The Mathematical Association of America, Washington, DC, 1993.

Charles, R., Lester, F., and O'Daffer, P. *How to Evaluate Progress in Problem Solving*, National Council of Teachers of Mathematics, Reston, Va, 1987.

Cohen, D., ed.*Crossroads in Mathematics: Standards for Introductory College Mathematics Before Calculus*, American Mathematical Association of Two-Year Colleges, Mcmphis, TN, 1995.

Cohen, D. and Henle, J. "The Pyramid Exam," *UME Trends,* July, 1995, pp. 2 and 15.

COMAP. *For All Practical Purposes*, W. H. Freeman and Company, New York, 1991.

Committee on the Undergraduate Program in Mathematics (CUPM). "Assessment of Student Learning for Improving the Undergraduate Major in Mathematics," *Focus: The Newsletter of the Mathematical Association of America*, 15 (3), June 1995, pp. 24–28.

Committee on the Undergraduate Program in Mathematics. *Quantitative Literacy for College Literacy*, MAA Reports 1 (New Series), Mathematical Association of America, Washington, DC, 1996.

Crannell, A. *A Guide to Writing in Mathematics Classes* (1993). Available upon request from the author, or from http://www.fandm.edu/Departments/Mathematics/Writing.html.

Crannell, A. "How to grade 300 math essays and survive to tell the tale," *PRIMUS* 4 (3), 1994.

Crosswhite, F.J. "Correlates of Attitudes toward Mathematics," *National Longitudinal Study of Mathematical Abilities*, Report No. 20, Stanford University Press, 1972.

Davidson, N.A. *Cooperative Learning in Mathematics: A Handbook for Teachers*, Addison-Wesley Publishing Company, Inc., Menlo Park, CA, 1990.

Davis, B.G. *Tools for Teaching*, Jossey-Bass Publishers, San Francisco, 1993.

Douglas, R., ed. *Toward a Lean and Lively Calculus: Report of the conference/workshop to develop curriculum and teaching methods for calculus at the college level* (MAA Notes Series No. 6), Mathematical Association of America, Washington, DC, 1986.

Dubinsky, E. "A Learning Theory Approach to Calculus," in Karian, Z., ed. *Symbolic Computation in Undergraduate Mathematics Education*, MAA Notes Number 24, The Mathematical Association of America, Washington, DC, 1992, pp. 48–55.

Dubinsky, E. "ISETL: A Programming Language for Learning Mathematics," *Comm. in Pure and Applied Mathematics*, 48, 1995, pp. 1–25.

Emert, J.W. and Parish, C. R. "Assessing Concept Attainment in Undergraduate Core Courses in Mathematics" in Banta, T.W., Lund, J.P., Black, K.E.,and Oblander, F.W. eds., *Assessment in Practice: Putting Principles to Work on College Campuses*, Jossey-Bass Publishers, San Francisco, 1996, pp. 104–107.

Ewell, P.T. "To capture the ineffable: New forms of assessment in higher education," Review of Research in Education, 17, 1991, pp. 75–125.

Farmer, D.W. *Enhancing Student Learning: Emphasizing Essential Competencies in Academic Programs*, King's College Press, Wilkes-Barre, PA, 1988.

Farmer, D.W. "Course-Embedded Assessment: A Teaching Strategy to Improve Student Learning," *Assessment Update*, 5 (1), 1993, pp. 8, 10–11.

Fennema, E. and Sherman, J. "Fennema-Sherman mathematics attitudes scales: Instruments designed to measure attitudes toward the learning of mathematics by females and males,"*JSAS Catalog of Selected Documents in Psychology* 6 (Ms. No. 1225), 1976, p. 31.

Fenton, W.E. and Dubinsky, E. *Introduction to Discrete Mathematics with ISETL*, Springer, 1996.

Ferrini-Mundy, J., *CCH Evaluation and Documentation Project*, University of New Hampshire, Durham, NH, 1994.

Frechtling, J.A. *Footprints: Strategies for Non-Traditional Program Evaluation.*, National Science Foundation, Washington, DC, 1995.

Fullan, M.G. *The New Meaning of Educational Change*, 2nd edition, Teachers College Press, New York, 1991.

Fullilove, R.E., & Treisman, P.U. "Mathematics achievement among African American undergraduates at the University of California, Berkeley: An evaluation of the mathematics workshop program," *Journal of Negro Education*, 59 (3), 1990, pp. 463–478.

Ganter, S.L. "Ten Years of Calculus Reform and its Impact on Student Learning and Attitudes," *Association for Women in Science Magazine*, 26(6), Association for Women in Science, Washington, DC, 1997.

Gillman, L. *Writing Mathematics Well*, Mathematical Association of America, Washington, DC 1987.

Glassick, C.E., et. al. *Scholarship Assessed: Evaluation of the Professoriate*, Carnegie Foundation for the Advancement of Teaching, Jossey-Bass, San Francisco, CA, 1997.

Goetz and LeCompte, *Ethnography and Qualitative Design in Educational Research* , 1984.

Grasha, A.F. *Teaching with Style: A practical guide to enhancing learning by understanding teaching and learning styles*, Alliance Publishers, Pittsburgh, 1996.

Greenberg, M.J. *Euclidean and Non-Euclidean Geometries* (3rd ed.). W. H. Freeman, New York, 1993.

Griffith, J.V. and Chapman, D.W. *LCQ: Learning Context Questionnaire*, Davidson College, Davidson, North Carolina, 1982.

Hackett, G. and Betz, N.E. "An Exploration of the Mathematics Self-Efficacy/Mathematics Performance Correspondence," *Journal for Research in Mathematics Education*, 20 (3), 1989, pp. 261–273.

Hagelgans, N.L. "Constructing the Concepts of Discrete Mathematics with DERIVE," *The International DERIVE Journal*, 2 (1), January 1995, pp. 115-136.

Hastings, N.B. and Laws, P. *Workshop Calculus: Guided Exploration with Review*, Springer-Verlag, New York, vol. 1, 1996; vol. 2, 1998.

Hastings, N.B. "The Workshop Mathematics Program: Abandoning Lectures," in D'Avanzo, C. and McNeal, A., *Student-Active Science: Models of Innovation in College Science Teaching*, Saunders Publishing Co., Philadelphia, 1997.

Heid, M.K. "Resequencing Skills and Concepts in Applied Calculus using the Computer as a Tool," *Journal for Research in Mathematics Education*, 19 (1), NCTM, Reston, VA, 1988, pp. 3–25.

Herman, J.L., Morris, L.L., Fitz-Gibbon, C.T. *Evaluators Handbook*, Sage Publications, Newbury Park, CA. 1987.

Hoaglin, D.C. and Moore, D.S., eds. *Perspectives on Contemporary Statistics*, Mathematical Association of America, Washington, DC, 1992.

Hofstadter, D. *Gödel, Escher, Bach: an eternal golden braid*. Basic Books, New York, 1979.

Houston, S.K., Haines, C.R., Kitchen, A., et. al. *Developing Rating Scales for Undergraduate Mathematics Projects*, University of Ulster, 1994.

Hughes Hallett, D., Gleason, A.M., Flath, D.E., Gordon, S.P., Lomen, D.O., Lovelock, D., McCallum, W.G., Osgood, B.G., Pasquale, A., Tecosky-Feldman, J., Thrash, J.B., Thrash, K.R., & Tucker, T.W. *Calculus*, John Wiley & Sons, Inc., New York, 1992.

Hutchings, P. *From Idea to Prototype: The Peer Review of Teaching*. American Association of Higher Education, 1995.

Hutchings, P. *Making Teaching Community Property: A Menu for Peer collaboration and Peer Review*. American Association for Higher Education, 1996.

Johnson, D. and R. *Cooperative Learning Series Facilitators Manual*. ASCD.

Johnson D.W. and Johnson, F.P. *Joining Together; Group Theory and Group Skills*, 3rd ed., Prentice Hall, Englewood Cliffs, N.J., 1987

Joint Policy Board for Mathematics. *Recognition and Rewards in the Mathematical Sciences*, American Mathematical Society, Providence, RI, 1994.

Jonassen, D.H., Beissneer K., and Yacci, M.A. *Structural Knowledge: Techniques for Conveying, Assessing, and Acquiring Structural Knowledge*. Lawrence Erlbaum Associates, Hillsdale, NJ, 1993.

Keith, S.Z. "Explorative Writing and Learning Mathematics," *Mathematics Magazine*, 81 (9), 1988, pp. 714–719.

Keith, S.Z. "Self-Assessment Materials for Use in Portfolios," *PRIMUS*, 6 (2), 1996, pp. 178–192.

Kloosterman, P. "Self-Confidence and Motivation in Mathematics," *Journal of Educational Psychology* 80, 1988, pp. 345–351.

Kohn, A. "Effects of rewards on prosocial behavior," *Cooperative Learning*, 10 (3), 1990, pp. 23–24.

Leron, U. and Dubinsky, E. "An Abstract Algebra Story," *American Mathematical Monthly*, 102 (3), 1995, pp. 227–242.

Lester, F. and Kroll, D. "Evaluation: A New Vision," *Mathematics Teacher* 84, 1991, pp. 276-283.

Levine, A. and Rosenstein, G. *Discovering Calculus*, McGraw-Hill, 1994.

Loftsgaarden, D.O., Rung, D.C., and Watkins, A.E. *Statistical Abstract of Undergraduate Programs in the Mathematical Sciences in the United States: Fall 1995 CBMS Survey*, The Mathematical Association of America, Washington, DC, 1997.

Lomen, D., and Lovelock, D. *Exploring Differential Equations via Graphics and Data*, John Wiley & Sons Inc., 1996.

Madison, B. "Assessment of Undergraduate Mathematics," in L.A. Steen, ed., *Heeding the Call for Change: Suggestions for Curricular Action*, Mathematical Association of America, Washington, DC, 1992, pp. 137–149.

Martin, W. O. "Assessment of students' quantitative needs and proficiencies," in Banta, T.W., Lund, J.P., Black, K.E., and Oblander, F.W., eds., *Assessment in Practice: Putting Principles to Work on College Campuses*, Jossey-Bass, San Francisco, 1996.

Mathematical Association of America,*Guidelines for Programs and Departments in Undergraduate Mathematical Sciences*. Mathematical Association of America, Washington, DC, 1993.

Mathematical Sciences Education Board. *Moving Beyond Myths: Revitalizing Undergraduate Mathematics,* National Research Council, Washington, DC, 1991.

Mathematical Sciences Education Board. *Measuring What Counts: A Conceptual Guide for Mathematics Assessment,* National Research Council, Washington, DC, 1993.

Mathews, D.M. "Time to Study: The C4L Experience," *UME Trends*, 7 (4), 1995.

Maurer, S., "Advice for Undergraduates on Special Aspects of Writing Mathematics," *PRIMUS*, 1 (1), 1990.

McGowen, M., and Ross, S., Contributed Talk, Fifth *International Conference on Technology in Collegiate Mathematics,* 1993.

Miller, C.D., Heeren, V.E., and Hornsby, Jr., E.J. *Mathematical Ideas* (Sixth Edition), HarperCollins Publisher, 1990.

Moore, R.C. "Making the transition to formal proof," *Educational Studies in Mathematics* 27, 1994, pp. 249–266.

National Center for Education Statistics (Project of the State Higher Education Executive Officers). *Network News, Bulletin of the SHEEO/NCES Communication Network*, 16 (2), June 1997.

National Council of Teachers of Mathematics. *Curriculum and Evaluation Standards for School Mathematics*, National Council of Teachers of Mathematics, Reston, VA, 1989.

National Council of Teachers of Mathematics. *Professional Standards for Teaching Mathematics*, National Council of Teachers of Mathematics, Reston, VA, 1991.

National Council of Teachers of Mathematics. *Assessment standards for school mathematics*, NCTM, Reston, VA, 1995.

Novak, J.D. "Clarify with Concept Maps: A tool for students and teachers alike," *The Science Teacher*, 58 (7), 1991, pp. 45–49.

National Science Board. *Undergraduate Science, Mathematics and Engineering Education: Role for the National Science Foundation and Recommendations for Action by Other Sectors to Strengthen Collegiate Education and Pursue Excellence in the NextGeneration of U.S. Leadership in Science and Technology,* Report of the Task Committee on Undergraduate Science and Engineering Education, Neal, H., Chair, Washington DC, 1986.

National Science Foundation. *Undergraduate Curriculum Development in Mathematics: Calculus.* Program announcement, Division of Mathematical Sciences, Washington, DC, 1987.

National Science Foundation. *Undergraduate Curriculum Development: Calculus,* Report of the Committee of Visitors, Treisman, P. Chair, Washington, DC, 1991.

O'Brien, J.P., Bressler, S.L., Ennis, J.F., and Michael, M. "The Sophomore-Junior Diagnostic Project," in Banta, T.W., et al., ed., *Assessment in Practice: Putting Principles to Work on College Campuses*, Jossey-Bass, San Francisco, 1996, pp. 89–99.

Ostebee, A., and Zorn, P. *Calculus from Graphical, Numerical, and Symbolic Points of View, Volume 1,* Saunders College Publishing, 1997.

Penn, H., "Comparisons of Test Scores in Calculus I at the Naval Academy," in *Focus on Calculus*, A Newsletter for the Calculus Consortium Based at Harvard University, 6, Spring 1994, John Wiley & Sons, Inc., p. 6.

Perry Jr., W.G. *Forms of Intellectual and Ethical Development in the College Years*, Holt, Rinehart and Winston, NY, 1970.

Rash, A.M. "An Alternate Method of Assessment, Using Student-Created Problems." *PRIMUS*, March, 1997, pp. 89–96.

Redmond, M.V. and Clark, D.J. "A practical approach to improving teaching," *AAHE Bulletin* 1(9–10), 1982.

Regan, H.B. And Brooks, G.H. *Out of Women's Experience: Creating Relational Experiences*. Thousand Oaks CA: Corwin Press, Inc., 1996.

Reynolds, B.E., Hagelgans, N.L., Schwingendorf, K.E., Vidakovic, D., Dubinsky, E., Shahin, M., and Wimbish, G.J., Jr. *A Practical Guide to Cooperative Learning in Collegiate Mathematics*, MAA Notes Number 37, The Mathematical Association of America, Washington, DC, 1995.

Roberts, A.W., Ed. *Calculus: The Dynamics of Change*, MAA Notes, Number 39, MAA, Washington, DC, 1996.

Roberts, C.A. "How to Get Started with Group Activities," *Creative Math Teaching*, 1 (1), April 1994.

Rogers, C.R. "The Necessary and Sufficient Conditions of Therapeutic Personality Change." *Journal of Consulting Psychology* 21(2), 1957, pp. 95–103.

Rogers, C.R. *Freedom to Learn*. Merrill Publishing, 1969.

Schoenfeld, A. *Student Assessment in Calculus*, Mathematical Association of America, Washington, DC, 1997.

Schwartz, R. "Improving Course Quality with Student Management Teams," *ASEE Prism*, January 1996, p. 19–23.

Schwingendorf, K.E., McCabe, G.P., and Kuhn, J. "A Longitudinal Study of the Purdue C4L Calculus Reform Program: Comparisons of C4L and Traditional Students," *Research in Collegiate Mathematics Education*, CBMS Issues in Mathematics Education, to appear.

Science (entire issue). "Minorities in science: The pipeline problem," 258, November 13, 1992.

Selden A. and Selden, J. "Collegiate mathematics education research: What would that be like?" *College Mathematics Journal*, 24, 1993, pp. 431–445.

Selvin, P. "Math education: Multiplying the meager numbers," *Science*, 258, 1992, pp. 1200–1201.

Silva, E.M. and Hom, C.L. "Personalized Teaching in Large Classes," *PRIMUS* 6, 1996, p. 325–336.

Slavin, R.E. "When does cooperative learning increase student achievement?" *Psychological Bulletin* 94, 1983, pp. 429–445.

Smith, D.A. and Moore, L.C. (1991). "Project CALC: An integrated laboratory course," in Leinbach, L.C., et al, eds., *The Laboratory Approach to Teaching Calculus*, MAA Notes Number 20, Mathematical Association of America, Washington, DC, 1991, pp. 81–92.

Solow, A., ed. *Preparing for a New Calculus*, Conference Proceedings, Mathematical Association of America, Washington, DC, 1994.

Stage, F. and Kloosterman, P. "Gender, Beliefs,and Achievement in Remedial College Level Mathematics." *Journal of Higher Education*, 66 (3), 1995, pp. 294–311.

Steen, L.A., ed. *Calculus for a New Century: A Pump, Not a Filter*, Mathematical Association of America, Washington, DC, 1988.

Steen, L.A., ed. *Reshaping College Mathematics*, MAA Notes Number 13, Mathematical Association of America, Washington, DC, 1989.

Stenmark, J.K., ed. *Mathematics Assessment: Myths, Models, Good Questions, and Practical Suggestions*, National Council of Teachers of Mathematics, Reston, VA, 1991.

Stevens, F., et al. *User-Friendly Handbook for Project Evaluation*, National Science Foundation, Washington, DC, 1993.

Tall, D. "Inconsistencies in the Learning of Calculus and Analysis," *Focus on Learning Problems in Mathematics*, 12 (3 and 4), Center for Teaching/Learning of Mathematics, Framingham, MA, 1990, pp. 49–63.

Tall, D. (Ed.). *Advanced Mathematical Thinking*. Kluwer Academic Publishers, Dordrecht, 1991.

Thompson, A.G. and Thompson, P.W. "A Cognitive Perspective on the Mathematical Preparation of Teachers: The Case of Algebra," in Lacampange, C.B., Blair, W., and Kaput, J. eds., *The Algebra Learning Initiative Colloquium*, U.S. Department of Education, Washington, DC, 1995, pp. 95–116.

Thompson, A.G., Philipp, R.A., Thompson, P.W., and Boyd, B.A. "Calculational and Conceptual Orientations in Teaching Mathematics," in Aichele, D.B. and Coxfords, A.F., eds., *Professional Development for Teachers of Mathematics: 1994 Yearbook*, The National Council of Teachers of Mathematics, Reston, VA, 1994, pp. 79–92.

Toppins, A.D. "Teaching by Testing: A Group Consensus Approach," *College Teaching*, 37 (3), pp. 96–99.

Treisman, P.U. *A study of the mathematics performance of black students at the University of California, Berkeley,* unpublished doctoral dissertation, Berkeley, CA, 1985.

Trowell, S. *The negotiation of social norms in a university mathematics problem solving class.* Unpublished doctoral dissertation, Florida State University, Tallahassee, FL, 1994.

Tucker, A. ed. *Recommendations for a General Mathematical Sciences Program: A Report of the Committee on the Undergraduate Program in Mathematics*, Mathematical Association of America Washington, DC, 1981.

Tucker, A.C. and Leitzel, J.R.C., eds. *Assessing Calculus Reform Efforts: A Report to the Community*, Mathematical Association of America, Washington, DC, 1995.

West, R.D. *Evaluating the Effects of Changing an Undergraduate Mathematics Core Curriculum which Supports Mathematics-Based Programs*, UMI, Ann Arbor, MI, 1996.

Wheatley, G.H. "Constructivist Perspectives on Science and Mathematics Learning," *Science Education* 75 (1), 1991, pp. 9–21.

Wiggins, G. "A True Test: Toward More Authentic and Equitable Assessment," *Phi Delta Kappan*, May 1989, pp. 703–713.

Wiggins, G. "The Truth May Make You Free, but the Test May Keep You Imprisoned: Toward Assessment Worthy of the Liberal Arts," *The AAHE Assessment Forum*, 1990, pp. 17–31. (Reprinted in Steen, L.A., ed., *Heeding the Call for Change: Suggestions for Curricular Action*, Mathematical Association of America, Washington, DC, 1992, pp. 150–162.)

Winkel, B.J. "In Plane View: An Exercise in Visualization," *International Journal of Mathematical Education in Science and Technology,* 28(4), 1997, pp. 599–607.

Winkel, B.J. and Rogers, G. "Integrated First-Year Curriculum in Science, Engineering, and Mathematics at Rose-Hulman Institute of Technology: Nature, Evolution, and Evaluation," *Proceedings of the 1993 ASEE Conference, June 1993,* pp. 186–191.

Addresses of Authors

Dwight Atkins, Mathematics Department, Surry Community College, Dobson, NC 27017; atkinsd@surry.cc.nc.us

Janet Heine Barnett, Department of Mathematics, University of Southern Colorado, 2200 Bonforte Boulevard, Pueblo, CO 81001-4901; jbarnett@meteor.uscolo.edu

Steven F. Bauman, William O. Martin, Department of Mathematics, North Dakota State University, PO Box 5075, Fargo, ND 58105-5075; WiMartin@Plains.NoDak.edu

Deborah Bergstrand, Department of Mathematics, Williams College, Williamstown, MA 01267; dbergstr@williams.edu

Dorothee Jane Blum, Mathematics Department, Millersville University, P.O. Box 1002, Millersville, PA 17551-0302; dblum@marauder.millersv.edu

William E. Bonnice, Department of Mathematics, Kingsbury Hall, University of New Hampshire, Durham, NH 03824-3591; Bill.Bonnice@unh.edu

Martin Vern Bonsangue, Department of Mathematics, California State University, Fullerton, Fullerton, CA 92634; mbonsangue@fullerton.edu

Jack Bookman, Mathematics Department, Box 90320, Duke University, Durham, NC 27708-0320; bookman@math.duke.edu; Charles P. Friedman, Mathematics Department, University of Pittsburgh, Pittsburgh, PA

Joann Bossenbroek, Mathematics Department, Columbus State Community College, P.O. Box 1609, Columbus, OH 43216-1609; jbossenb@cscc.edu

David M. Bressoud, Mathematics and Computer Science Department, Macalester College, 1600 Grand Avenue, Saint Paul, MN 55105; bressoud@macalester.edu

Regina Brunner, Assistant to the Provost for Institutional Research and Planning, Kutztown University, P.O. Box 730, Kutztown, PA 19530-0730; rbrunner@kutztown.edu

G. Daniel Callon, Department of Mathematical Sciences, Franklin College, Franklin, IN 46131; callond@franklincoll.edu

Judith N. Cederberg, Department of Mathematics, St. Olaf College, Northfield, MN 55057; cederj@stolaf.edu

John C. Chipman, Department of Mathematical Sciences, Oakland University, Rochester, MI 48309-4401; chipman@oakland.edu

Ellen Clay, Stockton State College, Jim Leeds Road, Pomona, NJ 08240; claye@vax003.stockton.edu

C. Patrick Collier, Department of Mathematics, University of Wisconsin, Oshkosh, Oshkosh, WI 54901; collier@uwosh.edu

Annalisa Crannell, Department of Mathematics, Franklin and Marshall College, Lancaster, PA 17604-3003; a_crannell@acad.fandm.edu

Steven A. Doblin, Wallace C. Pye, Department of Mathematics, University of Southern Mississippi, Box 5165, Hattiesburg, MS 39406-5165; wallace_pye@bull.cc.usm.edu

Ed Dubinsky, Mathematics and Computer Science Department, Georgia State University, Atlanta, GA 30303-3083; edd@odin.cs.gsu.edu

Steven R. Dunbar, Department of Mathematics and Statistics, University of Nebraska-Lincoln, P.O. Box 880323, Lincoln, NE 68588-0323; sdunbar@math.unl.edu

Charles E. Emenaker, Mathematics, Physics and Computer Science Department, Raymond Walters College, University of Cincinnati, 9555 Plainfield Road, Blue Ash, OH 45236-1096; emenakce@ucrwcu.rwc.uc.edu

John W. Emert, Charles R. Parish, Department of Mathematical Sciences, Department of Mathematical Sciences, Ball State University, Muncie, IN 47306; emert@bsu-cs.bsu.edu

Deborah A. Frantz, Department of Mathematics and Computer Science, Kutztown University, Kutztown, PA 19530; frantz@kutztown.edu

Michael D. Fried, Department of Mathematics, University of California at Irvine, Irvine, CA 92717; mfried@math.uci.edu

Susan L. Ganter, Director, Program for the Promotion of Institutional Change, American Association for Higher Education, One Dupont Circle, Suite 360, Washington, DC 20036; sganter@aahe.org

Jacqueline Brannon Giles, 13103 Balarama Drive, Houston, TX 77099; jbgiles@aol.com

Bonnie Gold, Department of Mathematics, Monmouth University, West Long Branch, NJ; bgold@mondec.monmouth.edu

Richard A. Groeneveld, W. Robert Stephenson, Department of Statistics, Iowa State University, Ames, IA 50011; rgroen@iastate.edu, wrstephe@iastate.edu

Nancy L. Hagelgans, Department of Mathematics, Ursinus College, P.O. Box 1000, Collegeville, PA 19426-0627; nhagelgans@acad.ursinus.edu

Joel David Hamkins, Mathematics Department 1S-215, CUNY-CSI, 2800 Victory Boulevard, Staten Island, NY 10314; hamkins@integral.math.csi.cuny.edu

Nancy Baxter Hastings, Department of Mathematics and Computer Science, Dickinson College, P.O. Box 1773, Carlisle, PA 17013-2896; baxter@dickinson.edu

M. Kathleen Heid, 271 Chambers Building, University Park, PA 16802; ik8@email.psu.edu

Laurie Hopkins, Department of Mathematics, Columbia College, Columbia, SC 29203; lhopkins@colacoll.edu

Pao-sheng Hsu, Department of Mathematics and Statistics, University of Maine, Orono, Orono, ME 04469; hsupao@maine.maine.edu

Philip Keith, English Department, St. Cloud State University, 720 Fourth Street, St. Cloud, MN 56301-4498; pkeith@stcloudstate.edu

Sandra Z. Keith , Mathematics Department, St. Cloud State University, 720 Fourth Street, St. Cloud, MN 56301; keith@stcloudstate.edu

Patricia Clark Kenschaft, Department of Mathematics and Computer Science, Montclair State University, Upper Montclair, NJ 07043; kenschaft@math.montclair.edu

Alan P. Knoerr, Michael A. McDonald, Department of Mathematics, Rae McCormick, Department of Education, Occidental College, Los Angeles, CA 90041; knoerr@oxy.edu, mickey@oxy.edu, mccor@oxy.edu

John Koker, Mathematics Department, University of Wisconsin-Oshkosh, Oshkosh, WI 54901; koker@vaxa.cis.uwosh.edu

A. Darien Lauten, Karen Graham, Joan Ferrini-Mundy, Department of Mathematics, Kingsbury Hall, University of New Hampshire, Durham, NH 03824-3591; kjgraham@hopper.unh.edu

David Lomen, Mathematics Department, University of Arizona, Tucson, AZ 85721; lomen@math.arizona.edu

William A. Marion, Department of Mathematics and Computer Science, Valparaiso University, Valparaiso, IN 46383; bmarion@exodus.valpo.edu

Mark Michael, Department of Mathematics, King's College, Wilkes-Barre, PA 18711; mmichael@gw02.kings.edu

Robert Olin, Lin Scruggs, Department of Mathematics, Virginia Polytechnic Institute & State University, Blacksburg, VA 24061-0123; Lscruggs@vt.edu

Albert D. Otto, Cheryl A. Lubinski, Carol T. Benson, Department of Mathematics, Illinois State University, Campus Box 4520, Normal, IL 61790-4520; cal@math.ilstu.edu

Judith A. Palagallo, William A. Blue, Department of Mathematical Sciences, The University of Akron, Akron, OH 44325; palagallo@uakron.edu

Charles Peltier, Department of Mathematics, Saint Mary's College, Notre Dame, IN 46556; cpeltier@ saintmarys.edu

Eileen L. Poiani, Department of Mathematics, St. Peter's College, Jersey City, NJ 07306-5997; poiani_e@spcvxa.spc.edu

Agnes M. Rash, Department of Mathematics and Computer Science, Saint Joseph's University, 5600 City Avenue, Philadelphia, PA 19131-1395; arash@sju.edu

Marilyn L. Repsher, Department of Mathematics; J. Rody Borg, Department of Economics, Jacksonville University, Jacksonville, FL 32211; lrepshe@ocean.st.usm.edu

Catherine A. Roberts, Mathematics Department, Northern Arizona University, Flagstaff, AZ 86011-5717; catherine.roberts@nau.edu

Sharon Cutler Ross, Department of Mathematics, DeKalb College, 555 N. Indian Creek Drive, Clarkston, GA 30021-2395; sross@dekalb.dc.peachnet.edu

Carolyn W. Rouviere, 377 Coreopsis Dr., Lancaster, PA 17606

Keith E. Schwingendorf, Purdue University North Central, SWRZ 109, 1401 South US 421, Westville, IN 46391-9528; ks@math.purdue.edu

Marie P. Sheckels, Department of Mathematics, Mary Washington College, Fredricksburg, VA 22401; msheckel@mwcgw.mwc.edu

Patricia Shure, Department of Mathematics, University of Michigan, Ann Arbor, MI 48109-1003; pshure@math.lsa.umich.edu

Joel Silverberg, Mathematics Department, Roger Williams University, Bristol, RI 02809; joels@alpha.rwu.edu

Linda R. Sons, Department of Mathematical Sciences, Northen Illinois University, DeKlab, IL 60115; sons@math.niu.edu

Elias Toubassi, Donna Krawczyk, Department of Mathematics, Building #89, University of Arizona, Tucson, AZ 85721; elias@math.arizona.edu

Sandra Davis Trowell, 2202 Glynndale Drive, Valdosta, GA 31602; strowell@surfsouth.com

Janice B. Walker, Department of Mathematics and Computer Science, Xavier University, Cincinnati, OH 45207; walker@xavier.xu.edu

Richard West, Department of Mathematical Sciences, US Military Academy, West Point, NY 10996-1786; ar7624@usma2.usma.edu

Alvin White, Mathematics Department, Harvey Mudd College, Claremont, CA 91711; awhite@hmc.edu

Brian J. Winkel, Department of Mathematical Sciences, United States Military Academy, West Point, NY 10996; ab3646@usma2.usma.edu